Pvt. Ltd.

Affectionately dedicated to

Kiran
Rohan
Deepa
Prashanth

Advances in Integrated Pest and Disease Management in Horticultural Crops

Vol. 2: Vegetable Crops

P. Parvatha Reddy

Former Director and Former Head,
Division of Entomology & Nematology,
Indian Institute of Horticultural Research
Bangalore – 560 089

2009

Studium Press (India) Pvt. Ltd.

Advances in Integrated Pest and Disease Management in Horticultural Crops

Vol. 2: Vegetable Crops

ISBN: 978-81-907577-8-2

Published by:

Studium Press (India) Pvt. Ltd.
4735 / 22, 2nd Floor, Prakash Deep Building
(Near Delhi Medical Association),
Ansari Road, Darya Ganj, New Delhi-110 002
Tel.: 23240257, 65150447; Fax: 91-11-23240273;
jngovil@gmail.com; jngovil@hotmail.com

Printed at:

Salasar Imaging Systems
C-7 / 5, Lawrence Road, Indl. Area, Delhi -110 035

Dr. M.V. Rao, MLC, Govt. of A.P.
President, Agri Biotech Foundation
Special Director-General, ICAR (Retd)
Special Secretary, DARE, GOI (Retd)
Vice-Chancellor, ANGRAU (Retd)
Ex-President, Andhra Pradesh Academy of Sciences

Plot No. 4, BHEL Enclave Akbar Road,
Bowenpally, Secunderabad 500 009,
Phone: (R) 040-27954748/27955795,
(O) 040 24002415 (Cell) 98494 77088;
Fax: 040-24002417

MVR/2008/ **10 September 2008**

FOREWORD

India is the second largest producer of vegetables next only to China contributing 11.9% of the total world production. The present area under vegetable crops is estimated to be about 7.163 million hectares with a production of 109.049 million tonnes. India occupies first position in peas; second in cabbage, cauliflower, onion and brinjal; third in potato; and fourth in tomato. India produces 14.7% of world onions. Vegetable crops contribute 58.93% share to the total production of horticultural crops in India.

The horticultural sector in 2005-2006 has contributed around 28% of GDP in agriculture. Vegetables, fruits, plantation crops and spices contributed 59.8, 30.9, 6.5 and 2.1%, respectively of total horticulture production. The tremendous potential for export of fresh and processed vegetables has made the horticulture sector as one of the thrust areas for export promotion. India exports vegetable seeds, fresh vegetables including onions, dried and preserved vegetables. Processed and value added vegetable products are other items of potential export. For this, focus should be more on the quality and phytosanitation aspects than on quantity.

Of the several factors responsible for lower productivity of vegetables, pests (insect pests, diseases and nematode parasites) are considered as important limiting factors. Some pest problems are of national importance which causes significant losses to vegetable crops. The annual losses due to pests to all the crops in India was estimated at Rs.60,000 million in 1983, which at today's prices could exceed Rs.200,000 million. Use of pest avoidance tactics, enhancement of biological pest suppression and adoption of other non-chemical methods of pest management would help to over come much of the pest problems. Use of pesticides when the pest density reaches ETL only needs to be advocated to reduce expenses to the grower.

The information on insect pests, diseases and nematode pests on vegetable crops (Solanaceous, bulbous, Malvaceous, Leguminous, Cucurbitaceous, root and leafy vegetable crops) is very much scattered. There is no book at present which comprehensively and exclusively deals with the above aspects on vegetable crops. The present book deals with integrated pest and disease management in fruit crops in detail using regulatory, physical, cultural, chemical, biological, host plant resistance and integrated methods. The book is extensively illustrated with excellent quality photographs enhancing the quality of publication. The utility of the book is further enhanced by chapters like pesticide residues, useful tips in plant protection and sources of critical inputs for IPM. The book is written in lucid style, easy to understand along with adoptable recommendations involving eco-friendly components of IPM. This book is a practical guide to growers of horticultural crops. Further, it is a useful reference to students, researchers and extension workers. The book can be used for teaching undergraduate and postgraduate courses as well.

In view of greater emphasis being given by the Government of India to horticulture by establishing horticulture mission and exclusive State Horticultural Universities, there is an urgent need for good text books to teach courses on different disciplines of horticulture. In this context this book written by Dr P Parvatha Reddy is of immense help to all those concerned with vegetable crops. With vast experience of more than three decades working with pests of horticultural crops, Dr Reddy has taken up this noble task single handedly and brought out an excellent publication which I am sure will be useful to all those who are interested in vegetable crops – the farmers, scientists, extension workers, students and policy makers. He deserves all appreciation and commendation.

(M.V. Rao)

PREFACE

India is the second largest producer of vegetables next only to China contributing 11.9% of the total world production. Vegetables occupy an area of 7.164 million ha with a production of 109.050 million MT during 2006. Vegetable crops contribute 58.93% share to the total production of horticultural crops. In vegetables, India occupies the first position in peas; second in cabbage, cauliflower, onion, brinjal; third position in potato and fourth in tomato. India produces 14.7% of the world onions. The per capita availability of vegetables has increased from 76 g in 1951 to 256 g in 2001 as against 280 g/day recommended by Indian Council of Medical Research for a balanced diet. Increase in production is mainly due to increase in productivity. Productivity is still lower in India (15 MT/ha) than the world average productivity (15.7 MT/ha). There is scope for increasing the productivity of vegetables in India further. The vegetable production target for 2020 has been fixed at 150 million MT.

Successful cultivation of vegetables is hampered due to the incidence of several insect pests, diseases and nematodes. Pest management in vegetables in South Asia is a very complex issue in view of the sub-tropical climate, intensive cultivation of high yielding varieties/hybrids under high fertigation, off-season production, lack of adequate knowledge on plant nutrition and protection, economic constraints to maximize qualitative returns and inability to afford risk. Diamondback moth (*Plutella xylostella*) in cabbage and cauliflower; fruit borer (*Helicoverpa armigera*) in tomato; shoot and fruit borer (*Leucinodes orbonalis*) in brinjal; serpentine leaf miner (*Liriomyza trifolii*), thrips (*Thrips tabaci*) and red spider mites (*Tetranychus urticae*) in many crops have become difficult to control. Disease intensity of enation leaf curl in okra, chilli leaf curl, gummy stem blight, *Didymella* leaf blight, *Alternaria* leaf blight in cole crops and viral diseases in cucurbits have increased throughout the country. A total of 65 million MT as pre-harvest losses in vegetables due to diseases alone in the world have been estimated that accounts for 10.5%. The destructive plant-parasitic nematodes are one of the major limiting factors in the production of vegetable crops throughout the world. Pesticides have offered a simple solution to farmers as they are cheap, readily available, easy to apply and non-specific. Thus, it is not surprising that with 3% of the area; nearly 17% of pesticide uptake is on vegetables in India.

However, lack of adequate control, pesticide resistance, resurgence of secondary pests and out-break of new pests have necessitated many growers to seek scientific information on management of pests and re-evaluate the role of pesticides in vegetable production. Many IPM technologies using trap crop, botanicals and bioagents have been developed in some vegetables and a few are validated in farmer's field.

The information on pests (insects, mites, nematodes and vertebrates) and diseases (fungal, bacterial, viral/mycoplasma/phytoplasma) on vegetable crops (Solanaceous, Bulbous, Malvaceous, Cruciferous, Leguminous, Leafy, Cucurbitaceous and Root vegetable crops) is very much scattered. There is no book at present which comprehensively and exclusively deals with the above aspects on vegetable crops. The present book deals with integrated pest, disease and nematode management in vegetable crops in detail using regulatory, physical, cultural, chemical, biological, host plant resistance and integrated methods. The book is extensively illustrated with excellent quality photographs enhancing the quality of publication. The utility of the book is further enhanced by chapters like pesticide residues, allied topics in pest management and sources of critical inputs for IPM. The book is written in lucid style, easy to understand language along with adoptable recommendations involving eco-friendly components of IPM. This book is a practical guide to growers of vegetable crops. Further, it is a useful reference to students, researchers, extension workers and policy makers. The book can also be used for teaching undergraduate and postgraduate courses as well. Suggestions to improve the contents of the book are most welcome (E-mail: reddy_parvatha@yahoo.com).

In view of greater emphasis being given by the Government of India to horticulture by establishing horticulture mission and exclusive State Horticultural Universities, there is an urgent need for good text books to teach courses on different disciplines of horticulture. In this context the present book is of immense help to teach courses on plant protection in vegetable crops.

I profusely thank Dr. K. Srinivasan, Chief Agricultural Economist, TAFE, Chennai; Dr. N.K. Krishna Kumar and Shri. P.N. Krishna Murthy, Division of Entomology and Nematology, Indian Institute of Horticultural Research, Bangalore who have kindly made available many of the photographs included in this book. The publisher, Studium Press LLC, USA deserves commendation for their professional contribution.

Bangalore
February 20, 2009

P. Parvatha Reddy
34, UAS Layout, Ist Main, 7th Cross
Sanjay Nagar, Bangalore-560 094

CONTENTS

1 VEGETABLE CROPS

1.1. INTERNATIONAL SCENARIO

China is the largest producer of vegetables contributing 47.3% of the total world production (Table 1.1). Vegetables occupy an area of 22.518 million ha in China with a production of 434.332 million MT. Productivity of vegetable crops is highest (33.8 MT/ha) in Spain. The major vegetable growing countries of the world include China, India, USA, Turkey, Egypt, Italy, Iran, Russia, Spain, Japan, Mexico and Korea.

Table 1.1. Major vegetables producing countries of the world

Country	Area (in '000 ha)	Production (in '000 MT)	Productivity (in MT/ha)	% Share of world production
China	22518	434332	19.3	47.3
India	7164	109050	15.2	11.9
USA	1307	36845	28.2	4.0
Turkey	1060	26290	24.8	2.9
Egypt	600	16165	26.9	1.8
Italy	593	15994	27.0	1.7
Iran	631	15760	25.0	1.7
Russia	928	15404	16.6	1.7
Spain	395	13356	33.8	1.5
Japan	424	11745	27.7	1.3
Mexico	687	11241	16.4	1.2
Korea	329	10965	33.3	1.2
Others	15984	200416	12.5	22.0
Total	**52620**	**917563**	**15.7**	—

Source: FAO (except Indian data) (Indian data source: Indian Horticulture Database 2006).

1.2. NATIONAL SCENARIO

More than 40 vegetable crops belonging to Solanaceous, Cucurbitaceous, Leguminous, Cruciferous, root crops and leafy vegetables are grown in Indian tropical, subtropical and temperate regions. Important vegetable crops grown in the country are onion, tomato, brinjal, peas, beans, okra, chilli, cabbage, cauliflower, pumpkin, bottle gourd, cucumber, musk melon, water melon, radish, carrot, palak and methi. West Bengal, Uttar Pradesh,

Bihar, Orissa, Tamil Nadu, Gujarat, Maharashtra, Karnataka, Andhra Pradesh, Jharkhand, Kerala, Madhya Pradesh, Haryana and Punjab are the major vegetable-growing states.

India is the second largest producer of vegetables next only to China contributing 11.9% of the total world production. Vegetables occupy an area of 7.164 million ha with a production of 109.050 million tonnes (Bijay Kumar, 2007) (Tables 1.1 and 1.2). Vegetable crops contribute 58.93% share to the total production of horticultural crops. India produces 14.7% of the world onions. In vegetables, India occupies the first position in peas; second in cabbage, cauliflower, onion, brinjal; third position in potato and fourth in tomato. The per capita availability of vegetables has increased from 76 g in 1951 to 256 g in 2001 as against 280 g/day recommended by ICMR for a balanced diet. Increase in production is mainly due to increase in productivity. Productivity is still lower in India (15 t/ha) than the world average productivity (15.7 t/ha). There is scope for increasing the productivity of vegetables in India further. The vegetable production target for 2020 has been fixed at 150 million tonnes.

Table 1.2. Area, production and productivity of vegetable crops in India (2005–06)

Vegetable crops	Area ('000 ha)	Production ('000 MT)	Productivity (MT/ha)	% Share of world production
Brinjal	553.9	9136.3	16.5	8.4
Cabbage	267.4	6921.6	22.1	5.4
Cauliflower	291.8	5260.1	18.0	4.8
Okra	376.1	3684.0	9.8	3.4
Onion	695.1	9248.4	13.1	8.5
Peas	314.0	2298.7	7.3	2.1
Potato	1550.2	29093.7	18.8	26.7
Sweet potato	129.4	1148.6	8.9	1.1
Tapioca	252.7	7924.9	31.4	7.3
Tomato	534.5	9361.8	17.5	8.6
Others	2199.5	25971.4	11.8	23.8
Total	**7163.9**	**109049.5**	**15.2**	

(*Source*: Indian Horticulture Database 2006).

Potato, tomato, brinjal, chilli, cabbage, cauliflower, onion, peas, gourds, okra and melons are now grown on commercial scale with increasing demand, better output, change in production system and development of suitable *check varieties*. These crops have high potential for foreign exchange earning. Today the emphasis on vegetable breeding is on value addition, disease resistance and better quality.

1.3. EXPORT

The vegetable crops which have export potential include onion, potato, chilli, okra, pea, tomato, brinjal, cucumber, cauliflower, bitter gourd, asparagus, celery, capsicum, sweet corn, French bean, lima bean, processed mushrooms, gherkins and organically grown vegetables. Low pungent chillies and low fibre content ginger have good export potential. Some of the new products introduced in export market are tomato paste in bulk aseptic packs, processed button mushrooms, freeze dried and instant frozen vegetables. Dehydrated onion and tomato products have great potential as export commodities.

The important vegetables exported include onions-7.1%, cabbage-4%, cauliflower-4%, okra-3%, peas-3%. Traditional vegetables like onion, potato, okra, bottle gourd and chilli are exported to Malaysia, Singapore, Gulf, Sri Lanka, Bangladesh, Pakistan and Nepal. Non-traditional vegetables like asparagus, cabbage, capsicum, sweet corn, baby corn, green peas, French bean, cucumber, gherkin, cherry tomato, brussels sprouts, broccoli, processed and frozen French bean and lima bean are exported to European countries, Australia, Gulf and S.E. Asian countries. French bean, lima bean, sweet corn and capsicum are exported to Gulf.

1.4. PEST SCENARIO

Successful cultivation of vegetables is hampered due to the incidence of several insect pests. Given the sub-tropical climate, intensive cultivation of high yielding varieties/hybrids under high fertigation, off-season production, lack of adequate knowledge on plant nutrition and protection, economic constraints to maximize qualitative returns and inability to afford risk, insect pest management in vegetables in South Asia is a very complex issue. Diamondback moth (*Plutella xylostella*) in cabbage and cauliflower; fruit borer (*Helicoverpa armigera*) in tomato; shoot and fruit borer (*Leucinodes orbonalis*) in brinjal; serpentine leaf miner (*Liriomyza trifolii*), thrips (*Thrips tabaci*) and red spider mites (*Tetranychus urticae*) in many crops have become difficult to control. Bean fly (*Ophiomyia phaseoli*) and pod borers in leguminous vegetables, and fruit fly in cucurbits are also becoming important. Insect control in vegetables is usually through a mixture/cocktail of insecticides that either can be under or over dosed. Pesticides have offered a simple solution to farmers as they are cheap, readily available, easy to apply and non-specific. Thus, it is not surprising that with 3% of the area, nearly 17% of pesticide uptake is on vegetables in India.

However, lack of adequate control, insecticide resistance, resurgence of secondary pests and out-break of new pests have necessitated many growers to seek scientific information on management of insect pests and re-evaluate the role of pesticides in vegetable production. Many IPM technologies using trap crop, botanicals and bioagents have been developed in some vegetables and a few are validated in farmer's field. Ecologically sound and economically viable IPM is possible only through dissemination of scientific and up to date information to farmers either by the extension agencies or those who train them.

1.5. DISEASE SCENARIO

Vegetable crops are more susceptible to diseases caused by fungi, bacteria, virus, viroid, mycoplasma and phytoplasma that result in reduction of yield and quality. Seed-borne diseases and their inoculum in vegetables have increased with adoption of OGL and GATT agreement. Diseases affecting from seedling stage to harvest and spoilage caused by pathogens during transit, storage and marketing are the main constraints in vegetable production. A total of 65 million tonnes as pre-harvest losses in vegetables due to diseases alone in the world have been estimated that accounts for 10.5%. Introduction of hybrids in vegetable crops has resulted in new disease problems like *Pseudocercospora* leaf blight in tomato and brinjal in India. Disease intensity of enation leaf curl in okra, chilli leaf curl, gummy stem blight, *Didymella* leaf blight, *Alternaria* leaf blight in cole crops and viral diseases in cucurbits have increased throughout the country. Intensive and monoculture of vegetables also increased the disease pressure of soil-borne pathogens such as *Sclerotium*, *Sclerotinia* and *Rhizoctonia*. The off-season cultivation of vegetables has prolonged the survival period of pathogens in the soil. The severity and incidence of major fungal diseases of various vegetable crops are presented in Table 1.3.

Table 1.3. Prevalence of fungal diseases of vegetable crops

Crop	Disease	Pathogen	Severity (%)	Incidence (%)
Brinjal	Phomopsis blight	*Phomopsis vexans*	10–70	32.5
	Collar rot	*Sclerotium rolfsii*	10–30	20.0
	Leaf blight	*Alternaria alternata*	5–35	25.2
Tomato	Alternaria blight	*Alternaria solani*	15–90	40.0
	Septoria leaf spot	*Septoria lycopersici*	5–20	10.0
	Collar rot	*Sclerotium rolfsii*	10–40	20.0
	Fruit rot	*Rhizoctonia, Cladosporium, Pythium*	5–20	15.0
Chilli/ Capsicum	Die-back	*Colletotrichum capsici*	15–60	35.0
	Anthracnose	*Colletotrichum capsici*	10–30	15.0
	Leaf blight & Fruit rot	*Alternaria alternata* *Alternaria solani*	5–20	3.7
	Leaf spot	*Cercospora capsici*	10–35	20.0
Cucurbits	Downy mildew	*Pseudoperonospora cubensis*	10–80	30.0
	Anthracnose	*Colletotrichum lagenarium*	30–90	55.0
	Cercospora leaf spot	*Cercospora citrulina*	20–50	30.0
	Vine & Fruit rot	*Sclerotium, Rhizoctonia, Pythium, Sclerotinia*	10–20	15.0
	Phytophthora blight	*Phytophthora cinnamoni*	25–60	35.0
Cowpea/Pea/ Beans	Leaf blight	*Cercospora cruenta, Pseudocercospora cruenta*	20–90	60.0
	Stem rot	*Macrophomina bataticola*	5–20	10.0
	Phoma leaf blight	*Phoma phaseolina*	3–15	12.0

Table 1.1. (Contd.)

Crop	Disease	Pathogen	Severity (%)	Incidence (%)
	Rust	*Uromyces phaseoli*	5–30	10.0
	Powdery mildew	*Erysiphe pisi*	10–90	50.0
	Downy mildew	*Peronospora pisi*	5–60	15.0
	White rot	*Sclerotinia sclerotiorum*	2–35	25.0
Cole crops	Leaf spot	*Alternaria brassicae, A. brassicola*	10–40	26.0
	Downy mildew	*Peronospora parasitica*	5–25	10.0
Okra	Leaf blight	*Cercospora abelomoschi, C. hibisina*	25–95	70.0
Onion/Garlic	Purple blotch	*Alternaria porri*	15–80	40.0
	Leaf blight	*Stemphylium vesicarium*	20–40	25.0
	Downy mildew	*Peronospora destructor*	10–25	10.0
Vegetable nursery	Damping off	*Rhizoctonia, Alternaria, Phytophthora, Fusarium, Colletotrichum, Pythium, Sclerotium, Phomopsis*	30–85	50.0

1.6. NEMATODE SCENARIO

The destructive plant-parasitic nematodes are one of the major limiting factors in the production of vegetable crops throughout the world. For centuries, man has been plagued by these microscopic organisms feeding on the roots of crop plants essential to his survival and well being. Roots damaged by the nematodes are not efficient in the utilization of available moisture and nutrients in the soil resulting in reduced functional metabolism. Visible symptoms of nematode attack often include reduced growth of individual plants, varying degrees of chlorosis, wilting of the foliage and sometimes death of plants. Furthermore, roots weakened and damaged by nematodes are easy prey to many types of fungi and bacteria which invade the roots and accelerate root decay. These deleterious effects on plant growth result in reduced yields (Table 1.4) and poor quality of crops. Nematode management is therefore, important for high yields and quality that are required by the high cost of modern vegetable production.

Table 1.4. Avoidable yield losses in vegetable crops due to plant parasitic nematodes

Crop	Nematode(s)	Yield loss (%)	Reference(s)
Potato	*Meloidogyne incognita*	42.50	Prasad, 1989
	Globodera rostochiensis	99.50	Prasad, 1989
Tomato	*M. incognita*	30.57–46.92	Bhatti & Jain, 1977; Reddy, 1985; Darekar & Mahse, 1988

Table 1.4. (Contd.)

Crop	Nematode(s)	Yield loss (%)	Reference(s)
	M. javanica	77.50	Anon, 1993b
	Rotylenchulus reniformis	42.25–49.02	Subramanyam *et al.*, 1990
Brinjal	*M. incognita*	27.30–48.55	Bhatti & Jain, 1977; Parvatha Reddy & Singh, 1981; Darekar & Mahse, 1988
Chilli	*Meloidogyne* sp.	24.54–28.00	Singh *et al.*, 2003
Okra	*M. incognita*	90.90 28.08	Bhatti & Jain, 1977; Parvatha Reddy & Singh, 1981
	M. javanica	20.20–41.20	Jain *et al.*, 1986
French bean	*M. incognita*	19.38–43.48	Das, 1994; Patel *et al.*, 2004; Parvatha Reddy & Singh, 1981
	M. javanica	30–40	Sharma *et al.*, 2002
Cowpea	*M. incognita*	28.60	Parvatha Reddy & Singh, 1981
	R. reniformis	13.20–32.00	Palanisamy & Sivakumar, 1981; Hasan & Jain, 1998
Peas	*M. incognita*	20.0–50.61	Parvatha Reddy, 1985; Upadhyay & Dwivedi, 1987; Sharma, 1989
	R. reniformis	15.8	Dalal & Vats, 1998
Carrot	*M. incognita*	56.64	Devi, 1993
Bitter gourd	*M. incognita*	36.72	Darekar & Mahse, 1988
Pointed gourd	*M. incognita*	30–40	Verma, 2001
Watermelon	*M. incognita*	18–33	Hasan & Jain, 1998
Mushroom	*Ahelenchoides sacchari*	40.6–100.0	Singh *et al.*, 2003
	A. composticola	35–60	Laqman Khan, 2001
	A. avenae	25.8–53.5	Bajaj & Jain, 2001
Colocasia	*M. incognita*	24	Anon, 1990b

2 SOLANACEOUS VEGETABLE CROPS

2.1. POTATO

Potato (*Solanum tuberosum*) is one of the important vegetable crops grown in India. It is being cultivated in 1.550 million hectares producing 29.094 million tonnes of potatoes with an average yield of 18.8 tonnes per hectare (Bijay Kumar, 2007) (Tables 2.1 and 2.2). It thrives best in cool climate. Therefore it is a summer crop in the hills and a winter crop in the plains. The major potato growing states include Uttar Pradesh, West Bengal, Bihar, Punjab, Gujarat, Madhya Pradesh, Assam, Uttarakhand, Haryana and Karnataka.

Table 2.1. Major potato producing countries of the world

Country	Area (in '000 ha)	Production (in '000 MT)	Productivity (in MT/ha)	% Share of world production
China	4884	73460	15.0	22.4
Russia	3071	37280	12.1	11.4
India	1550	29094	18.8	8.9
Ukraine	1516	19462	12.8	5.9
USA	439	19091	43.5	5.8
Germany	277	11624	42.0	3.6
Poland	588	10369	17.6	3.2
Belarus	462	8185	17.7	2.5
Netherlands	156	6777	43.4	2.1
France	156	6605	42.2	2.0
Others	6102	105368	17.3	32.2
Total	**19200**	**327314**	—	—

Source: FAO (except Indian data) (Indian data source: Indian Horticulture Database 2006).

Table 2.2. Area, production and productivity of leading potato growing states in India (2005–06)

State	Area (in '000 ha)	Production (in '000 MT)	Productivity (MT/ha)
Uttar Pradesh	443.2	9986.8	22.5
West Bengal	320.6	7106.6	22.2
Bihar	308.9	5702.5	18.5
Punjab	75.0	1223.5	16.3

Table 2.2. (Contd.)

State	Area (in '000 ha)	Production (in '000 MT)	Productivity (MT/ha)
Gujarat	41.0	1088.5	26.6
Madhya Pradesh	53.3	799.7	15.0
Assam	90.1	629.1	7.0
Uttarakhand	22.2	450.6	20.3
Haryana	19.0	374.4	19.7
Karnataka	57.5	368.4	6.4
Others	119.5	1363.6	11.4
Total	**1550.2**	**29093.7**	**18.8**

(*Source*: Indian Horticulture Database 2006).

2.1.1. Pests

2.1.1.1. Aphid, *Myzus persicae*

(i) Damage: Wingless green peach aphids tend to be pear shaped. They may have variable colors but usually are straw-colored or light-green (Fig 2.1). Hind cornicles slant so they converge. A key characteristic is tubercles at the base of the antennae. Potato aphids are more elongate and larger, about 1/6 to 1/8 inch, than the green peach aphid. They often are bright green, but pinkish forms occur. They are similar in general appearance to the English grain aphid. Both potato and green peach aphids produce winged forms. Winged aphids are most important in spread of viruses to plants since they move readily within and between fields. Wingless green peach aphids tend to occur on lower, older potato leaves. Potato aphids more commonly colonize the upper part of the potato plant. During the summer, aphids reproduce asexually (no males) and give live birth to young. Under optimal conditions, a generation may be completed in 10 days. Green peach aphids overwinter in the egg stage on various *Prunus* species, particularly plum and apricot, in and around towns. It may also be introduced on infested greenhouse plants.

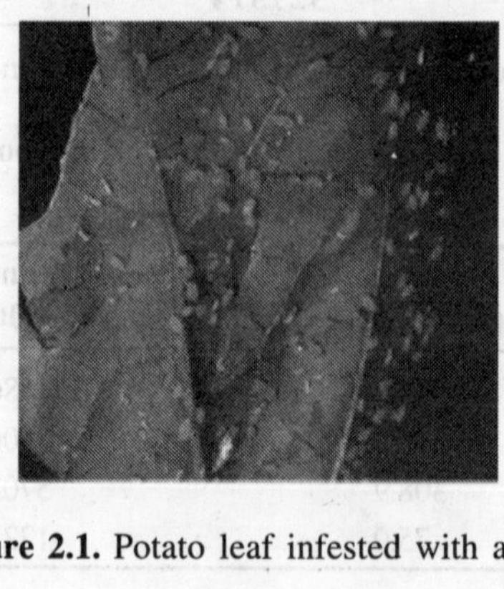

Figure 2.1. Potato leaf infested with aphids

(ii) Management

Cropping in aphid-free periods or when their population is low by adjusting the planting dates in the Indo-Gangetic plains, where about 90% of seed crop is grown *i.e.,* up to 15th October in North-western plains, up to 5th November in North Eastern Plains. Remove all weed hosts susceptible to viruses and for aphids (especially those having yellow flowers) from within and around the vicinity of seed potato plots. The haulm cutting of seed crop should be done as soon as the aphid number crosses the critical level *i.e.,* 20 aphids/ 100 leaves. Care must be taken to prevent any regrowth of stumps before harvest.

Monitor flights of winged aphids by using yellow traps (yellow dish pans with water and a bit of bleach). Sweep fields at least weekly to monitor aphids. Make sure to get the sweep net deep in the canopy since a lot of aphids are found on lower leaves. Pull up yellowish plants to check for aphid colonies. Apply pesticides only when necessary. Use of some insecticides can exacerbate aphid problems, because the control has more effect on natural enemies than on aphids. Insecticides can only provide limited control of aphid transmitted virus diseases. The green peach aphid can be highly resistant to insecticides registered on potatoes. Therefore, choose insecticides carefully for effective control. Protect natural predator populations that are important in aphid management. Trap crops around the edges of fields can help aphid control. Direct damage from aphid feeding in a strictly commercial field needs high populations. If aphid levels are approaching 50 aphids per plant, treat the field.

Several predators and parasites help control aphids. Among the more important insects are lady beetles, damsel bugs, green lacewings, minute pirate, and parasitic wasps.

2.1.1.2. Cut-worm, *Agrotis ipsilon, A. flammatra*

(i) Damage: Cutworms feed at night. During the day they can be found under clods of soil or in cracks in the ground near injured plants (Fig 2.2). Cutworms have a smooth appearance, three pairs of legs and five pairs of prolegs. Damage early in the season includes stems cut off at or below ground level. Later, chewed foliage is the most common symptom. Tubers that are exposed on the surface or in cracks in the soil can be chewed by some cutworms. Cutworms overwinter as immature larvae or eggs in the soil.

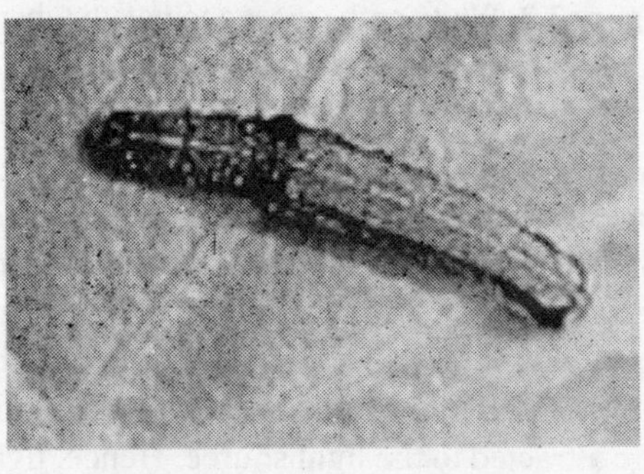

Figure 2.2. Cut worm of potato

(ii) Management

Hot weather ploughing in plains and autumn ploughing in hills reduce the population of immature stages. A number of birds such as crow, mynah, starling, etc. feed on insects that get exposed upon ploughing. Planting of potato on 25th November recorded least foliage damage (1.25%).

Some of the natural enemies which play important role in the management of cut worms are *Broscus punctatus, Liogryllus bimaculatus, Anpopus hypsipylae, Ichneumon* sp., and *Turanogonia chinensis*.

Bacillus thuringiensis is quite effective against cut worms. The percentage foliage damage in *Bacillus thuringiensis* treated plots ranged between 0.0 to 2.0% compared to 1.6 to 6.4% in control. Similarly, tuber damage recorded at harvest was also less in *B. thuringiensis* treated plots. Further, significantly higher yield of potatoes was recorded from *B. thuringiensis* treated plots.

2.1.1.3. White Grub, *Holotrichia consanguinea*

(i) Damage: White grubs can severely damage seed pieces and tubers (Fig 2.3). These insects feed on grass roots and should be considered as a serious threat when potatoes are to be grown in ground immediately following sod.

Figure 2.3. White grub damage to potato tuber

(ii) Management

Mechanical control by collection of beetles should be organized on community basis and started in late May or early June after the first shower by vigorous shaking of host trees/shrubs and then killed. Repeated ploughing before monsoon (April–May) exposes the grubs and pupae for predation by natural enemies (birds) or may be hand collected and destroyed. Flooding of the fields for 7–10 days, wherever possible may be undertaken. Adopt suitable crop rotations. Apply only well rotten FYM/compost in the fields.

A majority of beetles are attracted to the light source. Hence, light traps may be operated for mass collection.

2.1.1.4. Tuber Moth, *Phthorimaea operculella*

(i) Damage: Tuber moth is one of the most serious pests of potato in the field as well as in storage. On an average 105 eggs are laid by each female at an average temperature of 28°C with an incubation period of 4.93 days. The larval, pupal and adult stages last for 8.87, 5.10 and 24 days, respectively. The pest is observed during December to March in the field and April onwards in stored potatoes.

The tuber worm is a small caterpillar, the larva of an inconspicuous gray and black moth called the potato tuber moth. The larva is usually dull white with a brown head. The tuber worm can have several generations a year in warm areas. Tuber worms also feed on other solanaceous species like nightshade, tomato and eggplant. On potato tubers, the larva feed just below the surface and sometimes deep into the flesh. The tubers can appear black since the tuber worm tunnels are filled with larval feces (Fig 2.4). In any case, tuber worm feeding leaves the tuber un-marketable. Tuber worms will feed on the tubers. Varieties that have tubers well protected by soil can tolerate higher populations of tuber worms.

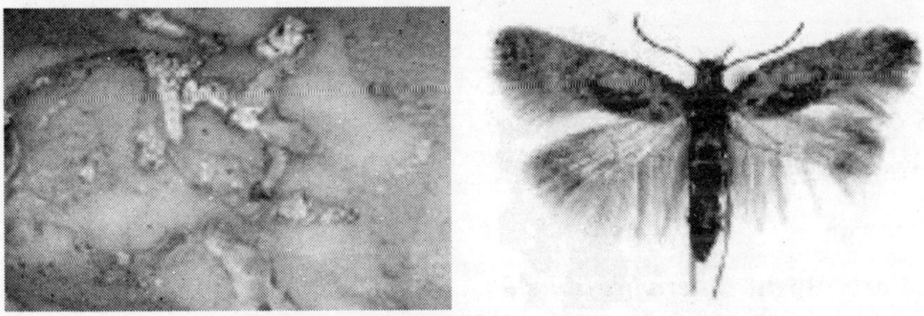

Figure 2.4. Tuber moth damage on potato

(ii) Management

The pest is controlled through deep planting, high ridging and cold storage of seed. Larvae failed to enter the potato tubers through wet soil layers. Keep the tubers in storage before sun set covered with 5 cm sand layer to prevent the adult moths from laying eggs on the eyes of the tubers (Patel, 1980). Use 2.5 cm thick layers of dried lantana or eucalyptus leaves below and on top of the potato heaps which checks tuber moth infestation. Planting varieties that set tubers deep in the soil and hilling the plants up well to protect the tubers from tuber worm can help reduce damage.

In country stores, use of sex pheromones on water traps at 4 traps/100 m^3 store space are effective for mass trapping of males. Pheromones are available to monitor the tuber moth populations. The trap consists of a dishpan of water with a little soap to break the surface tension of the water. Place a tent over the trap to help catch adult moths and place the pheromone inside to attract them. If you catch 10–20 per day you should treat the field.

Foliar application of carbaryl at 0.15% and chlorodane 5% dust at 20 kg/ha is found effective in controlling the pest under field conditions. Pesticide treatment must occur before vine kill.

Use of biocontrol agents like *Bacillus thuringiensis* and Granulosis Virus are also advocated.

2.1.1.5. Epilachna Beetle, *Epilachna vigintioctopunctata*

(i) Damage: Grubs scrape the chlorophyll from the leaves leaving only veins (Fig 2.5).

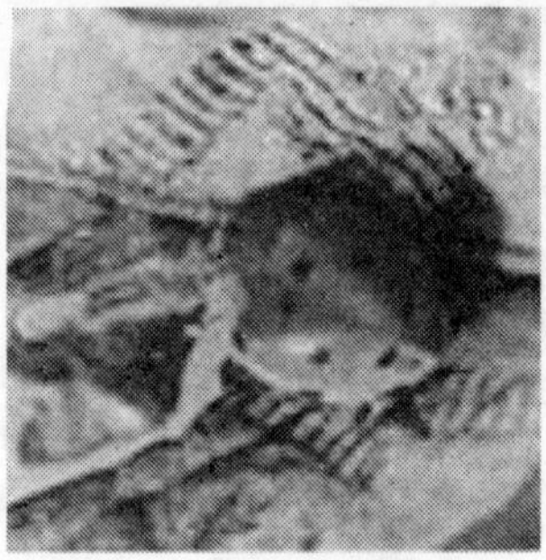

Figure 2.5. Epilachna beetles on potato leaves

(ii) Management

Apply 5% carbaryl dust at 25 kg/ha.

2.1.2. Diseases

2.1.2.1. Early Blight, *Alternaria solani*

Early blight of potato, caused by the fungus *Alternaria solani,* poses a significant risk to crop productivity in the field and to tuber quality in storage. This disease can occur over a wide range of climatic conditions and can be very destructive, sometimes resulting in complete defoliation of plants. Early blight is found in all potato growing regions and is a major disease in hot climates, especially where crops are irrigated. This disease is most severe on maturing or nutritionally deficient potato vines.

(i) Symptoms: Initial infection is most frequent on older leaflets on lower portions of plants. Lesions first appear as small, (1–2 mm) dark brown to black spots. The spots enlarge and develop concentric rings of raised and depressed necrotic tissue which gives the lesion a characteristic "target spot" appearance. Advanced lesions often have angular margins because of limitation by leaf veins and a narrow chlorotic halo frequently surrounds the spot (Fig 2.6). As the disease progresses, affected leaves turn yellow and senescent and either dry up or fall off.

Tuber infection occurs less frequently than leaf infection and are usually evident only after several months of storage. Tuber lesions are dark, sunken, irregularly shaped and often surrounded by a raised violet border (Fig 2.6). The underlying flesh is dry, leathery, and usually brown. In advanced decay, the tissue is often water-soaked and yellow to greenish yellow. Lesions can enlarge during storage, and tubers can become shriveled in advanced cases. Early blight tuber lesions are not as prone to invasion by secondary organisms as compared to other tuber rots.

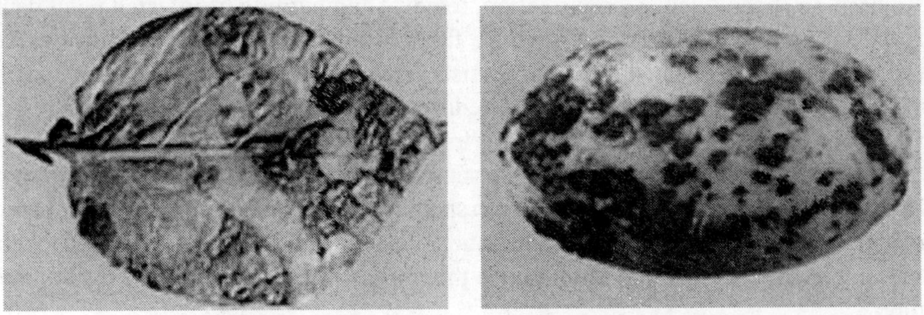

Figure 2.6. Early blight on potato leaves and tuber

(ii) Disease Cycle: *Alternaria solani,* can persist in crop debris, soil, infected tubers, or other solanaceous hosts. Spores are carried primarily by wind and infect potato leaves under ideal conditions of warm temperatures (68–86°F) and high moisture levels. Disease development is also favored by high relative humidity (about 95 percent). Spores are frequently produced after heavy dews and frequent rains and are blown in from mycelium growing on debris or on infected cultivated plants. The germinating spores penetrate susceptible tissue directly or through wounds and produce new conidia that are further spread by wind, splashing rain, etc. Periods of alternating wet and dry weather favor spore dispersal and disease spread. Early blight is often more prevalent on older, senescing tissues and particularly when plants have been predisposed by injury, poor nutrition, insect damage, or other type of stress. Because of the relative resistance of young to intermediate-aged potato plants, major outbreaks of disease do not occur until late in the season.

If sufficient moisture is present, spores can germinate and infect tubers. Infection of potato tubers occurs mostly through wounds inflicted during harvest. Immature tubers are particularly susceptible to infection by *A. solani.*

(iii) Management

Use disease-free seed tubers for planting. Avoid cultivation of solanaceous crops nearby potato fields. Spray 1% Bordeaux mixture to control the disease. Use disease resistant *cvs.* like Kufri Sindhuri, Kufri Jyoti, Kufri Lalima for planting.

The disease is controlled primarily through the use of resistant cultivars, cultural practices, and foliar fungicides such as chlorothalonil, maneb, triphenyl tin hydroxide and mancozeb. Cultivars with levels of field resistance are available, however no immunity to disease exists in commercial potato cultivars or in their wild parents. Highly susceptible cultivars such as Red Norland, Norchip, and Superior should be avoided in locations where early blight is prevalent and disease pressure is high. Field resistance to foliage infection is associated with plant maturity. Late maturing cultivars are usually more resistant than early maturing cultivars; therefore, one should avoid planting early and late varieties in the same or adjacent fields.

Cultural practices, such as crop rotation, removal and burning of infected plant debris, and eradication of weed hosts helps reduce the inoculum level for subsequent plantings and susceptible crops. Because *A. solani* persists in plant debris in the field from one season to the next, rotation with non-host crops reduces the amount of initial inoculum. Other cultural control measures include avoiding irrigation in cool, cloudy periods, harvesting fully mature tubers and avoiding bruising or other mechanical damage and adjacent potato and tomato fields planted the previous season should be ploughed immediately after harvest. Current recommendations include a combination of cultural and chemical measures to retard the development of early blight and reduce the impact of this disease on yield and tuber quality.

To minimize tuber infection after harvest, tubers should be stored under conditions that promote rapid suberization. Suberization of the wounded surface provides a significant degree of protection against the penetration of *A. solani.* The storage of tubers for three weeks at approximately 61°F with more than 90% relative humidity will facilitate rapid wound healing.

The most common and effective control method of early blight is the application of foliar fungicides. The protectant fungicides recommended for late blight control (*i.e.,* maneb, mancozeb, chlorothalonil, and triphenyl tin hydroxide) are also effective against early blight when applied at approximately 7–10 days intervals. The application of foliar fungicides is not needed in plants at the vegetative stage, when they are relatively resistant. Accordingly, spraying should commence at the first sign of disease or immediately after bloom. Frequency of subsequent sprays should be determined according to the genotype and age-related resistance of the cultivar. Protectant fungicides should be applied initially at relatively long intervals and subsequently at shorter intervals as the crop ages.

Early season applications of fungicides before secondary inoculum is produced often have minimal or no effect on the spread of the disease. Early blight can be adequately controlled by relatively few fungicide applications if the initial application is properly timed. Regular inspection of fields after plants reach 30 cm in height is recommended in order to detect early infections.

Disease forecasting models can be used to predict the onset of early blight and specify when the initial fungicide application of a season should be applied. For example, the computerized forecasting system, modified FAST, was developed to identify periods when environmental conditions are favorable for the development of *A. solani* on potato and to provide an efficient fungicide application schedule. This forecasting model calculates the increased susceptibility of potato plants according to their physiological age. It determines the number of days (P-days) after potato emergence at which plants become susceptible to the disease. In general, early maturing cultivars should be sprayed at approximately 250P-days, whereas fungicide applications for late maturing cultivars should commence at approximately 300P-days

2.1.2.2. Late Blight, *Phytophthora infestans*

The late blight of potato shot into prominence during Irish potato famine in 1844–45 which resulted in one million deaths due to starvation and another 1.5 million fled the country when entire potato crops rotted in the field or in storage because of infection by *Phytophthora infestans*. The losses are caused by the foliage infection which leads to premature death of the plant and consequently the reduction in tuber yield. The losses are also caused directly by tuber infection which leads to rottage of infected tubers in the field and stores (Bhattacharya *et al*., 1990). The disease is capable of wiping out entire potato commercial fields within a week if the conditions are wet. Farmers who grow potatoes are at serious risk of losing their entire income for the season. Losses are more in hilly regions where crop is grown under rain-fed conditions as compared to plains. In India losses range between 21–74% in Shimla hills, 11–38% in Mukteshwar hills, 33–65% in Darjeeling hills and 32–39% in Ooty hills.

(i) Symptoms: Late blight affects the foliage and tubers of potato (Fig 2.7). The characteristic foliar symptoms on any susceptible plant are irregular to circular lesions. Initially, in wet weather, the lesions may appear water-soaked; in dry weather, the centres tend to dry out. The dark centre is often surrounded by yellow, chlorotic tissue or brown, collapsed tissue. These spots often occur at the tips or edges of leaves. If moist weather persists, the leaf spots enlarge rapidly to blight larger areas of the foliage. Dark brown discolouration of leaf stalks and stems is also commonly encountered. The symptoms frequently develop first on lower leaves but can develop at any level in the canopy. The symptoms are usually most dramatic inside the canopy. During humid weather, a white fuzz or mildew-like growth appears on the leaves (upper or under sides) and on stems. This indicates that the fungus is producing spores.

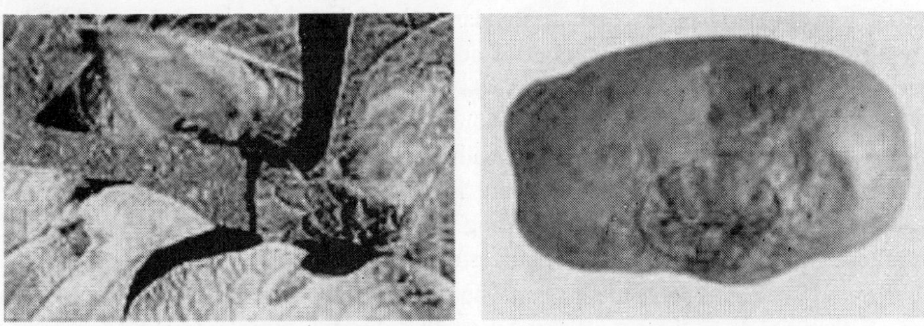

Figure 2.7. Late blight on potato leaves and tuber

Infected potato tubers are discolored by a reddish-brown, granular rot. The rot can be superficial or penetrate more deeply into the tubers (2.5 cm or more). Tubers must be cut open to observe this symptom. The rot continues to develop after the tubers are harvested. Further decay of the tubers is often hastened by secondary fungi and bacteria. The soft rot bacteria often turn tubers into a foul smelling liquid.

Tubers become infected when spores of the fungus wash down through the soil and come into contact with the fleshy organs.

(ii) Epidemiology: The infected seed tubers and refuse piles in the temperate regions and cold stored infected tubers in the sub-tropical regions serve as the primary source of the inoculum.

The fungal growth and spores produced could only survive in living tissues. Infected potato tubers used for seed, infected tubers left in cull piles, and unharvested, infected tubers left in the ground were the sources of inoculum, or spores, for the disease from one growing season to another.

The development of the disease is favored by daytime temperatures of 21–27°C (70–80°F) and night time temperatures of 10–16°C (50–60°F). The presence of free moisture (such as dew, rain or fog) on the foliage also favors the disease. Free moisture is a prerequisite for spore germination and infection and thus is essential for disease development. As the foliage becomes further blighted, more spores are produced and spread of the disease to adjacent plants becomes more likely. As the season progresses, potato tubers can become infected as spores from the blighted foliage are washed down through the soil and contact developing tubers.

(iii) Management

The effective management of late blight of potato can be achieved by reduction of both foliar and tuber infection by adopting three pronged attack *i.e.,* by employing chemicals, cultural practices and host resistance in an integrated manner.

(a) Cultural Methods: The most effective strategy for managing late blight is to avoid sources of inoculum (spores). Do not save potato seed from year to year. Use healthy certified potato seed. Many states that produce potato seed have seed certification programs to ensure that the seed meets certain standards for disease levels. Using certified seed also reduces the carry over of a number of potato viruses. Certified seed, however, is not a guarantee that late blight or other diseases will not be present. Examine your seed carefully before you plant, and plant only sound, blemish-free tubers. Examine tubers for late blight symptoms by cutting them open with a knife. Immediately sterilize or thoroughly wash the knife if late blight symptoms are found.

Use disease-free seed tubers for planting. Ridges should be made up high enough to cover all daughter tubers and reduce chances of their infection upon exposure.

Destroy any rejected seed tubers that you do not plant (bury them deeply in an area away from the field, burn them, or discard them in a plastic bag in the trash). Make sure any tubers put into compost piles are completely decomposed. Pull up and destroy potato plants that come up from tubers left in the field from the previous season.

If late blight becomes severe, remove diseased plants by digging them up. Destroy these plants immediately by one of the following: burying them deeply in an area away from the field, burning them, or by bagging them in a plastic bag and discarding the bag. These steps will help avoid production of a larger number of spores. Harvest all potato tubers in the field. If late blight occurs when the tubers have already 'sized up', harvest the crop as soon as possible to avoid post-harvest tuber rot. Again, destroy diseased foliage and stems.

It is best to avoid wetting the leaves when you irrigate; soak the ground around the plants instead of spraying the plants with water. If this is not possible, water your garden mid-day so that the foliage dries off quickly. Eliminate weeds around the field so as to maximize air circulation around the plants. Additionally, earthing up the soil around the plants to provide a more substantial barrier against tuber infection from spores washing down through the soil.

The following cultural practices may be adopted to reduce the disease incidence:

- Use of disease-free seed.
- Removing/rogueing of ground keepers.
- Adjustment in planting time.
- High ridges and proper earthing up to avoid exposure of seed tubers.
- Water management–light or no irrigation when the blight is likely to appear.
- Killing of haulms at 75% crop damage.

(b) Host Resistance: There are several potato varieties that show resistance to late blight. These varieties will slow down, but do not prevent, the development of the disease. Elba is currently the most resistant potato variety available. Potato varieties with moderate levels of resistance include: Kennebec, Sebago, Allegany, and Rosa.

In north Indian plains, use disease resistant *cvs.* like Kufri Sindhuri, Kufri Jyoti, Kufri Badshah, Kufri Chipsona 1 and 2, Kufri Jawahar and Kufri Sutlej. In hills, use disease resistant *cvs.* like Kufri Jyoti, Kufri Giriraj in Himachal Pradesh hills, Kufri Kanchan and Kufri Jyoti in Darjeeling hills, Kufri Swarna and Kufri Thenamalai in Nilgiri hills.

(c) Chemical Methods: Potatoes are susceptible to late blight at any time during the growing season. If the growing season is wet, late blight occurrence is very likely. Fungicides may be necessary to protect your plants from infection. Continuous fungicide coverage is necessary to protect plants from infection. Spray 1% Bordeaux mixture to control the disease.

When disease first appears, begin treatment with chlorothalonil, maneb or copper oxychloride. Additional pesticides that may be used later in rotation include Heritage or products containing the active ingredients *Bacillus subtilis* or basic cupric sulfate.

The following strategies can be adopted for chemical control:

- Application of prophylactic spray of contact fungicide and subsequently need based fungicide sprays.
- Application of 1 or 2 sprays of systemic fungicides mixtures (metalaxyl, ridomil, etc.) in areas where blight comes every year in epiphytotic form. First and last spray should be of contact fungicides (mancozeb, captafol, copper oxychloride, Bordeaux mixture) to avoid development of resistant strains of the fungus.

(d) Biological Methods: Biocontrol agents like *Pencillium* sp. and *Trichoderma* sp. have been found effective.

2.1.2.3. Dry Rot or Black Scurf, *Rhizoctonia solani*

Rhizoctonia stem and stolen canker, commonly called black scurf because of its tuber phase, is an occasionally serious disease of potato found in all production areas of the world. This disease affects potato development from emergence to harvest. Under cold, moist soil conditions following planting, *Rhizoctonia* can affect emergence and the developing plant. The development of resting structures (sclerotia) causing the black scurf or "dirt that won't wash off" phase of this disease occurs after vine desiccation. Most reports of this disease claim it will limit total yield of a potato crop only under extreme conditions. *Rhizoctonia* can seriously limit marketable yield, however, by causing infected plants to produce an abnormally high percentage of tubers that are misshapen, knobby and cracked while also adversely affecting size distribution. *Rhizoctonia* will invariably decrease the amount of bruise-free potatoes during harvest due to the increased percentages of these misshapen and knobby tubers.

(i) Symptoms: Perhaps the most readily apparent phase of *Rhizoctonia* disease is the black scurf or sclerotia present on tuber surfaces. These sclerotia can vary in size from very small, flat, superficial black specks to large, raised, irregularly shaped masses that can cover a major portion of the tuber. While black scurf can, under extreme conditions, affect the marketability of table stock/fresh pack potatoes, it is an extremely important seed-borne phase of the pathogen that deserves attention by all potato producers.

Lesions characteristic of *Rhizoctonia* on stems and stolons are brown to black and sunken. These cankers can continue to expand and are capable of girdling stems and stolons of young developing plants (Fig 2.8). *Rhizoctonia* infection of older plants very seldom lead to girdled stems that die. However, the health of these plants can be severely compromised and they can frequently become more susceptible to other diseases, particularly early blight.

Figure 2.8. Black scurf on collar region of potato

Potato plants are most severely affected in the spring when underground sprouts can be killed prior to emergence. The secondary sprouts that develop are generally less vigorous and emerge much later causing irregular, uneven stands. Sprouts of severely affected plants that do not die are frequently stunted and remain so for the rest of the growing season. Infections of stolons that occur early in the growing season frequently result in pruning to tuber formation or abortion of tubers early in their development.

Mid-season *Rhizoctonia* infections of potato plants will result in deep, sunken cankers on the main stem. The above ground portion of the plant will appear yellow with some purpling and upward curling of the foliage. These symptoms are frequently confused with leaf roll or purple top. A key in differentiating *Rhizoctonia* from these two diseases is the frequently formation of aerial tubers in the lower leaf axils and the brown, sunken lesions just below the soil surface.

(ii) Epidemiology: *Rhizoctonia solani* can be either a soil-borne or seed-borne pathogen. The fungus survives in soil as mycelium in decomposing plant tissues. It also survives as sclerotia on tuber surfaces (seed-borne) or in the soil for extended periods. Populations of *R. solani* decline in the absence of a susceptible host although the rate of decline is affected by soil type, rotational crops and possibly the amount of organic matter present in the soil.

Disease development on emerging sprouts is favored by cold, wet soil conditions. Although these conditions may not increase disease incidence, disease severity is generally much greater. These conditions slow sprout development but favor germination of sclerotia and infection causing cankers to develop on young, under-developed tissues. Tuber-borne inoculum is very important in this phase of the disease while soil-borne inoculum is believed to generally more important in stem and stolon infections. However, tuber-borne inoculum is also important in the pruning of the first stolons developed by the plant. When stolons become pruned, the secondary stolons are generally weaker and much shorter in length. Tubers produced on these shorter stolons are generally the ones that become misshapen because they tend to grow around the plant stem. Cool, moist conditions, with moisture being the most critical factor, also favor disease development on stems and stolons. Disease development is optimum at around 65°F and decreases as soil temperatures increase.

(iii) Management

Management of *Rhizoctonia* canker must be integrated since no single tactic is totally effective. First and foremost is crop rotation. In this regard it is important to consider both the number of years between potato crops and the rotational crops used in the system. Most research has shown that one potato crop in three or four is adequate to reduce soil inoculum to economic levels. However, soil type can potentially impact the rate of pathogen decline. Observations suggest that the rate of *Rhizoctonia* decline in sandy soils exceed that of loamy soils with more organic matter. Rotational crops such as sugar beet and dry beans, which are affected primarily by different AG's of *R. solani* tend to raise the general activity of *Rhizoctonia* in the soil thereby increasing the inoculum potential of soil-borne inoculum. Cereal crops such as oat and barley are generally regarded as being good rotations with potato for the reduction of soil-borne inoculum.

Since young developing potato crops are more susceptible to injury by *Rhizoctonia*, practices that encourage rapid emergence and plant development will reduce disease severity. Potatoes planted into soils with excessive amounts of residue from a previous crop, such as corn, tend to have serious *Rhizoctonia* problems. Soils with large amounts of residue remain cold for extended periods and more moist, thereby favoring disease development. Therefore, high amounts of crop residue should be dealt with prior to planting.

Cultural practices such as pre-cutting seed, shallow hilling or hill drag-off and delaying planting until soil temperatures are more than 60°F encourage emergence and will also reduce sprout pruning. It is also important to avoid high soil moisture during this period. Early irrigation, prior to emergence, should be kept to a minimum. Potato hills with steep sides and sharp peaks generally warm more rapidly than large, flat beds which can also reduce *Rhizoctonia* disease severity.

Seed selection is also important. Seed tubers free of obvious sclerotia is more desirable than seed with a significant amount of black scurf. Sclerotia coverage as little as 5% can significantly increase disease incidence and severity. In reality, however, much less black scurf sclerotia can significantly increase disease development. One black sclerotium per square inch, which is far less than 1% coverage, can significantly increase disease pressure under environmental conditions that favor disease development. For this reason, disease-free seed is the best option. Harvesting seed tubers within 2–3 weeks of vine desiccation will reduce the amount of sclerotial formation. Tubers left in the soil for more than 3 weeks after vine desiccation will generally have significantly higher levels of black scurf which can lead to deep discounts on the fresh market.

A number of fungicide seed piece treatments have been tested for their effectiveness in controlling the seed-borne *Rhizoctonia* inoculum. Of these, Maxim has consistently provided the best control of tuber-borne *Rhizoctonia*.

Use disease-free seed tubers for planting. Seed treatment with acetic acid +Zn, boric acid or *T. viride* at 3.5 g/kg was found very effective. Combination of soil solarization of black scurf infested fields and seed treatment with *T. viride* further improved the disease control.

Soil application of mustard/groundnut cake at 2.5 MT/ha significantly reduced the number of black scurf diseased tubers and intensity of the disease.

Tuber damage and injury must be avoided during harvest and storage. Tubers should be washed to remove contaminated soil adhering to the tubers, followed by drying under shade reduces the infection substantially. In plains, tubers should be stored in cold stores. In country stores, tubers must be examined periodically and rotting tubers sorted out.

2.1.2.4. Wart, *Synchytrium endobioticum*

(i) Symptoms: Tumors or warts appear on tubers (Fig 2.9), stems and stolons. The wart pathogen is readily seed and soil-borne. Warted tubers used as seed are the chief means of disease spread. The wart pathogen may also spread through seed of wart immune varieties grown in infested land, contaminated soil carried on the feet of workers, animals or farm implements and manure containing diseased material.

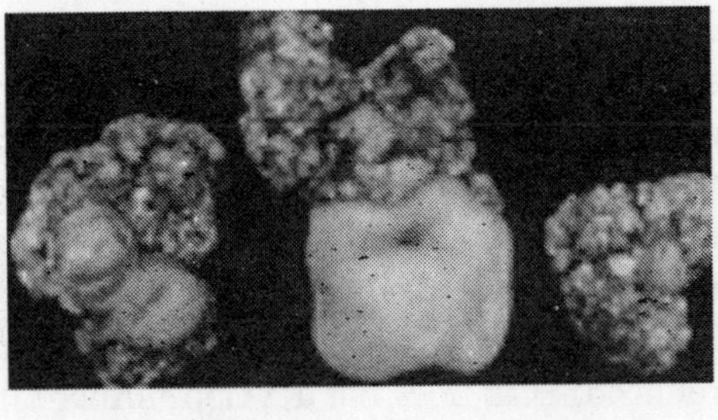

Figure 2.9. Wart on potato tubers

(ii) Management

Cultivate immune potato *cvs.* like Kufri Jyoti, Kufri Sherpa, Kufri Kanchan and Pimpernel. Practice long crop rotation (5 years or more). Rogue out infected plants along with tubers. Rogue out plants of susceptible varieties by burning and destroying warty tubers. Use disease-free seeds for planting.

2.1.2.5. Common Scab, *Streptomyces scabies*

Potato scab is caused by a soil-borne organism that is worldwide in distribution. The same organism also attacks many other root crops as well as the fleshy roots of weeds. It has been reported on beet, turnip, rutabaga, radish, carrot and parsnip. Beets are especially susceptible when soil conditions are favorable for the disease.

(i) Symptoms: Scab spots are more or less circular brown, roughened areas, with irregular margins. Sometimes the ridged portions are in broken concentric rings. The spots may be raised or warty in appearance, level with the surface, or somewhat sunken. Superficial russetting to deep pitting may occur (Fig 2.10). White grubs, wire worms, and millipedes are attracted to this affected tissue so that scab spots may be further enlarged or deepened. Scab begins when tubers start forming. Initially the spots may be so small that they are not noticed. As the tuber continues to grow, the areas of these reddish-brown spots also enlarge. An older tuber has too thick a protective layer on its surface to be invaded readily.

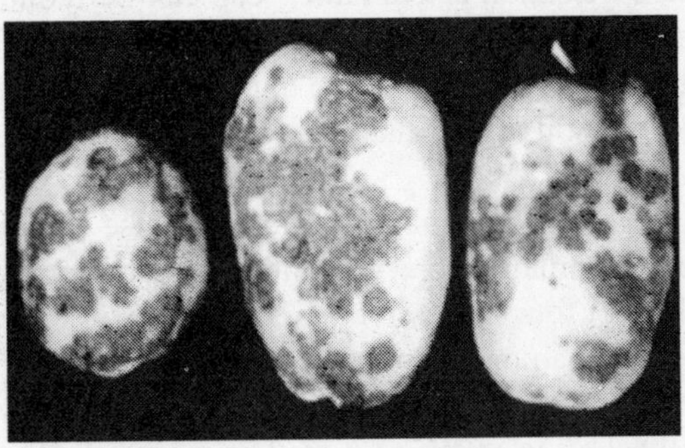

Figure 2.10. Common scab symptoms on tubers

(ii) Disease cycle: *Streptomyces scabies* can live on decomposing material in the soil and does not require a potato or root crop to remain alive. However, it does become more aggressive on any given crop, such as potatoes, if the same crop is grown year after year without rotation. The organism infects the tuber by means of a tiny thread that pushes directly through the tender skin of the forming potato. It slowly continues to grow until the crop is harvested. The infected areas on the potato respond at once by laying down a corky layer. As each layer is invaded by the scab organism, new cork formation takes place until a mature scab spot is produced. Scab does not develop further after the potatoes are dug, but the organism remains alive all winter in storage.

(iii) Management

Use disease-free seed tubers for planting. Irrigate the crop repeatedly to keep the moisture to the field capacity right from tuber initiation until the tubers measure 1 cm diameter. Follow crop rotation with wheat, peas, oat, barley, lupin, soybean, sorghum, bajra and adopt green manuring to keep the disease in check. Plough the potato fields in April and leave the soil exposed to high temperature during May-June. Grow disease resistant *cvs.* like Patna Red, Kufri Alankar, Kufri Sindhuri, Kufri Dewa and Kufri Lalima.

Management in the vegetable field involves correcting the conditions that are favorable for the scab organism. Some of the important points to be considered are given below:

- Rotations with other than root crops should be employed as the size of the field permits. This way the build up of aggressive scab strains can be avoided.
- Purchase certified potato seed pieces. Seed treatment will help to prevent the introduction of the organism into relatively scab-free soils but is not a replacement for using clean seed. Currently, thiophanate-methyl + mancozeb (Tops MZ) is recommended for seed piece treatment against scab.
- Plant resistant varieties. If scab has been a problem before, test one of the following varieties under local conditions to determine whether it is suitable: Belchip, Centennial Russet, Keuka Gold, Norchip, Norgold Russet, Norland, Onaway, Pike, Russet Burbank, Salem or Superior. Of these, Norchip, Norland, Superior and Russet Burbank are probably the most available as seed tubers.
- In a garden where irrigation is available, scab can be reduced by keeping the soil reasonably wet for several weeks while the young tubers are beginning to form.
- Great care should be taken to avoid the application of fertilizer or other materials that tend to make the soil highly alkaline. At the head of the list are lime, ashes, and farm yard manure.
- One of the best methods for combating scab is the use of acid producing fertilizers, especially those that contain liberal amounts of sulphate of ammonia. If this practice is repeated for three or four successive years, the soil will finally become acid enough (pH 5.2) to keep the scab organism from developing.
- Test the pH of the soil and apply elemental sulphur to lower the soil pH to 5.2. Sulphur is usually slow in action so it may be more effective if it is applied a year ahead of the potato crop.

2.1.2.6. Bacterial Wilt or Brown Rot, *Ralstonia solanacearum*

Bacterial wilt of potato continues to be a major disease since 19th century. In India, the disease was first recorded on potatoes by Cappel (1892). This was the first bacterial disease recorded in India. The disease is more severe in tropical, subtropical and warm climates. In Ranchi district of Jharkhand (India), cultivation of rainfed potato was abandoned due to this disease (Gufran and Chakravarti, 1960) and similar instances are recorded in Indonesia (Machmud, 1986) and in Peru (French, 1986a). In India, the losses reported are up to 55% in Kumaon hills (Paharia, 1963), 0.33–40% in Maharashtra (Paharia, 1963), 20–25% in Hyderabad (Nath *et al.*, 1958) and over 75% in some localities of Karnataka (Gadewar *et al.*, 1991).

(i) Symptoms: The disease affects both above and under ground parts of the plant. Infected plants show sudden wilting and complete collapse of plants. Vascular browning has been reported to be the characteristic symptom of the disease and the bacterial mass ooze out.

(ii) Epidemiology: Latently infected material (tubers, rhizomes and corms) is the single most important carrier of the bacterium from one place to another. Bacteria spread from plant to plant by root contact or through the water films or flowing water. In general, increase in the temperature enhances wilting. High soil moisture favours survival of the pathogen in the soil, infection and exit of the bacterium from the host and spread in the soils.

(iii) Management

The management practices ought to be directed towards reducing the inoculum in the field and avoiding introduction of fresh inoculum, growing the crop when the environmental conditions are not favourable for the pathogen and developing the resistant varieties or by biological means such as the use of avirulent strains of the pathogens and/or rhizobacteria.

(a) Cultural Methods: Seed production may be undertaken only in disease-free areas (higher hills of Himachal Pradesh and plains of Punjab). Selection of disease-free tubers for planting, soil application of bleaching powder, hot/cold weather cultivation, disinfection of cutting knives with mercurials, adopting crop rotation for 2–3 years with non-host crops (maize, sorghum, finger millet, cereals, garlic, onion, lupin, cabbage, green manuring crops like sunnhemp, etc.) and use of resistant cultivars have been recommended. Restrict post-emergence tillage to minimum. Practice full earthing up immediately after planting. Collect and burn all diseased plant material including tubers after harvest. The pathogen is susceptible to high temperature. Raising the soil temperature by covering the soil with polythene film has been found effective in reducing the wilt. Opening the deep furrows in infested fields at 50–60 cm intervals, spreading the straw and burning (Jhumming cultivation) reduced the wilt incidence by 100%.

In north-eastern and north-western hills, early planting of the summer crop in February and harvesting in June can avoid the wilt. In Shillong and Ooty, planting in last week of July reduced the bacterial wilt and the wilt was negligible to nil in crop planted in first week of September. These practices reduce the cropping period to only 90 days. Therefore short duration varieties can be used to minimize the losses.

Crop rotation with maize, wheat, barley, oat, sunnhemp, finger millet and vegetables like cabbage, onion, garlic reduced the wilt incidence to the extent of 94%. Studies undertaken in north-western hills using 6 cropping sequences revealed that the reduction in wilt incidence ranged between 25 to 81% by following 2 years rotations. Potato–finger millet–finger millet–potato gave maximum reduction in wilt incidence (81%) (Table 2.3). Roots of finger millet were free from *R. solanacearum* infection, whereas paddy, maize, wheat and beans carried the infection. However, rhizosphere population of the pathogen was found to be reduced.

Table 2.3. Effect of crop rotation on bacterial wilt incidence and yield of potato in north-western hills (Bhowali)

Crop rotation	% Wilt incidence	Yield loss	Population of *R. solanacearum* in cfu/root system of different crops
Potato-Potato-Potato-Potato	51.7 (+92)	56.9	9.0×10^6 (Potato)
Potato-Bean-Bean-Potato	19.7 (–25)	18.3	9.2×10^4 (Bean)
Potato-Paddy-Paddy-Potato	12.0 (–53)	29.7	1.0×10^5 (Paddy)
Potato-Wheat-Wheat-Potato	9.2 (–63)	18.8	4.2×10^4 (Wheat)
Potato-Maize-Maize-Potato	8.7 (–66)	13.1	1.0×10^6 (Maize)
Potato-Finger millet-Finger millet-Potato	4.7 (–81)	7.1	0.0 (Finger millet)

Intercropping of potato with maize and cowpea were the most effective in reducing the wilt incidence.

In non-endemic areas (Gujarat, Maharashtra, north-western and north-central plains and north-western high hills), use of disease-free seeds alone is adequate to control the disease completely. In mid hills (north-western mid hills up to 2200 MSL, north-eastern hills and Nilgiris hills), use of disease-free seeds, cold weather cultivation and application of bleaching powder at 12 kg/ha will be adequate for the disease control. Application of bleaching powder can be substituted with 2 years crop rotation with wheat, barley, finger millet, cabbage, cauliflower, knol-khol, carrot, onion, garlic. In eastern plains and Deccan plateau, the disease can be reduced to considerable extent by following cultural practices like use of disease-free seeds, adjusting the date of planting, application of bleaching powder, straw burning in infested soil, blind earthing, hot weather cultivation and crop rotation (Table 2.4).

(b) Biological Methods: Promising biocontrol agents include *Bacillus* spp., *B. polymyxa* and *Pseudomonas fluorescens* which have been found to reduce the wilt development and incidence. Root colonization by *P. fluorescens* has been reported. *Bacillus* spp. reduced bacterial wilt incidence up to 79%, while *P. fluorescens* by 43 to 75%. Summer (during May-June in Plateau and Plains where very hot and dry climate prevails) and winter ploughing (during December-January in hills where cold temperature prevails) alone reduced wilt incidence by 66% and 50%, respectively. Dipping of seed tubers in 0.25% solution of the formulation of *Bacillus cereus* (strain B4) and *B. subtilis* (strain B5) for 20 min. was effective in reducing the wilt by 59–72% and enhancement of tuber yield by 16.7–25.6%, respectively. Tuber treatment with *P. fluorescens/B. subtilis* reduced wilt incidence by 76.31 and 53.15% and increased tuber yield by 220 and 129%, respectively.

Table 2.4. Control of bacterial wilt and brown rot

Treatment	% Reduction		% Yield increase
	Wilt	Brown rot	
Healthy seed	50–76	40–46	18–27
Bleaching powder at 12 kg/ha + Blind earthing	70–94	50–85	22–28
Summer ploughing in plains	50–83	50–70	5–25
Winter ploughing in hills	60–75	62–78	35–40
Early planting in hills	77–80	75–85	44–68

Avirulent and bacteriophage producing strains of *R. solanacearum* induce resistance to bacterial wilt. Tuber bacterization followed by one soil application of avirulent strain of *Ralstonia solanacearum* significantly reduced the disease incidence to an extent of 80 to 82.5% and increased the tuber yield (15.08 to 15.83 t/ha) compared to control (6.04 to 6.94 t/ha).

(c) Host Resistance: Some of the resistant/tolerant *cvs*. under cultivation are Ontario, Sebago, Kathadin and Green Mountain in USA; Prisca in Madagascar; Myrylin and Greta in Mauritius and Caxamarca and Molinera in South America.

2.1.2.7. Soft Rot, *Erwinia carotovora*

(i) Symptoms: Fruits develop sub-epidermal soft-rot, which spreads rapidly reducing the tissue to a pulp in 4 to 6 days.

(ii) Management

Avoid excess irrigation. Provide proper drainage. Adjust planting time to avoid hot weather during plant emergence. Harvest the crop before the temperature raises to 28°C. Harvest the crop only when the tuber skin is fully cured. Avoid injury to tubers and sort out injured tubers. Treat tubers before storage with 3% boric acid for 30 min. and dry under shade. Store the produce in well ventilated cool stores or cold stores.

2.1.2.8. Potato Virus X (PVX)

(i) Symptoms: The characteristic symptoms of PVX include interveinal mosaic, mild or barely perceptible mosaic, light and dark green patches, mottling with stunting, crinkling (Fig 2.11) with severe strains and top necrosis in some varieties. An infection of PVX has the potential to reduce the yields up to 10–30%.

Figure 2.11. PVX on potato leaves

(ii) Transmission: Use of infected tubers is the main reason for spread of PVX from one place to another. PVX is contagious and readily spread upon contact of plant foliage and roots in the field as well as upon injury/cutting of the tubers.

PVX is readily transmitted mechanically. It is spread by foliar or sprout contact, seed cutters, pick-type planters, cultivators, spray equipment, and possibly by insects with chewing mouth parts.

(iii) Management

Strict sanitation should be practiced in the field and also in stores. Disinfect all field equipments with 3% Trisodium phosphate or 1% calcium hypochlorite solution. Use disease-free seed tubers for planting. Rouging of diseased plants along with their tubers at least twice in growing season helps in reducing the incidence of the disease. Dehaulm the crop before the aphids cross the critical level. Follow thermotherapy by meristem culture for eliminating the virus from seed tubers. Potato *cvs.* having resistance are Kufri Ashoka, Kufri Badshah, Kufri Jawahar, Kufri Jyoti, Kufri Lalima, Kufri Pukhraj and Kufri Sindhuri.

2.1.2.9. Potato Virus Y (PVY)

(i) Symptoms: The characteristic symptoms of PVY include mild to severe mosaic, veinal necrosis, leaf drop streaking, stunting of plants, rugosity, bunching or twisting of leaves, down-turning of leaflet margins and necrosis, dropping and collapse of older leaves (Fig 2.12). Symptoms of PVY may vary by cultivar. Necrosis and veinal necrosis of petioles often cause leaves to drop or cling to the stems. Infected plants have bunchy tops and leafless lower stems or clinging necrotic leaves. An infection of PVY has the potential to reduce the yields up to 60–80%.

(ii) Transmission: PVY is a potyvirus which is transmitted through aphids, *Myzus persicae* and *Aphis gossypii* in non-persistent manner.

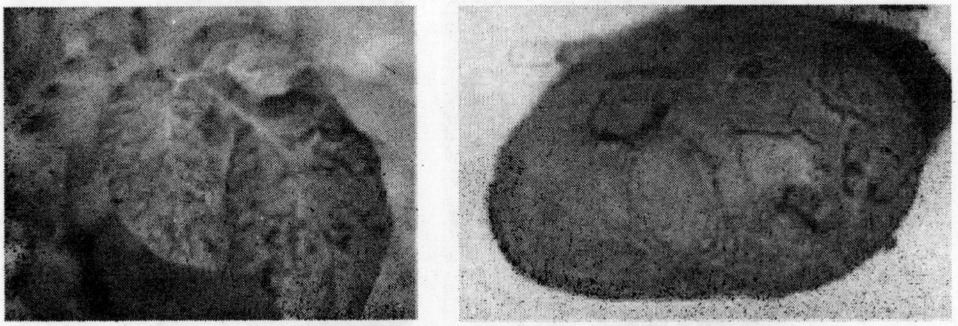

Figure 2.12. PVY on potato leaves and tuber

PVY is a stylet-borne virus transmitted by at least 30 species of aphids, particularly the green peach aphid and potato aphids. Cereal aphids in high numbers become important vectors of PVY. Aphids landing on infected plants acquire PVY within a few seconds of feeding and can transmit the virus immediately, in a non-persistent manner. While aphid transmission is the most important means of spread in the field, all PVY strains also are mechanically transmissible by leaf contact and may be carried in seed tubers. Current-season infection late in the growing season may not cause foliar symptoms.

(iii) Management

A combination of thermotherapy coupled with meristem culture eliminated the virus from seed tubers. Potato *cvs*. Kufri Jyoti, Kufri Kuber, Kufri Lalima, Kufri Sindhuri and Kufri Ashoka have in built resistance/tolerance to PVX, PVY and PLRV.

2.1.2.10. Potato Leaf Roll Virus (PLRV)

Potato leaf roll is caused by the potato leaf roll virus. The severity of symptoms depends on the strain of virus, susceptibility of the cultivar, time of infection, and environment. Yield may be substantially reduced in plants grown from infected tubers. An infection of PLRV has the potential to reduce the yields up to 60–80%. Potato leaf roll severity can be reduced as a result of a comprehensive aphid suppression and monitoring program.

(i) Symptoms: The characteristic symptoms of PLRV include rolling of upper leaves (Fig 2.13) in primary infections and of the lower leaves in the secondary infection, leaves turning leathery, brittle, pink or brown pigmented, stunting of plants and making rattling noise when shaken.

Lower leaves of plants grown from leaf roll-infected seed tubers become rolled and leathery and may turn light pink or yellow. Infected plants usually are stunted and more upright than healthy plants. Young leaves may become yellow or slightly pinkish and rolled and may stand upright in early current-season infections (Fig 2.13). Current-season infections that develop late in the season may cause no visible symptoms, but tubers usually become infected. Early season infections may reduce yield.

Figure 2.13. Leaf roll virus infected plants and tuber of potato

(ii) Transmission: The virus is transmitted by aphids, *Myzus persicae* and *Aphis gossypii* in persistent manner. The green peach aphid, *Myzus persicae* is the primary vector. Aphids acquire the virus after several hours of feeding and require an incubation period before being able to transmit the virus to non-infected plants. Once infected, the aphid usually retains the virus for life.

(iii) Epidemiology: Leaf roll is introduced by planting infected seed or by aphids carrying the virus to the field from an outside source. Leaf roll is transmitted by several species of aphids. The green peach aphid is the primary vector.

(iv) Management

Plant high quality, certified seed. Two hour daily treatment of tubers for 6 weeks at 40°C inactivates PLRV. Treatment of tubers in hot water at 55°C for 15–20 min has also been found effective in eradicating the virus. Control aphid populations in the field, and on other hosts. Rogue infected plants early in the season to control disease sources in the field. Consider early vine-kill when high aphid populations are present.

Potato *cvs*. Kufri Jyoti, Kufri Kuber, Kufri Lalima, Kufri Sindhuri and Kufri Ashoka have in built resistance/tolerance to PVX, PVY and PLRV.

Integrated Pest and Disease Management

- Deep summer ploughing to expose resting stages of pests to sunlight.
- Use disease-free seed material for planting.
- Grow resistant varieties (Table 2.5).
- Adjustment of planting and harvesting time for avoidance of pests and diseases. Optimum time for planting is 15th to 30th October in western and central plains and 1st to 15th November in eastern plains.
- Covering the exposed tubers by high ridging.
- Use yellow sticky traps at 10/ha for attracting aphids.
- Use sex pheromone traps at 20/ha for mass trapping of tuber moth.
- Rouging of diseased plants.

Table 2.5. Management practices for pests and diseases of potato at each crop stage

Crop stage	Action	Reason
Before planting	Clean up crop area and remove weeds and self-sown potatoes.	Weeds and self-sowns harbour viruses, insect pests and diseases which can be transferred to the new crop.
	Avoid continuous cropping with potatoes (minimum 2 years between crops).	Reduces the build-up of pests and diseases in the soil.
Planting	Plant certified seed.	Certified seed is grown under strict hygiene regulations and will reduce the risk of introducing seed-borne pests and diseases to the crop.
	Plant varieties which are less susceptible to powdery scab or common scab.	In paddocks where scabs have been present in the past, less susceptible varieties will reduce disease risk.
	Change planting depth according to relative risk of pests and diseases.	Deep-setting varieties are less susceptible to potato moth attack. Shallow planting will help reduce *Rhizoctonia* infection.
	Plant in optimum conditions of soil temperature (16–20ºC) and soil moisture.	High temperature and high moisture can lead to seed piece breakdown (especially in Coliban). Low temperature and high moisture favours infection by *Rhizoctonia.*
Hilling	Maintain and improve soil structure and create well-drained hills.	Poorly-drained soils which are easily saturated favour development of diseases like powdery scab, blackleg and phoma (gangrene).
	Maintain good soil cover throughout season.	Good soil cover acts as a barrier, preventing exposure of tubers to potato moth and greening.
	Monitor pests and diseases throughout season.	Control measures are only required when pest populations reach a critical level.
	Monitor beneficials and release if necessary.	
Irrigation/rain	Check soil coverage of tubers regularly throughout season.	Rolling and hilling prevents greening of tubers as well as protecting them from pests and diseases attack.
	Manage irrigation to avoid pests and diseases attack.	Paddocks with a history of powdery scab should not have saturated soil during tuber set. Paddocks with a history of common scab should have moist soil during tuber set.

Table. 2.5. (Contd.)

Crop stage	Action	Reason
	Avoid water stress.	Maintaining good crop health makes plants less susceptible to pests and diseases attack (especially black dot).
Die-down/slashing	Maintain soil barrier using irrigation/rolling.	Prevents greening, potato moth attack and late blight.
Harvesting	Ensure tubers are mature before digging.	Mature tubers have a protective skin which reduces the risk of infection.
	Harvest as soon as possible after maturity.	Increased time in the ground increases risk of exposure to pests and diseases (*e.g. Rhizoctonia*, silver scurf, black dot, and wireworm).
	Handle tubers carefully during harvest.	Damage to tubers creates an entry point for disease (*e.g. Phoma*, dry rot, soft rot).
	Harvest under cool (12–18ºC), moist conditions.	Hot and dry, or cold, conditions increase damage to tubers and the risk of rots.
Post harvest	If tubers were very wet at harvest, place them in a drying tent with circulating warm air until dry.	Free moisture on tubers increases the risk of rots and of seed piece breakdown.
	If tubers were cold at harvest or have just been removed from cool storage, warm them to at least 12ºC before grading or cutting for seed.	Handling cold tubers is likely to damage them and increase the risk of disease.
	Handle tubers carefully during grading and inspection.	Damage to tubers increases risk of infection and can make them unsalable.
	If tubers are to be cool stored, cure them beforehand by holding them at 14–16ºC and 85–95% relative humidity, with good airflow, for 10–14 days.	Curing heals skin wounds and reduces the risk of storage rots (*e.g. Phoma*, dry rot).
	Don't retain tubers for seed.	Non-certified seed increases chances of pests and diseases (especially viruses) developing in next season's crop.
	Store tubers at optimum temperature and humidity (3–4ºC for seed; 4–10ºC for ware; 10ºC for processing tubers; 85–95% relative humidity for all).	Storing tubers under optimum conditions minimizes disease and prolongs storage life.

2.1.3. Nematodes

Cyst, root-knot, dry rot and lesion nematodes have been recognized as the major nematode pests of potato.

2.1.3.1. Root-knot Nematode, *Meloidogyne* spp.

Thirumalachar (1951) noticed root-knot galls on potato tubers from Shimla during 1950 for the first time and since then it has been recorded from all the potato growing regions of the country. The root-knot nematodes, *Meloidogyne* species are by far the most important nematode pests of potatoes in India. Severe infestation of root-knot nematodes leading to crop failure have been noticed in Mahasu district of Himachal Pradesh and Hassan in Karnataka where farmers did not get even as much as planted by them.

An initial level of 2 root-knot larvae per gram of soil resulted in 100% tuber infection with an overall yield reduction of 42.5% (Prasad, 1989).

(i) Distribution: *M. incognita* has been recorded on potato in Himachal Pradesh, Uttar Pradesh, Assam and Karnataka; *M. javanica* in Himachal Pradesh, Uttar Pradesh, Punjab, Haryana, Delhi, Bihar and Maharashtra; and *M. hapla* in Gulmarg (Jammu and Kashmir) and Shimla (Himachal Pradesh).

(ii) Symptoms: Plants affected by root-knot nematodes are generally stunted, slightly yellow and may wilt during hot weather. The diagnostic symptoms of a *Meloidogyne* attack on potatoes are the galls found on both roots and tubers. Tuber surface becomes uneven and warty because of numerous blisters like galls (Fig 2.14). When an infested tuber is transversely cut, the white glistening swollen females of the size of pin head can be easily seen embedded in the potato tissue. Such tubers are invariably small, unfit for marketing, rot quickly and cause considerable yield losses. Galling on tubers may render them unsaleable.

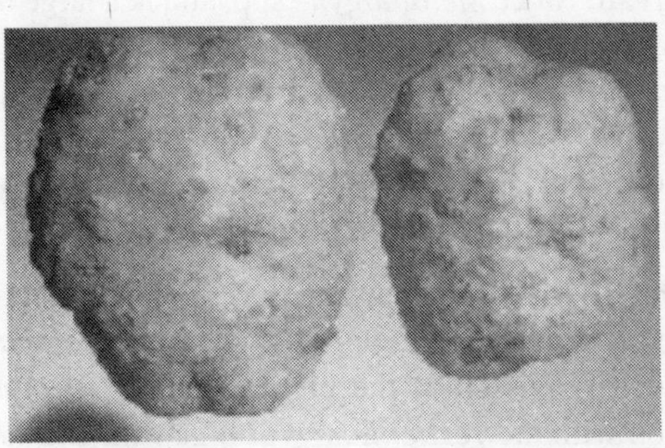

Figure 2.14. Potato tubers with blisters incited by *Meloidogyne incognita* (Courtesy: Sasser, 1971)

(iii) Life Cycle: There are up to five generations on the susceptible host under favourable conditions. Both roots and tubers are infected, however, the first generation occurs mainly on the root systems, while the succeeding generations attack tubers.

(iv) Interaction with Other Pathogens: *Meloidogyne* species are associated with the fungi such as *Rhizoctonia solani* and *Verticillium* species and the bacterium, *Ralstonia solanacearum* in forming disease complexes.

Jatala *et al*. (1990) reported the interactions between *R. solanacearum* and nematodes, especially *Meloidogyne* spp. in potato in relation to breeding for combined resistance.

(v) Ecology: Root-knot damage to potatoes is usually associated with light soil. Aggravated attack by *Meloidogyne* species on potatoes under irrigation is usually ascribed to spread of the nematodes in irrigation water. *M. hapla* requires cool climate, whereas *M. incognita* and *M. javanica* require higher temperatures for their development and reproduction.

(vi) Races: Four races in *M. incognita* and 2 races in *M. arenaria* have been identified based on the North Carolina Differential Host Test which employed 5 differential hosts for delineating races (Table 2.6).

Table 2.6. Races of *Meloidogyne incognita* and *M. arenaria*

Meloidogyne		Cotton	Tobacco	Pepper	Watermelon	Peanut	Tomato
M. incognita	1	–	–	+	+	–	+
	2	–	+	+	+	–	+
	3	+	–	+	+	–	+
	4	+	+	+	+	–	+
M. arenaria	1	–	+	–	+	–	+
	2	–	+	+	–	+	+

In India, all 4 races of *M. incognita* have been reported and the race 1 is dominant (Sharma and Gill, 1992).

(vii) Survival and Spread: Since *Meloidogyne* spp. attack a large number of plant species, their population can be maintained on weeds and volunteer crops. They over winter usually in the form of eggs, although the ability of juveniles to go through anhydrobiosis may contribute to the survival of some *Meloidogyne* spp. Infected tubers, plant parts and planting material, as well as movement of infested soil by farm machinery and irrigation water are the main avenues of disseminating *Meloidogyne* spp.

(viii) Interaction with Other Pathogens: The most important interaction of the root-knot nematodes on potatoes is their association with *Ralstonia solanacearum*. Resistance of potatoes to bacterial wilt is broken in the presence of *M. incognita*. Other interactions include their association with the *Verticillium* wilt organism and *Rhizoctonia solani*.

(ix) Management

(a) Physical Methods

Heat Treatment: Incubation of potato tubers at 45°C for 48 hr has been shown to kill about 98.9 per cent *M. incognita* without affecting tuber viability (Nirula and Bassi, 1965).

Hot water treatment of potato tubers at 46–47.5°C for 120 min gave effective control of root-knot nematode infection.

Rabbing: In small holdings, burning of trash in the field before taking up tuber planting not only helps in reducing the nematode population but also enriches the soil.

(b) Cultural Methods: Tubers from root-knot infested potato crops should not be used as seed, as even outwardly clean looking seed may be infested. The movement of soil and water from the infested fields should be avoided. The field should be kept free from weeds, since root-knot nematodes have a wide host range and most of the weeds help in the build up of nematodes. Thus, clean cultivation (free from weeds which act as hosts) reduces nematode population to a great extent.

Deep ploughing and drying of soil in the summer months facilitates the drying and death of infective larvae thereby reducing initial inoculum in the soil. Dry fallow with 2 or 3 deep ploughings during the hot summer months gives excellent control of root-knot nematodes.

Intercropping: Potatoes grown with white mustard in a pot of infested soil were less heavily attacked by root-knot nematodes than potatoes growing alone. Potato root diffusate was ineffective in the presence of leachings from the roots of mustard seedlings. Mustard oil increased the yield of potatoes by reducing the severity of nematode attack. The active principle involved in mustard is allyl isothiocyanate which is toxic to the nematodes. The interplanting of marigold with potato has resulted in a 7% increase in tuber yield.

Intercropping of onion and maize reduce galling by *M. incognita* on potato roots. Potatoes intercropped with maize have been found to be extremely vigorous with well developed root systems.

Time of Planting: In long season, warm climate regions, potatoes may be planted during the winter months and harvested before injury occurs in the spring without visible infection. Frequently potatoes are planted in January-February and harvested in April-May before the second generation of nematodes develop and attack the tubers. Planting of potato during third or fourth week of March in Shimla hills would reduce both root and tuber infestation of *M. incognita* without affecting the yield (Ahuja, 1983; Prasad *et al.,* 1983). The yields were maximum which was concomitant with the lowest tuber infestation and lowest larval population in the soil at harvest. Late planting of autumn crop and early planting of spring crop in north-western plains; while in hills, early planting of summer crop in fourth week of March is ideal. At Jalandhar, early planting of spring crop in first week of January and late planting of autumn crop in second and third week of October reduced the root-knot infection in potato.

Crop Rotation: In north-western hills, rotation of potato with French beans, leafy vegetables, maize and wheat helped in minimizing the nematode damage and increasing the productivity of the crop (Raj and Nirula, 1969). Crop rotation with non-host crops like maize, wheat, barley, beans, etc. reduces nematode population. Growing of trap crop like French marigold, *Tagetes patula* in alternate rows with potato reduced the root-knot nematode larval population in soil, root galling and tuber infestation while the yields increased up to 123% over control. Similarly, growing *Cassia* sp. and incorporation of saw dust at 2.6 t/ha also helped in reducing tuber infestations. Tomato crop can be followed by peanuts without any damage to peanuts and nematode populations will be reduced.

Botanicals: The oil cakes (neem, groundnut, mustard and castor) were effective in reducing parasitic nematode population (*Hoplolaimus*, *Tylenchorhynchus, Meloidogyne* and *Helicotylenchus* species) and in increasing yields of potato. Application of eucalyptus leaf waste from oil distillation plants at 2.5 t/ha helps in reducing the nematode population.

(c) Chemical Methods: Application of carbofuran at 3 kg *a.i.*/ha in heavily infested fields and 2 kg *a.i.*/ha in moderately infested fields in two equal splits, first at planting and second at the time of earthing up, is recommended. Application of carbofuran at 3 kg *a.i.*/ha or aldicarb at 2 kg *a.i.*/ha or ethoprop at 10 kg *a.i.*/ha in two equal splits, first at planting and second at the time of earthing up, increased the yields by 18 to 23% while accounting for 80% reduction in tuber infestation.

(d) Biological Methods: The bacterium, *Pasteuria penetrans* has offered possibilities of biocontrol. Arbuscular mycorrhizal fungi such as *Glomus fasciculatum* and *G. mosseae* have shown promise in reducing the root-knot nematode infestation.

(e) Host Resistance: Potato *cv*. Kufri Dewa was found resistant to root-knot nematodes.

(f) Integrated Methods: Application of neem cake/FYM/compost enriched with *Trichoderma harzianum/Paecilomyces lilacinus* gave effective control of root-knot nematodes.

Following 2 years crop rotational sequence of maize-wheat-potato-wheat coupled with summer fallow after 2 or 3 deep ploughings in north-western plains reduced root-knot damage significantly.

2.1.3.2. Cyst Nematodes, *Globodera rostochiensis*, *G. pallida*

The potato cyst nematode is established as one of the major crop protection problems of the world. The ability of this nematode to build up to damageable levels in a short span of 5–6 years, substantial yield reductions in the crop, lack of inexpensive nematicides for soil treatment capable of providing adequate level of control under field conditions, the relative ease with which the cysts are dispersed with soil adhering to the seed tubers and the long persistence of eggs within the cyst in the absence of the host makes this nematode as probably the most important pest problem on potatoes.

The avoidable yield loss in susceptible potato due to cyst nematodes was 99.5 to 99.8% in summer and autumn crops. Total failure of the crop has been reported under severe infestation conditions. An average loss of about 9% of global potato is accounted to the cyst nematodes amounting to about 45 million tonnes (Prasad, 1989).

(i) Distribution: In India, it was first reported by Jones (1961) from Nilgiri hills, Tamil Nadu. It was also reported from Kodai Hills, Tamil Nadu, and Munar hills of Kerala (Ramana and Mohandas, 1998). Out of 9,000 ha under potato, 3,000 ha are infected by this nematode in Nilgiris. In Kodai hills, about 200 ha are infected (Thangaraju, 1983). Tomato and brinjal are also attacked by this nematode.

(ii) Symptoms: The disease usually occurs in patches. The infested plants exhibit typical symptoms of patchy growth of weak and stunted plants. The patches increase in time with continuous potato cropping. Under conditions of severe infestation, the plant growth is stunted and wilting occurs during hot part of the day. Plants show tufting of leaves at the top as the outer leaves turn yellow and die. Root system is smothered, secondary roots are induced at the collar region and the plants can be easily pulled out. Tubers formed are less in number and reduced in size. In extreme cases, tuber formation is arrested. The total photosynthesis per plant is also significantly reduced as a result of reduced leaf area and this is reflected in reduced potato yield (Fig 2.15).

Figure 2.15. Cyst nematode on potato. *Left*-Infected; *Right*-Healthy

(iii) Life Cycle: *Globodera rostochiensis* is amphimictic, with sessile, globose females and motile, vermiform males preceded by four larval stages. The second stage larva is the resistant, dormant and infective stage remaining inside the egg within the cyst formed by tanning of the female's cuticle. The cysts remain in the soil after death of the host, each containing about 500 eggs. After hatching, the second stage larva enters the root of a host plant, just behind the root tip or at the site of a new lateral root and moves through the root away from the tip before starting to feed on a group of pericycle, cortex or endodermis cells which are modified to form a large syncytial transfer cells on which the nematode feeds throughout the rest of its development to the adult stage. Sex is determined by nutrition, a poor food supply resulting in the formation of a male, a good supply, a female. Thus with heavy infections males predominate, the reverse with light infections.

The adult females swells greatly, rupturing the root cortex and protrudes from the root, exposing the vulval region while the head and neck remain inserted in the root (Fig. 2.16). As they develop, males become coiled within the earlier larval cuticles and on maturity they rupture them, then leave the root and are attracted to females by a secretion. Both sexes mate several times. Males do not feed but remain active for up to 10 days; females normally mate soon after rupturing the root cortex and within 50 days of larval invasion.

Figure 2.16. Potato roots infected with *Globodera rostochiensis* (Courtesy: Sasser, 1971)

(iv) Host Range: All known varieties of potato, tomato and brinjal are severely attacked by the cyst nematode. A number of other Solanaceae are recorded as hosts, but the majority of hosts are species and varieties of tuberous *Solanum*. The only susceptible plant outside the family is snap dragon, *Antirrhinum majus*.

(v) Interaction with Other Pathogens: The potato cyst nematode and the black-scurf fungus *Rhizoctonia solani* together caused much greater reduction in yield than did the organisms separately. The cyst nematode also interacts with *Verticillium dahliae*, enhancing the severity of the wilt caused by the fungus. The leaf symptoms appeared earlier and the tuber yields were depressed in the presence of the nematode.

The potato cyst nematodes, *Globodera rostochiensis* and *G. pallida* in association with the wilt bacterium, *R. solanacearum* causes wilt complex in potatoes.

(vi) Ecology: Well-drained, aerated sands, silts and peat soils with a moisture content of 50–75% water capacity are suitable for survival and movement of the free-living stages of the potato cyst nematode. A minimum soil temperature of 10°C is necessary for larval activity.

(vii) Spread and Survival: The cysts can be spread by the movement of soil through seed material for a very long distance. They can also be spread through field tools, implements, seedlings grown in infested soil for transplanting elsewhere, labour's feet, etc. from one field to another. The cysts can remain viable in the soil without a host for 30 years.

(viii) Physiological Races: At present, five pathotypes in *G. rostochiensis* and four pathotypes in *G. pallida* are differentiated using the following differentials (Table 2.7):

Table 2.7. Pathotypes of *Globodera rostochiensis* and *G. pallida*

Differential host	Nematode pathotypes*				
	1	2	3	4	5
	Globodera rostochiensis				
Solanum tuberosum	+	+	+	+	+
ex *andigena* CPC 1673	−	+	+	−	+
ex *kurtzianum* 60.21.19	−	−	+	+	+
ex *vernei* 58.1642/4	−	−	−	+	+
ex *vernei* 62.33/3	−	−	−	−	+
ex *vernei* 65.346/19	−	−	−	−	−
	Globodera pallida				
Solanum tuberosum	+	+	+	+	
ex *multidissectum* P 55/7	−	+	+	+	
ex *vernei* 62.33/3	−	−	+	+	
ex *andigena* x *multidissectum* D 47/1	−	−	−	+	
ex *andigena* x *multidissectum* D 47/2	−	−	−	−	

* (+) indicates ability to produce females; (–) indicates inability to produce females

In India, potato cyst nematodes are restricted to Nilgiri and Palani hills in South India, where 5 pathotypes—Ro 1, Ro 2 of *G. rostochiensis*, and Pa 1, Pa 2 and Pa 3 of *G. pallida* have been detected (Amalraj and Rao, 1992).

(ix) Management

(a) Regulatory Methods: In India, there is a Quarantine Act against the cyst nematode of potato (*G. rostochiensis*) in Nilgiris. Infected potatoes are not allowed to be transported from Nilgiris to other parts of India for seed and table purposes in order to prevent the spread of potato cyst nematode from Tamil Nadu to other states and Union Territories.

Potato seed pieces free of the cyst nematode can be produced commercially by seed certification. Use of certified planting material is fundamental to the prevention of cyst nematodes spreading into uninfested regions.

(b) Physical Methods: Fassuliotis and Sparrow (1955) demonstrated that irradiation of potato tubers with *x*-rays inhibited sprouting and also development of *G. rostochiensis*. Cysts of this nematode exposed to 20,000*r* contained only brown and dead eggs, while at 40,000*r* the eggs lost their contents completely.

Careful washing of soil adhering to potato tubers can considerably reduce the risk of spreading cyst nematodes. Commercial potato tuber washing apparatus has been developed.

(c) Cultural Methods: The potato cyst nematode (*G. rostochiensis*) can be eliminated by selecting nematode-free planting material.

Early Maturing Varieties: Early maturing varieties of potato like Irish Cobbler usually suffer less than do late maturing varieties. These varieties start growth in spring at soil temperature too low for much nematode activity and may be harvested before many of the nematodes can reproduce.

Crop Rotation: Continuous cropping with potatoes increases the cyst population of *G. rostochiensis*. More than three year rotations with wheat, strawberry, cabbage, cauliflower, peas, maize and beans reduces the nematode population to a safe level. Growing non-host crops like radish, garlic, beet root, French bean, cruciferous vegetables, turnip or green manuring crops bring down the cyst nematode population by more than 50%. A crop rotation pattern (using potato-French bean-peas-potato), wherein potato is grown once after 3–4 other crops, decreased the nematode population by 98 to 99% and increased yields of potato by 90%. In Nilgiris, crop rotation with cabbage and carrot gave effective control of cyst nematodes. Growing of vegetables and resistant potato in a long term rotation trial brought down the cyst propagule population up to 92% and was effectively used for nematode management at Nilgiris.

Botanicals: Incorporation of eucalyptus litter at 10 t/ha at the time of preparation of land reduces the population of the potato cyst nematodes. Application of eucalyptus distillation waste at 2.5 t/ha helps to reduce the cyst nematode population in soil.

(d) Chemical Methods: Application of carbofuran at 3 kg *a.i.*/ha or aldicarb at 2 kg *a.i.*/ha or ethoprophos at 10 kg *a.i.*/ha was found to reduce the infestation of root-knot and cyst nematodes in potato crop. The yield increase ranged from 18 to 27%. The efficacy of these treatments further increased when applied in equal split doses.

Application of aldicarb at 2 kg *a.i.*/ha as spot treatment was the most economical dose which ensured reasonable level of nematode control while increasing yields (14 to 219% in summer and 15 to 115% in autumn), followed by carbofuran (18 to 206% in summer and 10 to 115% in autumn) and phenamiphos (15 to 105% in summer and 10 to 90% in autumn). The least build up of cyst nematode population was noticed in aldicarb and carbofuran (Rf = 0.5 to 1.2) followed by phenamiphos (Rf = 1.5 to 1.9) and control (Rf = 4.7 to 9.2).

(e) Biological Methods: Application of *Pseudomonas fluorescens* at 10 g/m^2 gives good control (Mani *et al.*, 1998). Application of neem cake at 1 t/ha or FYM at 20 t/ha or compost at 10 t/ha enriched with *Trichoderma harzianum/Paecilomyces lilacinus* gave effective control of cyst nematodes.

(f) Host Resistance: In the United States, both USDA and Cornell University scientists have evolved two commercial potato varieties–Peconic and Wauseon–which are resistant to the cyst nematode.

In India, Potato variety Kufri Swarna developed by the Central Potato Research Institute Substation at Ooty is resistant to both the species of cyst nematode and the late blight disease. The variety is also draught resistant and has shorter dormancy which is advantageous to Nilgiris region where potato is taken up mainly as a rain fed crop almost throughout the year and thus covers about 40% of potato area (Anon, 1985).

(g) Integrated Methods: The combination of disease escape (by planting early maturing varieties) and the use of nematicides gave good control of the potato cyst nematode. Hygiene in the form of seed certification, combined with crop rotation in seed growing areas is effective. Use of a resistant variety followed by a nematicide would kill 99% of the nematode population. The potato cyst nematode problem in Nilgiris is being managed by chemical treatments, crop rotations and utilizing the available sources of resistance (Kufri Swarna). The most effective management combines crop rotation, use of nematicides and resistant varieties to keep the nematode at an economically acceptable level (Jones, 1969) (Table 2.8).

Table 2.8. Integrated management of the potato cyst nematode

Sl. No.	Management method(s)	Resulting kill population (% initial population)	Kill (%)	Population after growing & harvesting a susceptible *cv.*, calculated at 2 assumed nematode multiplication rates (% initial population)	
				30x	70x
1.	4 years without potato	3	97	90	< 100
2.	1 year with resistant Potato	20	80	> 100	> 100
3.	Nematicide treatment (Fumigant)	25	75	> 100	> 100
4.	1 and 2	0.6	99.4	18	42
5.	1 and 3	0.75	99.25	22.5	52.5
6.	2 and 3	5	95	> 100	> 100
7.	All 3 methods	0.15	99.85	4.5	10.5

Better control of *G. rostochiensis* by *P. penetrans* was possible with soil solarization combined with application of FYM or karanj cake which resulted in greater kill of nematodes (Sitaramaiah and Naidu, 2003).

A well known integration of methods can control *G. rostochiensis* in potato in a 4-year rotation combining soil fumigation (causes 30% reduction in nematode population), cultivation of non-host (50% reduction), cultivation of resistant potato variety (30% reduction), cultivation of non-host (50% reduction) and then cultivation of susceptible variety of potato (Oostenbrink, 1972).

2.1.3.3. Potato Rot Nematode, *Ditylenchus destructor*

Ditylenchus destructor is the most important pest of potato tubers and is responsible for dry rot of tubers. High yield losses occur in the areas where climatological conditions favour establishment of the potato rot nematodes. The effect of nematodes will manifest itself at harvest or storage when infected tubers will rot.

(i) Symptoms: *Ditylenchus destructor* enters potato tubers through lenticels and initially causes small white mealy spots just below the surface that are only visible if the skin is removed. Infested areas enlarge and coalesce and light brown lesions, consisting of dry granular tissue, may be visible beneath the skin. As the infestation progresses, the tissues dry and shrink and the skin becomes cracked and papery (Fig 2.17). Internal tissues gradually darken and there are often secondary invasions of fungi, bacteria, mites, etc. If stored in moist conditions, a general rot may ensue and spread to neighbouring tubers.

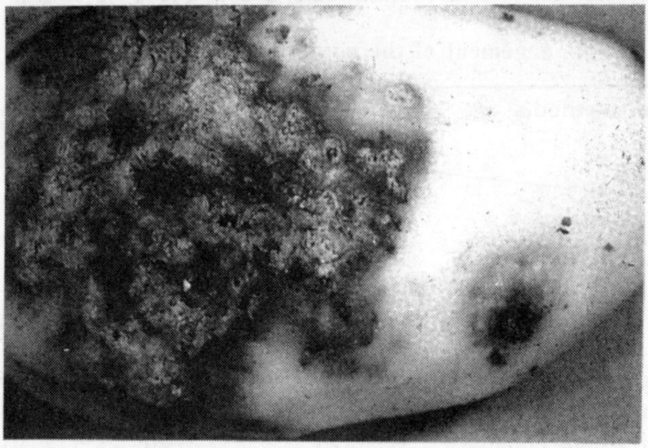

Figure 2.17. Dry rot of potato tuber incited by *Ditylenchus destructor* (Courtesy: Sasser, 1971)

(ii) Life Cycle: The life cycle is very similar to *D. dipsaci*. *D. destructor* does not seem to have a resistant stage, as it does not form 'nematode wool' or readily withstand desiccation but apparently survives in soil on alternative weed and fungal hosts. Thorne (1961) suggests that it may over winter as eggs.

(iii) Survival and Spread: *Ditylenchus destructor* has a wide host range, can survive on weeds and on a wide range of soil-inhabiting fungi. It can also survive on infected tubers left in the field. Spread occurs by introduction of infected tubers and in soil adhering to seed pieces. Irrigation water and cultivation by infested farm tools and machinery are other sources of inoculum dissemination.

(iv) Management

(a) Cultural Methods: The use of healthy seeds, destruction of infected plant parts in the field, control of weed hosts, late planting when soil moisture is inadequate for optimum invasion and early harvest reduces losses. Two to three years rotation of potato with beans, corn and fallow kept free of weeds gave a very marked reduction in nematode populations. Satisfactory control of *D. destructor* can also be obtained by crop rotation with small grains, vetch and lupine provided potatoes are grown once in 3–4 years. Rotation of potatoes with sugar beet and other non-host crops can reduce nematode populations.

The application of ammonium nitrate and ammonium sulphate (at 144 and 118 kg/ha, respectively) while planting potato reduced infection from 7% to 0.2–0.5%, and raised the yield from 10.6 t/ha to 16.3 and 25.2 t/ha, respectively.

(b) Chemical Methods: Darling (1959) found that heavily infested land treated with ethylene dibromide at 35 litres/ha in June followed by 17.5 litres/ha in August, gave clean potato crops in the following two years.

(c) Integrated Methods: The control of potato rot nematode was achieved by the combination of disease escape, hygiene, in the form of seed certification and crop rotation as follows (Winslow and Willis, 1972):

- Healthy seed, planted late, harvested early and stored as cool and dry as possible.
- Proper rotation of potatoes with non-host crops and growing of potatoes not more frequently than once in 3 or 4 years.
- Field hygiene in the form of removal of old infested tubers and weed control.

2.1.3.4. Lesion Nematode, *Pratylenchus penetrans*

(i) Symptoms: High populations of lesion nematodes cause areas of poor growth; plants are less vigorous, turn yellow and cease to grow. Damage is often caused by direct feeding and, usually, only cortical tissues are affected. Large nematode populations cause extensive lesion formation and cortex destruction of unsuberized feeder roots (Mai *et al.*, 1981). Tubers are often attacked and small lesions are formed on the surface.

(ii) Survival and Spread: Infected tubers are sources of nematode inoculum and aid in survival of the nematodes.

(iii) Management

(a) Physical Methods: Hot water treatment of infected tubers at 50°C for 45 to 60 min. aid in reducing nematode spread.

(b) Cultural Methods: In areas infested by *P. penetrans*, beet should precede the susceptible potato in the crop rotation system. The cultivation of legumes should be avoided since nematodes breed well in them.

(c) Host Resistance: Potato *cv.* Russet Burbank is tolerant, while Peconic appears to be less susceptible to the lesion nematode.

2.1.3.5. Other Nematodes, *Quinisulcius capitatus* and *Helicotylenchus dihystera*

The pathogenicity of the stunt nematode, *Quinisulcius capitatus* frequently occurring in hill tracts was established on potato *cv.* 'Kufri Jyoti' at Shimla (Prasad and Sharma, 1985). The nematode build up ranged from 5 to 8 times at initial inoculum levels of 10 to 1000 at 45 days. Besides, the plant growth parameters were also reduced which affected the tuber production. The tuber reduction ranged from 14 to 29% and was negatively correlated with nematode build up.

The spiral nematode, *Helicotylenchus dihystera* was also pathogenic to potato accounting for 9–27% reduction in yield in 90 days with 2–4 times nematode build up (Prasad, 1990).

2.1.3.6. Root-knot Nematode, *Meloidogyne* spp. and Wilt, *Verticillium dahliae* Disease Complex

The root-knot nematodes often interact with *V. dahliae* in development of disease complex in potato.

(i) Management

(a) Cultural Methods: In fields infested with root-knot nematodes and *V. dahliae,* Scholte (1990) found that a five-year rotation of maize-sugar beet-barley-barley-potato had significantly higher potato yields than continuous potato.

2.1.3.7. The Lesion Nematode, *Pratylenchus penetrans* and Wilt, *Verticillium dahliae* Disease Complex

(i) Management

(a) Chemical Methods: Methylisothiocyanate has been successfully used for soil fumigation to reduce the populations of soil-borne fungus, *V. dahliae* and alleviate the effects of disease complex involving wilt-inducing fungus (*V. dahliae*) and the lesion nematode (*P. penetrans*) on potato (Rowe *et al.*, 1987).

2.1.3.8. The Cyst Nematode, *Globodera rostochiensis* and Wilt, *Verticillium dahliae* Disease Complex

The association of fungi like *Rhizoctonia solani* and *V. dahliae* with *G. rostochiensis* has been reported on potato indicating that the cyst nematode is involved in the disease complex.

(i) Management

(a) Chemical Methods: Soil fumigation with methyl bromide resulted in yield increase of 68% for potato fields infested with *Globodera* spp. and *V. dahliae* (Hide and Corbett, 1974).

In *V. dahliae-G. rostochiensis* disease complex on potato, a mixture of methyl bromide and chloropicrin was found to have comparatively more fungicidal and nematicidal efficacy (Hide and Corbett, 1974).

2.2. TOMATO

Tomato (*Lycopersicon esculentum*) is the most important and remunerative vegetable crop in India. It is being cultivated in 0.535 million hectares producing 9.362 million tonnes of tomatoes with an average yield of 17.5 tonnes per hectare (Bijay Kumar, 2007) (Tables 2.9 and 2.10). Andhra Pradesh, Orissa, Karnataka, Maharashtra, West Bengal, Bihar, Gujarat, Chattishgarh, Madhya Pradesh, Tamil Nadu and Haryana are the major tomato-growing states in India. A rich source of minerals, vitamins and organic acids, tomato fruits provides 3–4% total solids, 15–30 mg/100 g ascorbic acid, 7.5–10.0 mg/ 100 ml titratable acidity and 20–50 mg/100 g fruit weight of lycopene. Tomato fruits contain 0.4% protein, 2.5% carbohydrates, 1.5 mg iron and 2 mg vitamin C in 100 g of fresh weight. Fruits are good for people suffering from constipation, jaundice and indigestion.

Table 2.9. Major tomato producing countries of the world

Country	Area (in '000 ha)	Production (in '000 MT)	Productivity (in MT/ha)	% Share of world production
China	1305	31626	24.2	24.7
USA	167	11043	66.3	8.6
Turkey	260	10050	38.7	7.9
India	535	9362	17.5	7.3
Egypt	195	7600	39.0	5.9
Italy	139	7187	51.8	5.6
Spain	72	4810	66.5	3.8
Iran	139	4781	34.4	3.7
Brazil	61	3453	57.0	2.7
Mexico	119	2800	23.6	2.2
Others	1624	35281	21.7	27.6
Total	**4615**	**127993**		

Source: FAO (except Indian data) (Indian data source: Indian Horticulture Database 2006)

Table 2.10. Area, production and productivity of leading tomato growing states in India (2005–06)

State	Area (in '000 ha)	Production (in '000 MT)	Productivity (MT/ha)
Andhra Pradesh	76.5	1453.5	19.0
Orissa	100.4	1332.2	13.3
Karnataka	44.5	1188.1	26.7
Maharashtra	35.0	987.0	28.2
West Bengal	50.0	857.2	17.2
Bihar	46.0	727.2	15.8
Gujarat	29.3	650.0	22.2
Chattishgarh	29.2	365.8	12.5
Madhya Pradesh	20.4	306.7	15.0
Tamil Nadu	22.0	277.7	12.6
Haryana	17.1	257.3	15.0
Others	64.1	959.1	15.0
Total	**534.5**	**9361.8**	**17.5**

(*Source*: Indian Horticulture Database 2006)

2.2.1. Pests

2.2.1.1. Fruit Borer, *Helicoverpa armigera*

(i) Damage: Tomato fruit borer is a polyphagous pest infesting tomato, reducing marketable yield and market value of the crop. This pest has already developed resistance to a number of insecticides in different crops, and is thus difficult to manage. Losses up to 50% in Tamil Nadu (Srinivasan, 1958); 65% in Punjab (Singh and Singh, 1975) and 22 to 38% in Karnataka (Tewari and Krishna Murthy, 1984) have been reported due to this pest attack.

The adult of this pest is stout and medium sized moth and has a dark circular spot in the center on the forewing. They lay small, single, and whitish, round eggs on the trifoliate leaves beneath the top most flower cluster. Eggs hatch in about 3–4 days and the first instar larvae initially feed on the leaves and later migrate to the developing green fruits. Later the larvae bore into the fruits with the posterior end outside the hole (Fig 2.18). Full-grown caterpillars show characteristic whitish and dark brown longitudinal stripes. Pupation takes place in the soil. Eight or more eggs/30 trifoliate leaves below the top most flower cluster was observed to cause economic loss.

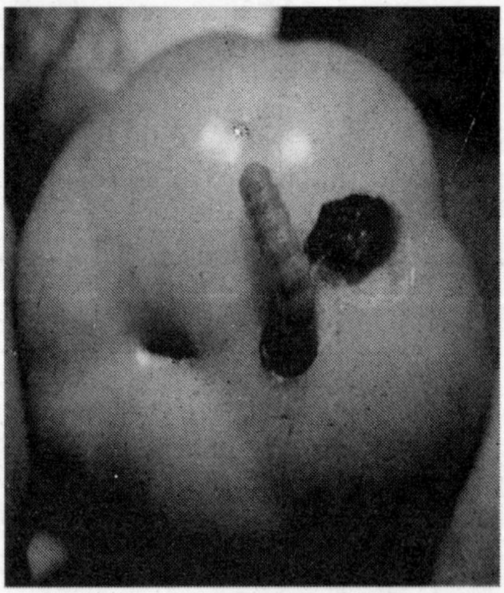

Figure 2.18. Fruit borer damage on tomato

(ii) Management

(a) Cultural Methods: Mechanical collection and destruction of bored fruit at periodic intervals (3–4 times) brings down the borer incidence to less than 2%.

Soil application of neem cake at 250 kg/ha at flowering reduced the borer incidence to 13.2% as compared to 33.2% in control. Hence, soil application of neem cake at flowering may be used as a component of IPM against tomato fruit borer.

Foliar sprays of neem soap and pongamia soap both at 1% starting from 25 DAP and repeated four more times at 10 days interval reduced the borer incidence to less than 7% as compared to 33% in unsprayed control plots (Krishna Murthy and Krishna Kumar, 2002) (Table 2.11). The soap sprays were also effective in reducing whitefly, red spider mite and leaf miner incidence.

Table 2.11. Efficacy of neem and pongamia soaps on tomato fruit borer

Treatment	Dose	Mean fruits bored (%)
Neem cake	250 kg/ha	13.21
NSKE	4%	11.12
Neem oil	1%	13.24
Pongamia oil	1%	13.76
Soluneem	600 ppm	7.97
Neem soap	1%	6.64
Pongamia soap	1%	6.96
Control	—	33.23

(b) Biological Methods: The egg parasitoid, *Trichogramma brasiliensis* gave as high as 90% parasitism when 2.5 lakh/ha were released, while *T. pretiosum* also proved effective at 2.5 lakh/ha and resulted in less than 2% fruit damage with 60–80% parasitism (Krishnamoorthy and Mani, 1996). In Solan, Himachal Pradesh region, *T. chilonis* gave 90–100% parasitism following releases. In Gujarat, fruit borer was checked by releasing *T. chilonis* at 2.5 lakh/ha at weekly interval during Dec-Feb.

Five releases of *T. brasiliensis* at 50,000/ha commencing the first release at the time of flower initiation and repeated at weekly intervals resulted in egg parasitism up to 78.4% as compared to 12.3% in control. Percentage bored fruits in parasitoid released plots was 19% in comparison with 42% in control plots. Releases of *T. brasiliensis* and *T. pretiosum* each at 2.5 lakhs/ha resulted in significant reduction in borer damage to 8.92 and 7.20, respectively as compared to 23.06 and 13.72% in control (Table 2.12).

Table 2.12. Efficacy of *Trichogramma* spp. against *Helicoverpa armigera* on tomato

Trichogramma spp.	Dosage/ha (in lakhs)	Parasitism (%)
T. brasiliensis	5.0	100
	2.5	20–71
	3.0	78.4
T. chilonis	2.5	20–96
	0.63–10.88	76.3
T. exiguum	5.0	100
T. pretiosum	5.0	100
	0.55–10.58	92.4
	2.5	31.3

Application of 5 rounds of *Ha* NPV at 250 LE/ha at weekly intervals commencing first spray on flower initiation is needed to check the borer very effectively.

Larval parasitoids like *Campoletis chloridae, Eriborus argentiopilosus* and *Senometopia illota* and the nematode, *Hexamermis* sp. cause 10 to 40% mortality of fruit borer in tomato fields.

The tomato fruit borer can be effectively controlled by sprays of *Bacillus thuringiensis* (Dipel) at 0.5 kg/ha at 10 days interval. The borer infestation was found to be significantly low (7.41%) in 'Dipel' sprayed plots as against 21 to 30% in control (Krishnaiah *et al.*, 1981).

Five sprays of *Ha* NPV at 250 LE/ha gave 80.62% mortality of fruit borer as compared to 1.14% in control. The fruit damage was 4% in NPV treated plots as against 15% in the control plots (Anon., 1983).

Five rounds of sprays of *Ha* NPV at 250 LE/ha in combination with adjuvants and feeding stimulants at weekly intervals commencing the first spray on flower initiation were effective in increasing the mortality of tomato fruit borer, *H. armigera* to 75–85% in the treated plots as against 0 to 5.7% in control plots. The fruit damage was 4% in NPV treated plots as against 15% in control plots. Application of 3 rounds of NPV at 250 LE/ha along with adjuvants *viz.,* 1% soya flour and 0.01% soya oil applied during evening hours at weekly intervals, right from the initial observation of few eggs per plant, recorded significant reduction in tomato fruit borer damage (2.2%) and increase in yield (11.34 kg) as compared to control (21.7% fruit damage and 7.64 kg yield) (Narayanan, 1985).

Field applications of 5 rounds of *Nomuraea rileyi* at 3.2 × 10^8 conidia/ml along with Triton X-100 (0.01%) at weekly intervals during evening hours effectively controlled the fruit borer. The treatment reduced the fruit borer damage to 6.76% when compared to control (41.16%) and insecticide (8.02%) (Gopalakrishnan and Mohan, 1996).

(c) Integrated Methods: Use of African marigold (*Tagetes erecta*) as a trap crop for the management of fruit borer on tomato involves planting one row of 45 day-old marigold seedlings after every 16 rows of 25 day-old tomato seedlings (Fig 2.19) and spraying of *Ha* NPV at 250 LE/ha or 4% NSKE or 4% pulverized NSPE, 28 and 45 DAP coinciding with peak flowering.

Figure 2.19. Integrated management of tomato fruit borer using African marigold as a trap crop

For effective control of *H. armigera*, the egg parasitoid, *T. pretiosum* and *Ha* NPV could be integrated to tackle different stages of the pest. *T. pretiosum* was released at 2.5 lakh/ha and *Ha* NPV was sprayed 2 times at 250 LE/ha for effective control of the pest (4.18% fruit borer damage as compared to 21.79% fruit borer damage in control). The increase in yield was 65.5% over control and the reduction in fruit borer damage was 80.8% over control.

Three releases of *T. pretiosum* + 3 sprays of *Bacillus thuringiensis* at 1 kg/ha was found highly effective against fruit borer in Himachal Pradesh. The release of *T. pretiosum* (2.5 lakh/ha) + 2 sprays of *Ha* NPV (250 LE/ha) was found effective for the management of fruit borer.

Use nylon nets (40 gauge) to avoid insect vectors. *Ha* NPV was sprayed twice at 28 and 35 DAP. Spray of pongamia soap at 1% was given at 40 DAP for the management of fruit borer and leaf miner. The bored fruits were removed mechanically once 40 DAP. The fruit borer and leaf miner incidence were 2% and 0.7–2.7 mines/leaf, respectively.

2.2.1.2. Serpentine Leaf Miner, *Liriomyza trifolii*

(i) Damage: This is a polyphagous pest with a wide host range affecting many vegetable crops like tomato, beans, cowpea, okra, cucurbits *etc*. Damage by *L. trifolii* is severe during summer and rainy months in South India. Damage is more on hybrid tomatoes than open pollinated varieties. A computerized software on serpentine leaf miner called 'SLMINFO' has been developed at the Indian Institute of Horticultural Research, Bangalore, which gives information on host plants, insecticides used, resistance to insecticides, natural enemies, etc.

Serpentine leaf miner is a tiny, metallic fly which punctures the leaf lamina and feed on the oozing sap. Adult is 2–3 mm in size. Damage is severe on cotyledons and young leaves. It lays eggs on the outer margin of leaves. Within 2–3 days, whitish maggots hatch out of these eggs and start mining the leaves and pupate in 6–10 days. Pupation takes place in the soil and occasionally on the leaf surface itself. Pupal period is about 7–8 days. Typical serpentine shaped tunnels are formed in the leaf lamina indicating the path of feeding by the maggots (Fig 2.20).

(ii) Host Range: Economically important crop plants include cucumber, snake gourd, bottle gourd, pumpkin, watermelon, French bean, cowpea, okra, palak, gerbera, marigold, etc. (Virakthmath *et al*., 1993).

(iii) Management

As this pest tends to develop pesticide resistance, judicious use of pesticides is recommended.

- Often the incidence starts from nursery itself. Hence, remove infected leaves at the time of planting or within a week of transplanting.
- Apply neem cake to furrows at 250 kg/ha while planting and repeat after 25 DAP at flowering.

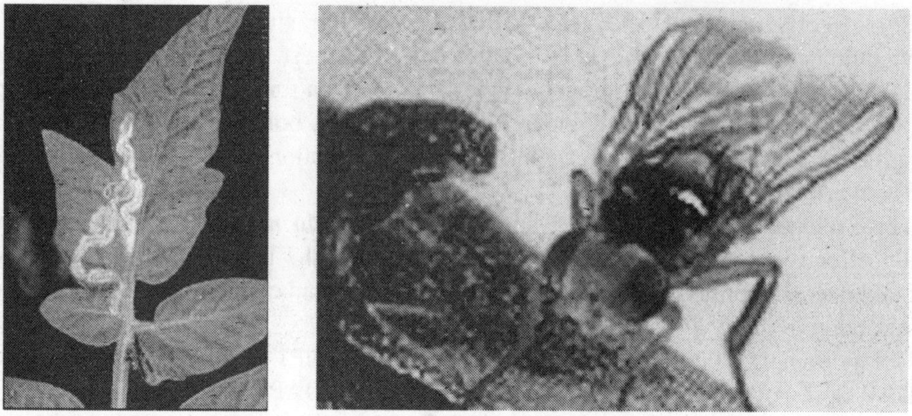

Figure 2.20. Serpentine leaf miner on tomato leaf and adult

- Spray 4% neem seed powder extract or 1% neem soap at 15–20 DAP.
- When the incidence is high, remove infected leaves first and mix 0.05% triazophos or 0.015% decamethrin with 7.5 g neem/litre and spray.
- Avoid indiscriminate use of insecticides so as to facilitate build up of the natural enemies.
- Avoiding excess application of N-fertilizers.
- Use of yellow sticky traps help in trapping adult flies.
- Tomato variety Sheetal was less attacked followed by Rupali, Rashmi and Naveen.
- *Hemiptarsenus varicornis* is the most predominant indigenous parasitoid with 5–15% parasitism (Krishna Kumar *et al.*, 1998).

Among the hosts, bittergourd (*Memordica charantia*) is least preferred and this information can potentially be used in the management of *L. trifolii* as a repellant crop around most susceptible tomato plants (Krishna Kumar *et al.*, 1997).

2.2.1.3. Red Spider Mites, *Tetranychus urticae*

(i) Damage: Tiny red mites feed mainly on the lower surface of the leaves by scrapping the epidermis causing yellowing of leaves followed by foliage drying. This pest was not at all considered as a pest of tomato during the least century. However, it has become a major pest of many vegetables like, tomato, brinjal, beans, okra, cucurbits etc. in recent times most probably due to indiscriminate use of pesticides and cultivation of vegetables during off season, particularly during summer.

(ii) Management

- IPM plots (using marigold as a trap crop and spraying 4% NSKE) were found to have no or very low incidence of this pest. Therefore, following IPM is ideal for preventing the outbreak of mite incidence.

- Generally incidence will be noticed in some patches in a plot. Remove the affected leaves immediately and spray dicofol 18.5 EC at 2.5 ml or any other acaricide like wettable sulphur 80 WP at 3 g/l.
- Spray lower leaves and lower leaf surface thoroughly as mites are generally observed there.
- As an alternative to the chemical acaricides spray neem oil/neem soap/pongamia soap each at 1%.
- When incidence is severe, first remove and destroy all severely infected leaves and then mix an acaricide with botanical and give a thorough spray.

2.2.1.4. Whitefly, *Bemicia tabaci*

(i) Damage: White fly is a well known vector which transmits tomato leaf curl virus and causing severe loss in yield. Sastry and Singh (1971) reported loss in yield ranging from 38 to 93% during different months of the year.

They have piercing and sucking mouthparts and both nymphs and adults feed on lower surface of the leaves causing deformation of young leaves (Fig 2.21). Severe cases leads to chlorotic leaves accompanied by wilting. Whiteflies also excrete honeydew, causing sooty mold.

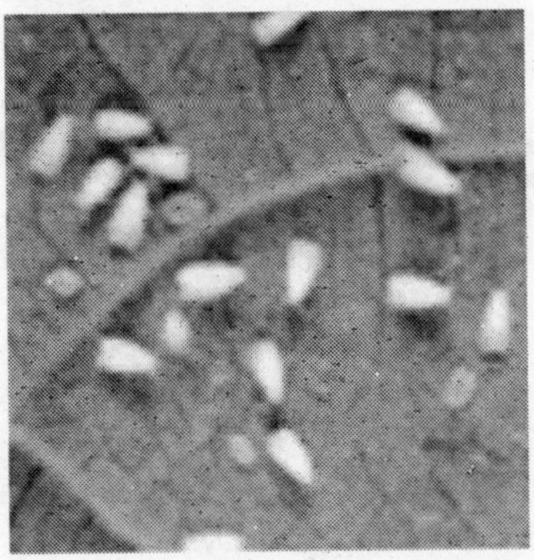

Figure 2.21. Whieflies on tomato leaf

(ii) Management

- Use virus resistant hybrids.
- Raise nurseries in nylon nets/polyhouses.
- Dip the roots with a systemic insecticide like imidacloprid (0.3 ml/l) or thiomethoxam (0.3 g/l) for 5 min before planting. If portrays are used for raising nursery, drench the portrays with the chemicals one day before transplanting.
- Fifteen days after planting, spray with the systemic insecticides used for root dipping.
- Rogue out the virus affected plants as soon as the symptoms are observed.

2.2.1.5. Thrips, *Thrips tabaci, Frankliniella occidentalis*

(i) Damage: The characteristic symptoms of thrips (Fig 2.22) damage are crinkling, cupping and upward curling of leaves. Flowers drop at later stage. The pest transmits tomato spotted wilt virus.

Figure 2.22. Thrips on tomato

(ii) Management

Spraying endosulfan or monocrotophos each at 2 ml/litre of water gives effective control of thrips.

IPM

The IPM package given below will take care of fruit borer, leaf miner, mite and whitefly (insect vector).

Nursery

- Raise marigold (tall African variety 'Golden Age' bearing yellow and orange flowers) nursery 15–20 days before tomato nursery.
- One week after germination of seeds, spray the seedlings with imidacloprid 200 SL at 0.3 ml/l or thiomethoxam 25 WP at 0.3 g/l.

Before transplanting

- Apply neem cake 250 kg/ha on ridges at the time of preparing land.
- Dip the roots of seedlings (do not dip the foliage as it may cause burning of leaves) with imidacloprid 200 SL at 0.3 ml/l or thiomethoxam 25 WP at 0.3 g/l for 5 min.

Main field

- Transplant 20–25 days old tomato and 45–50 days old marigold simultaneously in a pattern of one row of marigold for every 16 rows of tomato. However, the first and last row of the plots should be of marigold. Simultaneous flowering of both the crops ensures attraction of fruit borers to marigold flowers.

- Fifteen days after planting, spray imidacloprid 200 SL at 0.4ml/l or thiomethoxam 25 WP at 0.3g/l for leaf curl vector (whitefly) control.
- Apply neem cake at 250 kg/ha to ridges at 20–25 DAP (at flowering) to reduce nematode, fruit borer and leaf miner incidence.
- Spray *Ha* NPV at 250 LE/ha with 1% jaggery as sun screen at 28, 35 and 42 DAP in the evening.
- Spray marigold flowers with *Ha* NPV or destroy fruit borer larvae in them.
- As an alternative to *Ha* NPV spray, the egg parasitoids, *Trichogramma chilonis, T. braziliensis* and *T. pretiosum* at 2.5 lakhs/ha can be released (five releases at 50,000/ha/release). The first release has to be done at the flower initiation of the crop.
- If red spider mite incidence is noticed, spray neem soap 1% or neem oil 1% or any synthetic acaricide like dicofol 18.5 EC (at 1.5 ml/l), or ethion 50 EC (at 1.5 ml/l) or sulphur 80 WP (at 3 g/l) etc. Spray lower surface of the leaves.
- Mechanical collection and destruc n of bored fruits at periodic intervals (3–4 times after fruit set) to minimize fruit borer incidence to the minimum.
- Destroy leaf curl and other virus affected plants as soon as the symptoms appear in a few plants to minimize their spread.

Pesticide Residues

The persistence of terminal pesticide residues and waiting periods for tomato are presented in Table 2.13. (Awasthi, 2001).

Table 2.13. Terminal pesticide residues and their waiting periods for tomato

Pesticide schedule	Dosage/ Concentration (%)	Terminal residues (ppm/days)	Waiting period (days)
Monocrotophos spray at fruiting stage	0.05	0.14/15	10
Malathion spray at fruiting stage	0.10	0.08/7	3
Endosulfan spray at fruiting stage	0.07	0.12/20	7
Cypermethrin spray at fruiting stage	0.01	0.22/10	5
Fenvalerate spray at fruiting stage	0.01	0.08/15	10
Carbofuran soil application at planting	0.5 kg *a.i.*/ha	0.09/60	Nil
Phorate soil application at planting	1.0 kg *a.i.*/ha	0.85/75	Safe at 2nd harvest

2.2.2. Diseases

2.2.2.1. Damping-off, *Pythium aphanidermatum*

(i) Symptoms: In the pre-emergence phase of the disease, the young seedlings are killed even before they emerge out of the soil surface. However, this disease mostly occurs at post-emergence stage which is characterized by the toppling over of the infected seedlings anytime after they emerge from the soil until the stem has hardened sufficiently to resist invasion by the pathogen. Infestation usually occurs at ground level or through roots. The infected tissue appears soft and water-soaked. As the disease advances, the stem becomes constricted at the base of the plants that collapse later. Seedlings that are healthy looking one day may have collapsed the next morning.

(ii) Management

Use disease-free seeds for raising seedlings. Treat the seeds with chlorothalanil or captan or metalaxyl at 2 g/kg. Pick rotten fruits and destroy. Spray 0.2% metalaxyl or mancozeb just before onset of monsoon and 1% Bordeaux mixture or 0.3% copper oxychloride at 8–10 days interval. Spraying of 0.2% cymoxanil, propineb mancozeb or chlorothalanil is also helpful. Application of neem cake along with *Trichoderma* spp. may reduce the soil inoculum.

Cultural practices like rabbing, thin sowing, use of light soil, light but frequent irrigation and use of well decomposed FYM should be adopted. *Streptomyces distaticus* was effective in suppressing damping-off of tomato. Success in suppression of damping-off in tomato was achieved through soil and seed application of *Trichoderma virens* and *T. longibrachiatum*.

Seed treatment with *Trichoderma viride* and *Pseudomonas fluorescens* and addition of neem cake to the nursery beds enhanced seed germination and seedling stand. Soil drenching with 1% Bordeaux mixture at 1 litre/m^2 helps in reducing the disease incidence.

Soil solarization using polythene sheet for 30 days during peak summer was most effective and resulted in least occurrence of damping-off (3.9%) when compared to control (30.64%). Seed treatment with *Trichoderma viride* resulted in lowest incidence of damping off (5.7%).

Delivery of *T. harzianum* through solid matrix priming reduced the incidence of damping off in nursery. SMP is a process in which moistened seeds are mixed with an organic carrier and the moisture content is brought to a level just below that required for seed sprouting. SMP can further enhance the efficacy of the antagonist in treated seeds. *Trichoderma* spp. grow on seed surface during priming process and increase in numbers and other microbes may not easily dislodge it.

2.2.2.2. Early Blight, *Alternaria solani*

(i) Symptoms: It not only affects leaves but also the fruits. Symptoms appear on all above ground parts of the plant causing pre and post emergence damping off in seed beds. Symptoms appear as small, isolated scattered pale-brown spots on leaves. They enlarge and develop into concentric rings in the necrotic tissue and induce yellowing leading to leaf blight (Fig 2.23). Girdling of the stem at the base of the plant is also observed. The stem lesions are usually restricted to one side of a stem and become elongated and sunken. Similar symptoms also appear on the fruit causing fruit rot or stem end rot or blossom end rot. Under favourable conditions to the pathogen, losses may go up to 70%. It is more severe in June-July sown crop. In severe attacks, the entire plant may be defoliated, exposing the fruits to sun scald.

Figure 2.23. Early blight on tomato leaves

(ii) Management

The pathogen is soil as well as seed-borne. Use disease-free seed for raising seedlings. Healthy seeds should be collected only from disease-free fruits. Treat seeds with thiram 75 WP at 2.5 g/kg seed.

The crop debris and infected plant parts should be destroyed immediately after harvest by burning. Summer ploughing, avoiding susceptible crops adjacent to tomato, maintaining plant vigour, minimizing injury to plant are essential for checking the disease.

Give wider spacing of 90 × 60 cm in order to minimize the relative humidity in plant canopy. Remove the lower diseased leaves periodically and follow field sanitation.

Spray 0.2% chlorothalanil or 0.2% propineb or zineb or 0.3% copper oxychloride or 0.2% iprodion + carbendazim at 8–10 days interval starting at flowering stage.

2.2.2.3. Late Blight, *Phytophthora infestans*

The disease is more common in north-western India but also appears in Uttar Pradesh and Bihar in the months of December-January.

(i) Symptoms: Symptoms on leaves, twigs and fruits are more common. Blight appears on foliage as light-pale green water-soaked area. The lesions are wet in the morning and dried, shriveled in daylight. It enlarges rapidly until the entire leaflets are killed. In moist and cloudy weather, a downy white fungus develops near the outer margin of the lesion on underside of leaves. Green and ripe fruits affected by the blight show olivacious, greasy appearing spots which may expand until the entire fruit is covered.

(ii) Management

Certified seeds should always be used. Infected crop debris and fruits should be collected from the field and burnt.

When environmental conditions are favourable for late blight incidence, preventive sprays of mancozeb at 0.25% at 5–7 days interval provide good control. Spraying of metalaxyl + mancozeb applied within 2 days of infection is very effective.

2.2.2.4. Buckeye Rot, *Phytophthora nicotianae* var. *parasitica*

(i) Symptoms: This is a very serious disease and causes losses up to 92%. Spots with pale and dark brown concentric rings appear on fruits. These may be small or may cover a major portion of the fruit surface causing fruit rot (Fig 2.24), thereby lowering the yield considerably. One infected fruit with latent infection rots the entire basket overnight.

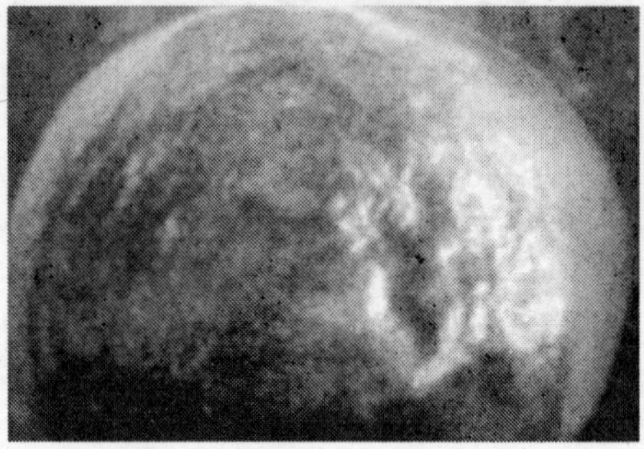

Figure 2.24. Buckeye rot on tomato fruit

(ii) Management

Ridge and furrow system of planting, staking plants at flowering stage and removal of basal leaves up to 15–20 cm height from ground at the start of monsoon reduce the disease incidence. The rotten fruits should be collected and destroyed. Spray 1% Bordeaux mixture or 0.2% propineb or dithionon or metalaxyl + mancozeb to reduce the disease incidence.

2.2.2.5. Septoria Leaf Blight, *Septoria lycopersici*

(i) Symptoms: The plant may be attacked at any stage of its growth. Small, circular, water-soaked grey spots having dark border first appear on under surface of lower leaves. On enlargement, they develop dark brown margins and sunken white-grey centres (Fig 2.25). Similar spots may develop on floral calyx and stem (rarely on fruits) causing severe defoliation resulting in yield losses and sun scalding of fruits.

Figure 2.25. *Septoria* leaf spot on tomato leaves

(ii) Epidemiology: The pathogen survives in the soil on infected plant debris and also in seeds. Low temperature, high moisture and humidity favour infection of the pathogen.

(iii) Management

Destroy infected plant material in the field. Treat the seeds with thiram 75 WP at 2 g/kg of seed. Spray chlorothalanil, propineb or mancozeb M at 0.2% or iprodion + carbendazim or mancozeb at 0.2% at 10–15 days intervals. Spraying of 0.4% copper oxychloride or 0.2% mancozeb or 0.1% carbendazim also gives effective control of the disease.

2.2.2.6. Wilt, *Fusarium oxysporum* f. sp. *lycopersici*

(i) Symptoms: The wilt disease is becoming a major disease of concern in many parts of the country. The disease is characterized by yellowing and wilting of leaves and finally the entire plant wilts and dies prematurely (Fig 2.26). Stem tissue often is discoloured throughout the plant. Vascular browning takes place in the root.

Figure 2.26. Tomato plant infected by *Fusarium* wilt

The fungus can persist in the soil for many years and is virulent at moderate temperature (26–28°C).

(ii) Management

The infection could be minimized by crop rotation and planting resistant varieties. Incidence of wilt can be significantly reduced by the application of 0.01% carbendazim or 0.2% benomyl as soil drench (Sen and Kapoor, 1974).

2.2.2.7. Collar Rot, *Sclerotium rolfsii*

It has wide host range and causes severe problem on tomato and other vegetable crops.

(i) Symptoms: The first symptom of the disease is observed as soft tissue necrosis of bark of the stem near soil line. White growth of cottony mycelium is clearly visible on the affected portion just below the soil surface. Later, dense silvery fungus growth along with white to light brown mustard-like sclerotia is observed on the same portion. Progressive drooping, yellowing or wilting of the entire plant is observed. Sometimes the plant is collapsed soon after infection. The disease is soil-borne where the pathogen survives in the form of sclerotia.

(ii) Management

Long rotations with cereals *viz.*, corn, sorghum and millets will reduce the disease incidence. All the weeds from the field should be removed. Use of infested FYM, soil, compost should be avoided in the field. Summer ploughing for 3 times and irrigating the field destroys the germinating sclerotia. Avoid dense planting and maintain proper aeration near collar region.

Soil drenching in the evening with 0.3% copper oxychloride or 0.25% captafol/captan near the collar region is effective to check the disease.

The disease was effectively managed under greenhouse and field conditions by soil + root-dip treatment and seed + soil treatment with *T. harzianum;* and seed + soil treatment with *B. subtilis.*

2.2.2.8. Anthracnose, *Colletotrichum gloeosporoides*

(i) Symptoms: Anthracnose is the disease of ripe and over ripe tomato. Lightly depressed circular lesions appear on fruits. Lesions enlarge and become sunken with concentric ring markings. Centre of the lesion is usually tan which later becomes dotted with small black specks. Lesions may appear on stems, leaves and fruits (Fig 2.27). Small circular brown lesions surrounded by yellow halo appear on leaves.

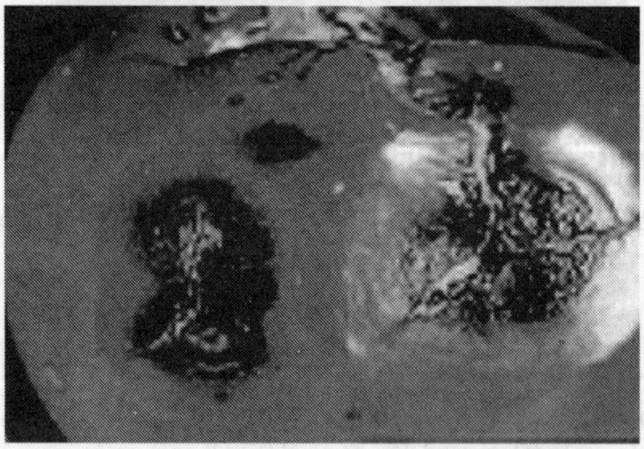

Figure 2.27. Anthracnose on tomato fruits

2.2.2.9. Bacterial Wilt, *Ralstonia solanacearum*

(i) Symptoms: This is the most serious bacterial disease of solanaceous crops. Characteristic symptoms of the disease are: wilting without leaf yellowing and collapse of the entire plant (Fig. 2.28). The vascular system shows browning. If a segment of the stem is cut and squeezed, bacterial ooze is visible (Fig 2.28). It causes sudden wilt and ultimately, the plant is killed resulting in total loss.

Figure 2.28. Bacterial wilt affected tomato plant

(ii) Management

(a) Cultural Methods: Soil amendment with green manuring crops, transplanting disease-free seedlings and following 3 years rotation with cereals/crucifers are effective for the management of wilt. Crop rotation with cowpea-maize-cabbage, okra-cowpea-maize, maize-cowpea-maize and finger millet-brinjal (Pusa Purple Cluster)-French bean are reported effective in reducing bacterial wilt. Intercropping of tomato with sorghum, maize, onion, garlic and marigold resulted in better crop stand. Grafting tomato on brinjal rootstocks such as Pusa Purple Cluster and Dingra Multiple Purple helps in reducing the disease incidence.

(b) Chemical Methods: Application of bleaching powder at 15 kg/ha and groundnut cake before planting was found effective.

(c) Biological Methods: Bacterial wilt of tomato was successfully controlled by incorporation of *Pseudomonas fluorescens*. Prior inoculation of tomato roots with *T. viride* and *P. fluorescens* protected the plants. Application of paste formulation of *P. fluorescens* in vermicompost exhibited lower wilt incidence with maximum reduction of wilt pathogen in the rhizosphere and significant yield enhancement.

Dipping of seedling roots in avirulent strain of *Ralstonia solanacearum* (Av 10) before transplanting gave least wilt incidence (6.66%) as compared to control (75%). Seed treatment with *P. fluorescens, P. aeruginosa* and *B. subtilis* resulted in least seedling mortality (< 1%, 2.66 and 4%, respectively) as compared to control (27.58%). Seedling root dip in *P. fluorescens/B. subtilis* followed by application of antagonists cell suspension to soil in root zone was found effective in reducing the disease incidence (16.66 and 25%, respectively) as compared to control (74.5%). Soil application of *Glomus mosseae* reduced the incidence of wilt.

(d) Host Resistance: Tomato *cvs*. Arka Abha, Arka Alok, Arka Shreshta, Arka Abhijit, Megha, Shakthi, Sonali, Sun 7610, Sun 7611, BT-1, BT-10, LE 79-5 are resistant to bacterial wilt.

(e) Integrated Methods: Integration of growing and incorporation of sunnhemp (green manure crop) into the soil and seed treatment with *Pseudomonas fluorescens* (10^{10} cfu/ml) was highly effective in reducing the bacterial wilt incidence to 6.75% and significantly increasing the fruit yield to 25.5 MT/ha as compared to high wilt incidence (55.6%) and low yield (7.8 MT/ha) in control. Thus, the efficacy of *Pseudomonas fluorescens* is increased under green manure incorporated soil against bacterial wilt of tomato.

2.2.2.10. Bacterial Spot, *Xanthomonas campestris* pv. *vesicatoria*

(i) Symptoms: Numerous small, dark spots appear on leaflets and stems of seedlings and cause severe damage. Prominent yellow halo is observed around the spots (Fig 2.29). The most conspicuous and damaging phase is infection of immature fruits. Small water-soaked spots appear on exposed surfaces of the green fruit (Fig 2.29). The disease appears in rainy season and continues up to fruit initiation stage. The disease is prevalent in warm humid weather when RH is more than 90% and temperature is 30–35°C.

Figure 2.29. Bacterial spot on leaves and fruits

(ii) Management

Seeds should be collected from the disease-free plants. Use of disease-free seeds, clean cultivation and 3 sprays of 1% Bordeaux mixture reduced the disease incidence. One spray of 150 ppm streptocycline should be given in the infected field to check the disease incidence. Use of tolerant varieties like LE-5 and spraying with 0.3% copper oxychloride provides satisfactory control (Pandey and Pandey, 2002).

2.2.2.11. Bacterial Stem and Fruit Canker, *Clavibacter michiganensis* pv. *michiganensis*

This disease of tomato was observed first during 1909 in the USA. In India, the disease was reported from Nagpur market during 1949 (Jain, 1951).

(i) Symptoms: Temporary and later on permanent wilting of leaflets of tomato plants are the first symptoms of the disease. Light streaks appear at the juncture of petiole and stem extending down the internode and up to the petiole. As these streaks become darker, epidermal tissues may rift and longitudinal canker like opening may appear in stem, petiole and midrib. When the stem of diseased plants cut longitudinally, a creamy white, yellow or brown line follows the phloem. Discoloration of vascular system can be traced up to the fruit, placenta and seeds. Secondary infection appears on green fruits as water-soaked spots with a white halo. The spots remain less than 3 mm in diameter exclusive of halo. Halo is the distinguishing character of bacterial canker of tomato (Fig 2.30).

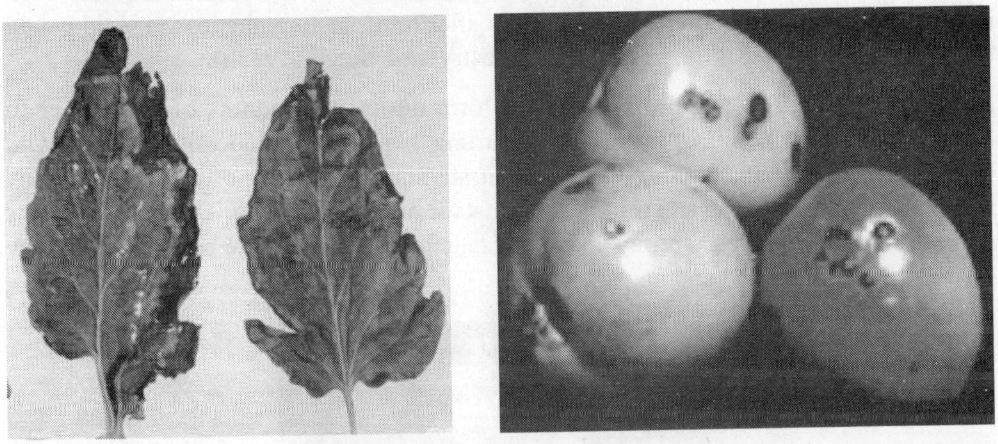

Figure 2.30. Bacterial canker on leaves and fruits of tomato

(ii) Disease Cycle: The disease is seed-borne and up to 70% seed-borne infection has been reported (Nedumaran and Vidyasekaran, 1982). Bacterium also survives in soil and the disease spreads to new area through infected seeds.

(iii) Management

- Seed treatment with hot water or organomercurials or streptomycin is beneficial.
- Three years crop rotation with non-host crops is effective.

2.2.2.12. Bacterial Speck, *Pseudomonas syringae* pv. *tomato*

Bacterial speck of tomato was first reported from Formosa in 1933. In India, the disease was observed during November 1960 to February 1961 at South Arcot, Tamil Nadu (Rangaswami and Gowda, 1963) and in August 1978 at Kanpur, Uttar Pradesh (Singh *et al.*, 1982).

(i) Symptoms: The disease starts as small, round, water-soaked spots, which gradually become yellowish brown and finally blackish brown with sunken ashy centre and pale yellowish green halo. Later on, spots become irregular to circular, measuring 1 mm in diameter, distributed over the entire leaf lamina and in severe cases forming large blotches. Spots on petioles are more or less oval.

(ii) Management

Seed treatment with 100 ppm Agrimycin-100 or 200 ppm Sandox or 100 ppm streptocycline is effective.

2.2.2.13. Tomato Leaf Curl Virus (TLCV)

This is one of the most important and devastating diseases of tomato which causes yield losses up to 92.4% if the plants get infection within 20 days after transplanting. However, if the plants are infected after 35 and 50 days after transplanting, the loss in yield would be reduced to 74 and 28%, respectively (Sastry and Singh, 1973b).

(i) Symptoms: Disease appears in severe form during the months of September to November. Leaves show downward rolling, curling, twisting and chlorosis (Fig 2.31). Old leaves turn leathery and brittle. Plants exhibit stunting, shortening of internodes, bushy appearance, non-flowering and fruiting at later stage of virus infection. The diseased plants induce partial or complete sterility. In case of late infection, plants do bear fruits but they are very small, deformed and unmarketable.

Figure 2.31. Leaf curl virus on tomato leaves

(ii) Epidemiology: High temperature and low humidity favours the disease incidence. The disease progresses during February-June with the prevalence of dry hot season with low humidity which favour whitefly multiplication and disease spread. The incidence may reach even up to 100% during summer months (Sastry *et al.*, 1978). The virus infects a number of cultivated hosts *viz.*, papaya, tobacco, bell pepper, sunnhemp, hibiscus, zinnia, petunia and weeds which serve as a source of inoculum.

(iii) Transmission: The virus is transmitted by the whitefly, *Bemisia tabaci*. Even a single whitefly can transmit the disease.

(iv) Management

Seed treatment with hot water maintained at 50°C or 2% trisodium phosphate solution for 25 min is recommended.

The nursery should be raised under nylon net to prevent entry of whiteflies which transmits leaf curl (Fig 2.32). Field sanitation should be given top priority as many weeds are alternate hosts for the whitefly vector. Crop should be frequently irrigated which invites less oviposition by whiteflies. Leaf curl affected plants should be uprooted and destroyed. When tomato is grown along with French bean and brinjal, the incidence of TLCV is low as they act as trap crops for whiteflies.

Figure 2.32. Tomato nursery covered with nylon net to prevent entry of whiteflies

The disease spread can be minimized by growing border or barrier crops like maize, sorghum and bajra (5 to 6 rows of these crops should be sown alround the tomato plot 50 to 60 days before transplanting tomato) which prevents incoming viruliferous whiteflies from entering into the tomato crop (Sastry *et al.*, 1977). Sowing of 5 rows of maize (border crop) about 5 weeks before transplanting tomato resulted in low incidence of leaf curl disease (14.42%) followed by sunnhemp (18.27%) and jowar (23.94%) as compared to 48.09% in control. Hence growing maize as barrier crop around tomato crop is effective in preventing further spread of the leaf curl disease.

Mulching tomato fields with saw dust, straw or yellow polythene sheets during early crop stages shall attract whiteflies which stay there long enough to be killed by reflected heat. Yellow sticky traps can also be used for controlling whitefly population (Smith *et al.*, 1979). Use of biocontrol agent like *Encarsia* has shown great promise in reducing whitefly population. In case of greenhouse whitefly, *Trialeurodes vaporarium* gave effective control of whiteflies.

Soil application of carbofuran at 1.5 kg *a.i.*/ha in nursery bed and also during transplanting of the seedlings is effective. Two to three foliar sprays of monocrotophos or dimethoate both at 0.1% at 10 days interval are also effective (Singh and Sastry, 1976). Singh *et al.* (1973) found that krishi oil (power oil) at 2% was very effective in reducing the spread of

TLCV. The use of oil prevents the acquisition and inoculation of the virus by whiteflies. It is advisable to spray krishi oil after fruit formation and spraying of insectcides should be discontinued later (Singh and Sastry, 1976).

Grow leaf curl resistant/tolerant varieties such as Avinash-2 (T), Hisar Anmol (H-24), Hisar Gourav, JK Asha, Nandi, Abhinav, Vaibhav, Sankranti, Mruthyunjaya-3.

2.2.2.14. Tomato Spotted Wilt Virus (TSWV)

This disease was first reported from India by Todd *et al.* (1975) from Nilgiri hills.

(i) Symptoms: The most characteristic symptoms are sudden browning of leaves followed by cessation of growth of the plants. Later on, the leaves become distorted and areas of necrosis are common. Leaves exhibit bronze colouring. The intensity of bronzing varies from inconspicuous green brown glaze to a distinct dark brown to an almost black glazed area. Bronzed areas usually roll inward and affected areas often become necrotic. The fruits on infected plants show numerous spots of about 1.2 cm in diameter with concentric circular markings. On ripe fruits, these markings consist of alternate bands of red and yellow which are diagnostic for the virus (Fig 2.33).

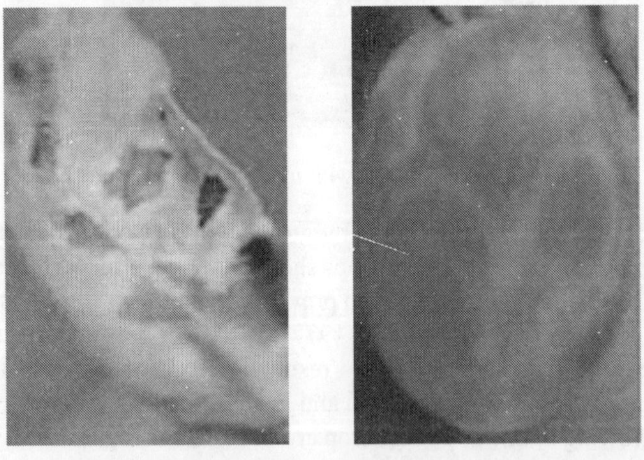

Figure 2.33. Spotted wilt virus on leaf and fruit

(ii) Transmission: The virus is transmitted by mechanical means. The important means of natural transmission is through thrips, *Thrips tabaci, Frankliniella schultzi* and *F. occidentalis*. The virus is acquired first by the larval form, and then subsequent adult can then transmit the virus. The minimum acquisition period is 15 min.

(iii) Host Range: In addition to tomato, this virus infects potato, brinjal, bell pepper, groundnut, pea, French bean, cowpea, lettuce, pineapple, marigold, *Chrysanthemum, Zinnia elegans*, *Aster* sp., *Gerbera* sp. and *Dahlia variablis*. These hosts are important source from which the disease can spread each year.

(iv) Management

- Raise the seedlings under nylon net cover.
- Uproot and burn the affected plants as soon as the disease is noticed.
- The disease spread can be controlled by spraying systemic insecticides.
- Three to four foliar sprays of 4% NSKE or 0.1% monocrotophos or dimethoate at 10 days interval are effective.
- Efficient weed control and avoiding cultivation of other hosts of the virus in the vicinity of susceptible crops may help in disease control.
- Physical barriers that limit the movement of thrips have been shown to reduce the disease incidence.
- Protected cultivation of tomato crop in greenhouses.
- Mulching of soil with aluminium-surfaced plastic mulch significantly reduced the disease incidence by repelling thrip vectors (Black *et al.*, 1986).

Spraying of Adathoda leaf extract (10%) was found to be equally effective as that of the existing practice of spraying Monocrotophos (0.15%) for the management of TSWV. Both these treatments recorded the least incidence of 10% and maximum yield of 3.2 and 3.9 kg/plot, respectively.

2.2.2.15. Tomato Mosaic Virus

The disease is caused by a strain of tobacco mosaic virus. The virus infects many solanaceous hosts. The loss in yield may vary from 20–80% depending upon the time of infection and age of the plant.

(i) Symptoms: The disease is characterized by light and dark green mottling on the leaves often accompanied by wilting of young leaves on sunny days when plants first get infected. The pathogen occasionally causes yellow-green mottling on fruits. The infected leaflets are usually distorted, puckered and smaller than normal ones (Fig 2.34). Sometimes the leaflets become indented and a 'fern leaf' symptom results.

Figure 2.34. Mosaic virus on tomato

(ii) Transmission: The virus is easily sap transmitted. TMV is a highly contagious disease and can be transmitted by clothes, tools, farm machinery, soil debris, etc. The virus is also seed-borne up to 80% in tomato.

(iii) Management

The following practices may be adopted for effective management of the disease:

- Select seeds from healthy plants and as an added precaution, soak them in 0.3% solution of trisodium phosphate for 48 hr before sowing.
- Dip the seedlings in solution of trisodium phosphate or monocrotophos both at 0.2% concentration for 10 min before transplanting.
- Carefully remove and destroy all infected plants in the nursery or in main field. Do not use TMV infected seedlings for transplanting.
- Crop rotation with non-solanaceous crops helps in managing the disease.

2.2.2.16. Tomato Big Bud

The disease is caused by phytoplasma. This disease causes heavy losses in yield of tomato fruits.

(i) Symptoms: The important symptoms of the disease are phyllody and virescence of floral parts, proliferation of auxiliary buds and shortening of leaves. The whole calyx is enlarged and forms a bladder-like structure with a toothed opening at the tip (known as big bud). Thickening of leaves, stem, petiole and peduncle is also common.

(ii) Transmission: The disease is transmitted by leaf hopper vector, *Orosius argentatus* and by grafting.

(iii) Management

- Strict rouging of infected plants from the field.
- Application of furadan at 1.5 kg *a.i.*/ha at transplanting followed by 3 to 4 foliar sprays of 0.1% monocrotophos at an interval of 10 days.
- Spraying of 200 ppm oxytetracycline hydrochloride solution at weekly interval.

IDM

- Crop rotation with French beans reduces bacterial wilt.
- Crop rotation with cereals, sesame and mustard, so also intercropping with marigold, onion, garlic reduce nematode infestation.
- Soil solarization of nursery beds for 2–3 weeks using transparent polythene sheets (100–160 gauge) kills soil borne pathogens.
- Raising nursery in raised seed beds (10–15 cm) provides good drainage and thereby prevent damping off.
- Use of bacterial wilt resistant varieties (Arka Abha, Arka Alok, Arka Shreshta, Arka Abhijit, Megha, Shakthi, Sonali).
- Seed treatment with *Trichoderma virde/T. harzianum/Pseudomonas fluorescens* at 20 g/kg of seeds to prevent damping off and nematodes.

- Deep summer ploughing helps in exposing pathogens and their resting stages to sun's heat.
- Use of barrier crops like pearl millet, maize and sorghum to reduce the incidence of tomato leaf curl.
- Collection and destruction of diseased plants/plant parts from the field from time to time.
- Soil application of neem cake at 250 kg/ha as basal dose to control nematodes.
- Soil application of FYM enriched with *Trichoderma virde* at 1 tonne/ha.
- Installation of yellow water-pan/sticky trap to monitor whitefly population which transmits leaf curl virus.

Integrated Pest and Disease Management

Nursery

- Deep off-season/summer ploughing helps in exposing resting stages of the pests and nematodes.
- Green manuring with sunnhemp in July–August.
- Raising nursery on raised beds (10 cm above ground level).
- Soil solarization using transparent polythene sheets on nursery beds for 2–3 weeks which helps in killing nematodes, soil-borne pathogens, weed seeds and resting stages of insect pests.
- Seed treatment with *T. viride* at 4 g/kg of seed to prevent soil-borne infection of fungi and nematodes.
- Soil application of *T. harzianum* at 50 g/m^2.
- Use leaf curl tolerant hybrids.
- Use nylon net (40 gauge) to avoid insect vectors (whitefly).
- Raising marigold nursery 20 days before tomato nursery.

Main Field

- Transplanting 1 row of 45 day-old marigold seedlings for every 16 rows of 25 day-old tomato seedlings. First and last row of field should be marigold.
- Giving wider spacing of 90 × 60 cm to reduce the incidence of early blight and powdery mildew.
- Soil application of fresh neem cake at 250 kg/ha at 20 DAP to reduce fruit borer and leaf miner incidence.
- Spraying 4% NSKE at 15 DAP to reduce the leaf miner incidence.
- Setting up of yellow sticky traps for whitefly at 10/ha.
- Installing pheromone traps at 13/ha to trap fruit borer adults.
- Release of *Trichogramma pretiosum* at 1 lakh/ha 6 times from flower initiation stage at weekly intervals brings about significant reduction in fruit borer damage.
- Spraying good quality *Ha* NPV at 250 LE/ha at 28 DAP and repeat 2 more times at weekly interval. Mix 2% jaggery and spray in the evening to prevent fast degradation of *Ha* NPV due to UV light.

- Spray chlorothalonil at 2 g/litre of water to control early and late blight diseases.
- Collection and destruction of borer infested fruits and leaf curl/bacterial wilt disease affected plants from time to time.
- Crop rotation with cereals such as wheat, sorghum and/or sesame, mustard or marigold to reduce nematode problem.

2.2.3. Nematodes

Root-knot, reniform and stubby root nematodes are the major limiting factors in the successful production of tomato crop.

2.2.3.1. Root-knot Nematodes, *Meloidogyne* spp.

Lal and Ansari (1960) reported *M. arenaria* for the first time on tomato from Bihar, India (Fig 2.35). *M. incognita* was responsible for 30.57 to 46.92% loss in fruit yield of tomato (Bhatti and Jain, 1977; Reddy, 1985; Darekar and Mahse, 1988), while *M. javanica* caused 77.5% loss in yield (Anon, 1993).

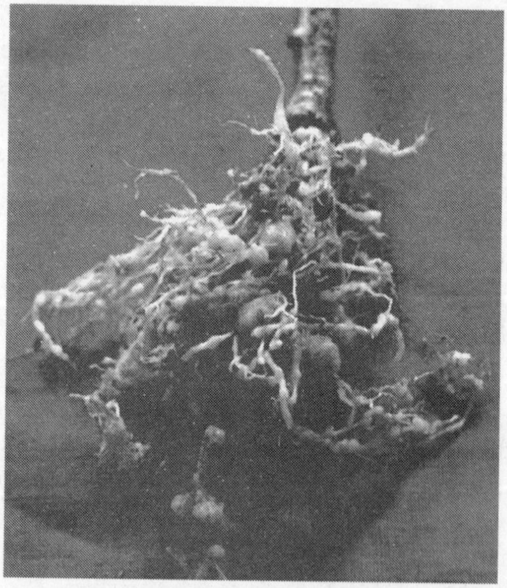

Figure 2.35. Heavy galling of tomato roots with *Meloidogyne incognita*

(i) Interaction with Other Pathogens: *M. javanica* increased the extent of damage by pre and post-emergence phases of damping off caused initially by *R. solani* and *Pythium debaryanum* in tomato (Nath *et al.*, 1984). Pani and Das (1972) have reported the association of root-knot nematode with bacterial wilt of tomato. The growth of tomato plants was significantly reduced in simultaneous inoculations with *M. incognita* and tobacco mosaic virus or when nematode preceded the virus inoculation (Goswami and Chenulu, 1974).

(ii) Management

(a) Physical Methods

Heat Treatment: The most important nematode pests controlled by greenhouse steaming are cyst nematode of potato attacking tomatoes and root-knot nematodes on tomatoes.

Rabbing: Rabbing of root-knot infested tomato nursery with bajra husk or paddy husk or saw dust at 7 kg/sq.m. a week prior to seeding is recommended for economic and effective management of root-knot disease (ICBR 1: 5.12).

Soil Solarization: Soil solarization of nursery beds with 100 gauge LLDPE clear plastic film during summer (in between April 15th and June 15th) for 15 days is recommended (Fig 2.36). It increased the number of transplantable seedlings by 615% giving net profit of Rs. 7,661 from 1000 sq.m. (ICBR 1: 5.65) and decreased the root-knot disease and weeds by 66% and 93%, respectively (Table 2.14).

Figure 2.36. Soil solarization for the management of root-knot nematodes in tomato Nursery

Table 2.14. Effect of soil solarization of nursery beds on root galling and number of transplantable seedlings of tomato in 1000 sq. m

Treatment	No. of seedlings	Weight of 100 seedlings	Root-knot index	Net profit (Rs.)	Cost: benefit ratio
With plastic	2,60,400	36 g	1.37	7,661	1 : 5.65
Without plastic	36,400	26 g	3.97	—	—
% Increase (+)/decrease (–) over control	(+) 615	(+) 38	(–) 66	—	—

Red plastic mulch suppresses root-knot nematode damage in tomatoes by diverting resources away from the roots (and nematodes) and into foliage and fruits (Adams, 1997).

(b) Cultural Methods: Several cultural methods like selection of nematode-free nursery sites, destruction of infested roots after crop harvest, crop rotation with non-hosts (maize, wheat, sorghum) or antagonistic crops (marigold, mustard, sesame), flooding, fallowing, deep summer ploughing and harrowing either alone or in combination proved to be reasonably effective and economical to check multiplication of root-knot nematodes on tomato.

Jain and Bhatti (1987) reported that planting of nematode-free seedlings in an infested field which received three summer ploughings increased the yield of tomato by 55%.

Decrease in *M. incognita* population on tomato occurs when the field was left fallow or following crop rotation with marigold, spinach and bottle gourd (Khan *et al.*, 1975).

Summer Ploughing: Ploughing not only leads to disturbance and instability in nematode community but also causes their mortality by exposing them to solar heat and desiccation. Three summer ploughings each at 10 days interval during June (average atmospheric temperature ranging between 40–46°C) at Hisar (Haryana) led to 96.5% reduction in *M. javanica* population, while fallowing itself during the same period registered 44.5% reduction (Jain and Bhatti, 1987). Additional use of plastic sheets for covering soil either in nursery beds or in field further enhanced nematode reduction. Such an approach also helps in reducing the intensity of weeds, fungi and bacteria in the soil.

Crop Rotation: To control *M. incognita* and *M. javanica*, a tomato crop can be followed by peanut without risk of damage to the peanuts. While the peanuts are growing, the nematodes can not reproduce. Instead many of the larvae in the soil die or become non-infective because of starvation and the attacks of predators, fungi and diseases. If the population is reduced sufficiently, tomatoes can be grown again after peanuts without serious injury (Taylor and Sasser, 1978). Tomato rotated with cotton-wheat-cotton resulted in low gall index due to *M. javanica* (root-knot index 1–2) and gave maximum yield (38% increase over continuous tomato) (Hashmi and Hashmi, 1990).

Sesamum–tomato sequence produced the least reproduction rate (0.47), least root-knot index (1.33) and highest yield (86 q/ha) of tomato which was at par with niger-tomato sequence that had reproduction rate of 0.56, root-knot index of 1.66 and highest yield of 75.5 q/ha of tomato (Sahoo *et al.*, 2004a).

In cropping sequence studies for the management of *M. javanica* on tomato, nematode population decreased to low levels when carrot, capsicum and onion were grown (Kanwar and Bhatti, 1992). The population of *M. incognita* was reduced considerably under groundnut-mustard-tomato cropping sequence (Sharma *et al.*, 1980).

Tomato nursery beds previously planted with trap crop (marigold) gave maximum reduction in root-knot nematode population in soil which also increased the germination of tomato seeds and production of more healthy (nematode-free) seedlings (Rangaswamy *et al.*, 1999).

Inclusion of rice, maize, groundnut and *Stylosanthes* in rotation sequences controlled root-knot nematodes and increased yields of tomato. In crop rotation trials, it was found that a crop of groundnut or strawberry grown prior to tomato in soil heavily infested with root-knot nematodes greatly increased yields of tomato. Rotation of tomato with leek (*Allium porrum*), onion, groundnut and sunnhemp reduced the population of root-knot nematodes.

Intercropping: Intercropping with marigold or mustard with tomato reduced the damage of root-knot and reniform nematodes. The root-knot development on tomato was low when interplanted with *Tagetes erecta*. The population of *Tylenchorhynchus, Helicotylenchus,*

Hoplolaimus, Rotylenchulus and *Pratylenchus* was also markedly reduced (Alam *et al.*, 1977; Khan *et al.*, 1971a). ∝-Terthienyl is the active principle in *Tagetes* spp. which is toxic to these nematodes.

Marigold intercropped with tomato at 1 : 4 and 1 : 6 ratios and mustard at 1 : 2 ratio was found to be effective in reducing root galls, egg masses and nematode population at harvest and gave cost: benefit ratio of 1 : 8.36, 1 : 7.88 and 1 : 3.31, respectively (Rangaswamy *et al.*, 1999).

Interculture of onion with tomato reduced gall index due to *M. javanica* significantly and improved tomato fruit yield. Intercropping with 2 rows of onion with one row of tomato was the most effective treatment in reducing the nematode infection and population and enhancing tomato yield (Ram and Gupta, 2001).

In pots having *Zinnia elegans* with tomato, the population of *M. incognita* and *R. reniformis* declined. The root-knot index on tomato was reduced to 1.08 compared to 3.75 in control. The nematode reproduction factor of *M. incognita* and *R. reniformis* was reduced to 0.88 and 0.80, respectively compared to 3.86 and 2.43, respectively in control (Tiyagi *et al.*, 1986).

Cultivation of tomato with castor results in considerable reduction in number of root-knot galls on tomato (Hackney and Dickerson, 1975).

Influence of Fertilizers: Increased levels of potash have significantly reduced the number of galls by *M. javanica* in tomato (Gupta and Mukhopadhyaya, 1971).

The growth of tomato increased (35%) and root galling (root-knot index 1 as against 4 in control) due to *M. incognita* , nematode population in soil (80% reduction) and egg mass production reduced (egg mass index 1 as against 4 in control) in soil amended with 20% fly ash (Haq *et al.*, 1985). The increase in plant growth may be ascribed to increased nutrient availability (N, K, Ca, Mg, Na, B, SO_4) and the reduction in nematode population might be due to toxic compounds (polycyclic aromatic hydrocarbons, dibenzofuran and dibenzo-p-dioxin mixtures) present in fly ash. Row application of fly ash at 0.6 kg/sq. m. enhanced the yield of tomato by 90.4% (Khan and Ghadipur, 2004).

Botanicals: Soil application of neem and subabul leaves to nursery beds at 0.5 t/ha gave better seedling growth and reduced root galling in tomato (Jain *et al.*, 1988).

Singh and Sitaramaiah (1966) applied finely divided oil cakes to root-knot infested soil and noted a reduction in disease incidence in tomato. The oil cakes (neem, groundnut, mustard and castor) were effective in reducing parasitic nematode population (*Hoplolaimus*, *Tylenchorhynchus, Meloidogyne* and *Helicotylenchus* species) and in increasing yields of tomato.

Application of neem cake at 15 g/spot or 100 g/m furrow, three weeks prior to transplanting of tomato has been reported to give 36–450% enhanced yield with corresponding reduction in gall index to 2.3 as against 4.3 in control (Anon., 1992).

Best results in respect of root-knot reduction due to *M. javanica* and increase in yield of tomato were obtained by amending the soil with saw dust at 2.5 tonnes per ha 3 weeks before planting and then applying N through urea at 120 kg per ha (Singh and Sitaramaiah, 1971). Rice hull ash at 2.5 tonnes per ha increased the yield of tomato by 133 to 317 per cent and reduced root-knot incidence by 46 to 100 per cent (Sen and Dasgupta, 1981). Application of poultry manure at 2.0, 2.5 and 3.0 tonnes/ha 15 days prior to tomato seeding is recommended for effective management of root-knot disease and production of more transplantable seedlings in tomato nursery giving ICBR of 1: 4.54, 1: 5.63 and 1: 6.68, respectively.

Green manuring with naffatia/besharmi or neem leaves (ICBR 1: 2.87) or water hyacinth (ICBR 1: 2.527) or calotropis (ICBR 1:2.26) or congress grass (ICBR 1: 2.09) each at 3 kg/sq.m. 15 days prior to seeding is recommended for economic and effective management of root-knot disease in tomato nursery. Incorporation of green materials of naffatia and neem leaves each at 2 kg/1.44 sq.m. (ICBR 1: 15.7) or congress grass and neem leaves each at 2 kg/1.44 sq.m. (ICBR 1: 15.6) or calotropis, naffatia, congress grass and neem leaves each at 1 kg/1.44 sq.m. (ICBR 1: 14.7) 15 to 20 days prior to tomato seeding gave effective management of root-knot nematodes and production of more healthy seedlings in tomato nursery during kharif season.

(c) Chemical Methods

Nursery Bed Treatment: Nursery bed treatment with aldicarb at 4 g *a.i.*/m^2 and carbofuran at 2g *a.i.*/m^2 were effective in increasing seedling growth and reducing root-knot nematode population of tomato (Jain and Bhatti, 1983; Ramakrishnan and Balasubramanian, 1981). In the main field, the above treatments were also effective and increased the fruit yields (88.2 and 55.9% increase in yield of tomato, respectively).

Bare Root-dip Treatment: Carbofuran and oxamyl at 1,000 ppm for 15 to 30 min. (Parvatha Reddy and Singh, 1979); phenamiphos at 250 to 750 ppm for 30 min (Thakur and Patel, 1985); dimethoate at 500 ppm for 6 hr, phosphamidon and dichlofenthion both at 1000 ppm for 8 hr (Jain and Bhatti, 1978) and thionazin at 500 ppm for 15 min (Reddy and Seshadri, 1975) gave effective control of root-knot nematodes on tomato when used as bare-root dips.

Main Field Treatment: Aldicarb, carbofuran, ethoprophos and phenamiphos each at 1 to 2 kg *a.i.* per ha were found effective in reducing the root-knot nematode population and in increasing fruit yields of tomato (Ahuja, 1983).

Dichlofenthion was found effective against *M. incognita* in tomato (Alam *et al.*, 1973; Saxena *et al.*, 1974).

(d) Biological Methods

Antagonistic Bacteria: Two bacteria *Bacillus licheniformis* and *Pseudomonas mindocina* and 2 fungi *Acrophialophora fusispora* and *Aspergillus flavus* were evaluated for their efficacy as bio-control agents individually and in various combinations against *M. incognita* infecting tomato. Individually, *B. licheniformis* gave best control although when all 4 agents were used, nematode multiplication was reduced by 82.06% (Siddiqui and Husain, 1991).

Application of *Bacillus thuringiensis* causes 95% mortality of *M. javanica* juveniles due to β-exotoxin production and suppression of gall formation, egg mass production and nematode population in soil. Field application of a nematicidal *B. thuringiensis* strain to tomato in Puerto Rico reduced galling in roots due to *M. incognita* and increased the yield significantly (Zuckermann *et al*., 1993).

Root material containing *Pasteuria penetrans* spores at 212–600 mg powder/kg soil, when broadcast in soil, showed reduction in galling on tomato roots and final nematode population. Tomato yield increased by 20% over untreated check, when *P. penetrans* was introduced to the nursery soil at 1×10^5 spores/g soil. The persistence of *P. penetrans* in the field soil was evident since about 50% of the final juvenile population was encumbered with the bacterial spores (Walia and Dalal, 1994). *P. penetrans* is highly resistant to heat, desiccation and has ability to survive for more than 2 years in soil, qualifies as the most potential biocontrol agent against root-knot nematodes infesting vegetable crops.

Application of *P. fluorescens* at 10 g/m^2 in nursery beds gave good control of root-knot nematodes (Shanthi and Sivakumar, 1995).

Avermectins B_1 and B_{2a} (applied to soil through drip irrigation systems) at rates ranging from 0.093 to 0.34 kg *a.i.*/ha applied as a single dose or 0.24 kg *a.i.*/ha applied as three doses each at 0.08 kg on tomatoes against *M. incognita* were as effective as oxamyl and aldicarb at 3.36 kg *a.i.*/ha (Garabedian and Van Gundy, 1983). The aqueous solution of avermectins (250 ml of 0.001%/m^2) significantly reduced root galls of tomato seedlings raised in root-knot infested nursery beds which yielded robust and healthy seedlings with no root infection (Parvatha Reddy and Nagesh, 2002). In a root-knot infested micro-plot study, avermectins at 0.015 kg/ha effectively controlled *M. incognita* on tomato and increased yield by 11 and 8% compared to the untreated and carbofuran treated plots, respectively. (Parvatha Reddy and Nagesh, 2002). Seedling bare root-dip in avermectin 100% gave highest reduction in root-knot index (1.3), soil nematode population (132.8), number of females/g root (12.8), egg masses/g root (5.4) and eggs/egg mass (125.2) in tomato (Jayakumar *et al*., 2005).

Antagonistic Fungi: Application of 'Royal 350' (*Arthrobotrys irregularis* cultured on oat seed medium) at 140 g/m^2 a month before transplantation of tomato resulted in good protection against root-knot nematodes. Among the egg parasites, efficacy of *Paecilomyces lilacinus* (commercially formulated as 'Biocon' in Philippines) has been found to be comparatively higher in suppressing the population of *Meloidogyne* spp. and *R. reniformis* on tomato. Bare root-dip treatment of tomato seedlings in a spore suspension of *P. lilacinus* at 4×10^5 spores/ml reduces *M. incognita* infestation in roots by 50% and increases the growth of seedlings by 20%. Application of organic amendments increases the effectiveness of *P. lilacinus* against *M. incognita*.

Arbuscular Mycorrhizal Fungi: Arbuscular mycorrhizal fungus (AMF), *Glomus fasciculatum* was effective against root-knot nematodes and *G. mosseae* against reniform nematodes on tomato. Organic amendments (oil cakes, calotropis leaves) in combination with AMF enhance the colonization of AMF on tomato roots which further increased plant growth and reduced gall index. Sundaram and Arangarasan (1996) observed that inoculation of *G. fasciculatum* at 10 g/plant was effective in increasing the uptake of nutrients and yield and reducing root galling (10) compared to 135.3 in control, followed by *G. mosseae, Gigaspora margarita* and *Acaulospora laevis*. Application of *G. fasciculatum* (Shimoga banana isolate) at 50 g/sq.m. (200 spores/g of soil) gave minimum number of galls per root system (39.2) at harvest and recorded maximum yield (26.47 t/ha) with an increased cost: benefit ratio (1 : 3.34) and decreased nematode population (Kiran Kumar *et al.*, 2004).

(e) Host Resistance: Tomato *cvs.* All Round, Anahu, Anahu R, Atkinson, Beefeater, Beefmaster, Better Boy, Big Seven, Cavalier, Eurocross, Nematox, SL-120, NTR-1, SL-12, Patriot, VFN-8, VFN Bush, Piersol, Radiant, Nemared, Ronita, Anahu, Bresch, Healani, Kewalo, Campbell-25, Punuui, Arka Vardan (Fig 2.37), Pelican, Hawaii-7746, Hawaii-7747 and Hisar Lalit have been reported to be resistant to root-knot nematodes.

Figure 2.37. F_1 Tomato hybrid 'Arka Vardan' resistant to root-knot nematodes

In a field with resistant tomato (Hisar Lalit) grown for a year, the yield was 27% higher than susceptible variety (HS-101) (Kanwar and Bhatti, 1990).

(f) Integrated Methods

Using Bio-agents and Botanicals: In nursery, integration of *Pasteuria penetrans* (at 28×10^4 spores/m^2), *P. lilacinus* (at 10 g/m^2 with 19×10^9 spores/g) and neem cake (at 0.5 kg/m^2) gave maximum increase in plant growth and number of seedlings/bed. Parasitization of *M. incognita* females was highest when neem cake was integrated with *P. penetrans*, while parasitization of eggs was highest when neem cake was integrated with *P. lilacinus*. In field, planting of tomato seedlings (raised in nursery beds amended with neem cake + *P. penetrans*) in pits incorporated with *P. lilacinus* (at 0.5 g/plant) gave least root galling and nematode multiplication rate and increased fruit weight and yield of tomato (Parvatha Reddy *et al.*, 1997).

Soil drenching with 5% Wellgro solution (at 200 ml/seed pan) along with *P. lilacinus* (at 2×10^4 spores/ml) gave maximum increase in seedling weight and highest reduction in root galling. Roots of tomato seedlings raised in the above treatment dipped in 5% Wellgro solution mixed with *P. lilacinus* spores (at 2×10^4 spores/ml) for 20 min when transplanted in pots gave maximum increase in plant growth, root colonization and parasitization of egg masses of *M. incognita* by the bioagent and highest reduction in root galling and final nematode population both in soil and roots (Rao and Parvatha Reddy, 1993a).

Bare root-dip treatment of tomato seedlings in 10% castor leaf extract mixed with *P. lilacinus* spores (at 1.5×10^8 spores/ml) for 20 min. significantly increased the plant growth and reduced root galling and final nematode population. The above treatment also gave significant increase in parasitization of eggs and egg masses and propagule density of *P. lilacinus* in roots (Rao *et al.*, 1999). Similarly, root-dip treatment of tomato seedlings in 5 and 10% neem leaf extract mixed with *P. lilacinus* spores (at 6×10^4 spores/ml) for 30 min gave significant reduction in root galling and final nematode population. Significant increases in root colonization, parasitization of eggs and egg masses and propagule density of the bioagent in soil were also noticed in the above treatment (Parvatha Reddy *et al.*, 1998).

Application of *Pochonia chlamydosporia* (100 ml/seed pan containing 4×10^5 spores/ml) with neem leaves (150 g/seed pan) increased seedling weight and colonization of roots with the bioagent. Tomato seedlings raised in the above treatment transplanted in pots resulted in maximum increase in plant growth, least root galling, nematode population both in soil and roots and highest parasitization of eggs, egg masses and propagule density of *P. chlamydosporia* in roots (Parvatha Reddy *et al.*, 1999). *P. lilacinus* in combination with castor leaves reduced the nematode population up to 89% and increased plant growth and yield in tomato (Zaki and Bhatti, 1991a).

Incorporation of neem cake (20 g/pot) along with *P. chlamydosporia* (10 ml/pot containing 4×10^5 spores/ml) gave maximum increase in plant growth and significant reduction in root galling and nematode population both in soil and roots. The above treatment also gave highest parasitization of eggs and egg masses of *M. incognita* and maximum propagule density of the bioagent in soil and roots (Rao *et al.*, 1998b).

Integration of neem cake (40 g/plant) with *T. harzianum* (at 5 g/plant with 4×10^8 spores/g) was effective in increasing the plant growth and reducing root galling and final population of *M. incognita* infecting tomato. The above treatment also gave maximum reduction in number of eggs/egg mass and increased root colonization, spore density in soil and parasitization of adult females with *T. harzianum* (Rao *et al.*, 1997c).

Application of *Aspergillus niger* and *P. lilacinus* along with mustard cake gave the maximum reduction in nematode population both in root and soil with enhanced plant vigour. *A. niger* being toxic agent kills the second stage juveniles present in the rhizosphere, while *P. lilacinus* being the egg parasite invade the eggs of *M. incognita* which escaped from the toxicity of *A. niger*. As a result, there is an overall reduction in root and soil population. Further, addition of mustard cake also helps to maintain the general plant health in addition to possessing nematicidal properties (Goswami *et al.*, 1998) (Table 2.15).

Table 2.15. Effect of *Aspergillus niger, Paecilomyces lilacinus* and mustard cake on biomass and multiplication of *Meloidogyne incognita* infecting tomato

Treatment	Biomass	No. of galls/plant	No. of egg masses/ plant	No. of eggs/ egg mass	Nematode population/ 500 g soil
Aspergillus niger	9.8	34	18	272	470
Paecilomyces lilacinus	10.2	42	23	262	580
A. niger + *P. lilacinus*	12.2	26	16	264	270
A. niger + *P. lilacinus* + Mustard cake	12.6	22	10	255	290
Nematodes alone	9.0	98	72	320	2880
Control	14.0	—	—	—	—
CD at 5%	**2.33**	**7.24**	**3.24**	**27.18**	**162.80**

Combined application of *P. penetrans* + *T. viride* + neem/castor cake at 1/3 doses was significantly superior compared to their individual applications in terms of increased plant growth and reduced root galling, egg mass production and final populations of *M. incognita*. Similarly, total parasitization by bioagents was higher in plants treated with these bioagents in terms of number of juveniles encumbered and adult females infected with *P. penetrans* and egg masses parasitized by *T. viride*, compared to their individual applications (Rangaswamy *et al.*, 1999).

Nursery bed treatment with *P. fluorescens* (with 1 × 10^9 spores/g) and *Pochonia chlamydosporia* (10 ml/pot containing 4 × 10^6 spores/ml) each at 20 g/m^2 and field application of 5 tonnes of FYM enriched with the above bioagents each at 5 kg, significantly reduced root-knot nematodes in tomato by 76% over untreated check. The yield increase was up to 21.7% with cost: benefit ratio of 1 : 4.9.

Using Arbuscular Mycorrhizal Fungi and Botanicals: In nursery, integration of neem cake (at 500 g/m^2) with *Glomus mosseae* (at 250 g/m^2 containing 16 chlamydospores/g) significantly reduced *M. incognita* population in soil, root galling, egg mass production, fecundity and produced vigorous tomato seedlings with increased root colonization with *G. mosseae*. The seedlings raised in the above treatment when planted in the main field significantly reduced root galling and increased fruit yield, root colonization with *G. mosseae* and chlamydospore population in soil (Parvatha Reddy *et al.*, 1998; Rao *et al.*, 1995).

Application of calotropis leaves at 400 g/m^2 along with *G. fasciculatum* (250 g/m^2 containing 25–30 chlamydospores/g) in nursery beds gave significant reduction in root galling and fecundity of *M. incognita*. The above treatment also gave maximum increase in plant growth and root colonization with *G. fasciculatum* (Rao *et al.*, 1996).

Using Bio-agents and Arbuscular Mycorrhizal Fungi: Interestingly, integration of bio-agent (*P. lilacinus*) and endomycorrhizae *(G. mosseae/G. fasciculatum)* have culminated in the successful management of *M. incognita* infecting tomato. This phenomenon facilitated standardization of a strategy wherein inoculation of mycorrhizae and bio-agent in the nursery beds protected the seedlings of tomato from the attack of *M. incognita*. Further, these mycorrhizal seedlings (colonized either with *G. mosseae* or *G. fasciculatum*) can be given a root-dip treatment with spore suspension of *P. lilacinus* for 5–10 min for the effective management of these nematodes in the main field after transplanting (Rao *et al.*, 1993).

Integration of *G. deserticola* with *P. chlamydosporia* gave effective management of *M. incognita* in tomato, increased seedling weight, root colonization with *G. deserticola* and *P. chlamydosporia* and parasitization of eggs of *M. incognita* with *P. chlamydosporia* and reduced root galling, egg mass production and fecundity of root-knot nematodes (Rao *et al.*, 1997).

Using Bio-agents, Arbuscular Mycorrhizal Fungi and Botanicals: Strategies to integrate botanicals, bio-agents and endomycorrhizae components for the management of root-knot nematodes infecting tomato have been developed. These strategies include inoculation of *G. mosseae* in neem leaf/neem cake amended nursery beds followed by the root-dip treatment of mycorrhizal seedlings of tomato in spore suspension of *P. lilacinus* for effective management of *M. incognita* under field conditions (Rao *et al.*, 1995).

In nursery, integration of neem cake (at 500 g/m^2) with *G. fasciculatum* (200 g/m^2 containing 15 spores/g) gave maximum increase in plant growth and highest root colonization with *G. fasciculatum* and least root galling. Integration of neem cake (at 500 g/m^2) with *T. harzianum* (at 100 g/m^2 with 4×10^8 spores/g) gave maximum parasitization of eggs with *T. harzianum*. In field, planting of tomato seedlings raised in nursery beds treated with neem cake + *G. fasciculatum* in pits incorporated with *T. harzianum* (at 0.5 g/plant with 4×10^8 spores/g) was effective in increasing tomato fruit yield, root colonization with *G. fasciculatum* and egg parasitization with *T. harzianum* and significant reduction in root galling and multiplication rate of *M. incognita* (Parvatha Reddy *et al.*, 1998).

Using Bio-agents and Chemicals: Maheshwari *et al.* (1987) reported that application of *P. penetrans* in combination with carbofuran, aldicarb, miral, sebufos and phorate significantly improved plant growth of tomato by greatly reducing galling due to *M. javanica*.

Integration of *Pochonia chlamydosopria* with carbofuran recorded maximum plant growth parameters and minimum gall index and nematode population. The above treatment also recorded maximum number of fruits/plant and yield/plant. Maximum parasitization of *M. incognita* with *P. chlamydosopria* was also noticed when the bioagent and the chemical were integrated (Gopinatha *et al.*, 2002).

Using Two Bio-agents: Integration of two bio-agents, *P. lilacinus* and *P. chlamydosopria* resulted in combined and complimentary effects for the successful management of *M. incognita* infecting tomato (Rao and Parvatha Reddy, 1992c).

In nursery, integration of *P. chlamydosopria* (200 g/m^2 containing 16 chlamydospores/g) with *P. penetrans* at 1 g mixed with 100 g of sand/m^2 (containing 10^5 spores/10 mg) was effective in increasing the seedling weight and root colonization with *P. chlamydosopria* and adult females of *M. incognita* adequately infected with *P. penetrans*. The tomato seedlings raised in the above treatment when transplanted in the field gave maximum reduction in root galling, fecundity, nematode population both in soil and roots and increased root colonization with *P. chlamydosporia*, infection of *M. incognita* females with *P. penetrans* and tomato fruit yield (Rao *et al.*, 1998a). Maheshwari and Mani (1988) also observed that population of *M. incognita* and *M. javanica* were suppressed effectively and yields of tomato were greater when *P. penetrans* and *P. lilacinus* were applied together.

Combined soil application of *P. lilacinus* and *Aspergillus niger* at the time of transplanting tomato is very effective in reducing root-knot nematodes.

The combination treatment with *T. harzianum* + *T. viride* each at 50 g (4×10^8 cfu/g) considerably increased the plant growth, yield and reduced the root galls and soil nematode population (Hassan and Sobita Devi, 2004).

Using Cultural and Chemical Methods: Combinations of deep ploughing (up to 20 cm) and nursery bed treatment with aldicarb at 0.4 g per m^2 and main field treatment with aldicarb/carbofuran at 1 kg *a.i.*/ha proved effective in the control of root-knot nematodes in tomato which also registered maximum yield (102.6% higher yield over control) (Jain and Bhatti, 1985).

In tomato, application of aldicarb and carbofuran each at 1 kg *a.i.* per ha in combination with neem cake and urea each at 10 kg N per ha at transplanting, produced maximum yield with lowest gall index (2.5) and nematode population, 90 days after planting (Routaray and Sahoo, 1985).

Using Botanicals and Physical Methods: Integration of soil solarization of nursery beds (with LLDPE transparent film of 25¼ for 15 days) with soil incorporation of *Calotropis* leaves at 4 kg/1.44 m^2 resulted in increased plant height (29.9 cm compared to 20 cm in control), fresh shoot weight (176.49 g compared to 100 g in control), transplantable seedlings (546/m^2 compared to 297 in control) with minimum root-knot index (1.83 compared to 3.85 in control) (Patel *et al.*, 2006). Soil solarization with clear LLDPE film (25μ) for 15 days in hot summer in combination with poultry manure at 2 t/ha proved effective in the management of nematodes and higher production of transplants in tomato nursery with cost benefit ratio of 1 : 3.10.

Using Botanicals and Chemical Methods: Application of carbofuran at 1 kg *a.i.*/ha in the nursery beds followed by neem cake at 400 kg/ha in the main field increased yield and reduced the gall index (Singh and Gill, 1998) (Table 2.16).

Table 2.16. Integrated management of root-knot nematodes in tomato

Treatment/Dose	% Reduction in root galling	% Increase in yield
Carbofuran (1 kg *a.i.*/ha) + Neem cake at 400 kg/ha	77.00	61.50
Carbofuran (1 kg *a.i.*/ha) + Urea (23.8 kg/ha) + Neem cake at 200 kg/ha	67.90	49.20
Phenamiphos (1 kg *a.i.*/ha) + Neem cake at 400 kg/ha	74.90	39.70
Phenamiphos (1 kg *a.i.*/ha) + Urea (23.8 kg/ha) + Neem cake at 200 kg/ha	67.90	31.43

Integration of chopped castor leaves (40 to 60 g/kg soil) 1 or 2 weeks before transplanting with application of aldicarb or carbofuran each at 2 kg *a.i.*/ha at transplanting, significantly reduced number of galls due to *M. javanica* and enhanced the growth of tomato (Ravi Dutt and Bhatti, 1986).

Integration of soil application of castor leaves with inorganic fertilizer (75 kg N/ha) enhanced plant growth of tomato and reduced *M. javanica* infestation (95% and 78% reduction in root galls and egg mass production, respectively) (Zaki and Bhatti, 1989).

2.2.3.2. Reniform Nematode, *Rotylenchulus reniformis*

Rotylenchulus reniformis was responsible for 42.25 to 49.02% loss in fruit yield of tomato (Subramanyam *et al.*, 1990) (Fig 2.38).

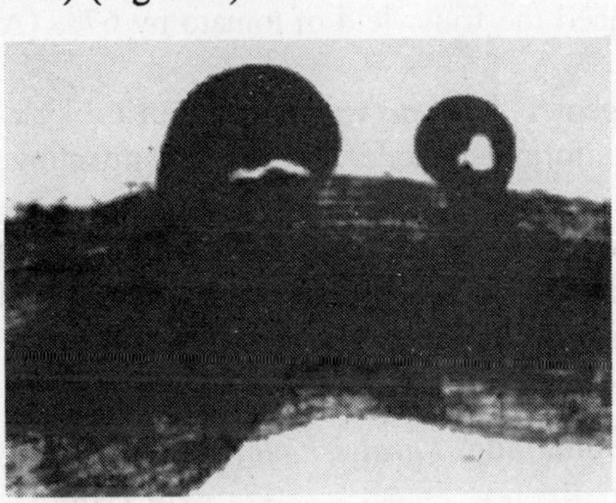

Figure 2.38. Tomato root infected with *Rotylenchulus reniformis*

(i) Management

(a) Cultural Methods: Soil amendment with fresh glyricidia leaves, sawdust and chicken manure controlled the root-knot and reniform nematode population on tomato (Castillo, 1985). The yield increases and the nematode suppressive effect of fresh chicken manure surpassed even phenamiphos (Table 2.17).

Table 2.17. Effect of soil amendment with organic materials on *M. incognita* and *R. reniformis* population and tomato fruit yield

Treatment	No. of nematodes/300 ml soil & 1 g roots			Yield (t/ha)
	No. of weeks after treatment			
	2	6	11	
Phenamiphos at 10 kg *a.i.*/ha	96	366	465	31.1
Rice straw at 10 t/ha	215	500	1921	29.1
Glyricidia leaves at 10 t/ha	201	533	1885	27.3
Saw dust at 10 t/ha	342	859	958	25.3
Chicken manure at 10 t/ha	17	169	1078	32.6
Control	552	816	2515	22.2

(b) Chemical Methods: Application of DBCP (50 l/ha), metham sodium (250 l/ha) or methomyl/aldicarb/thionazin (4–16 kg/ha) was effective in reducing the reniform nematode population and in increasing tomato yields (Sivakumar *et al.*, 1979).

Reddy and Seshadri (1972) reported that thionazin at 4 kg *a.i.* per ha proved effective against *R. reniformis* infecting tomato.

(c) Host Resistance: Balasubramanian and Ramakrishnan (1983) observed that Kalyanpur Selection-I, Kalyanpur Selection-III, CA-121 exhibited resistant reaction.

(d) Integrated Methods: Application of neem cake in the nursery at 100 g/m^2 followed by carbofuran at 1 kg *a.i.*/ha in the main field significantly reduced the soil population of *R. reniformis* and enhanced the fruit yield of tomato by 67% (Anitha and Subramanian, 1998).

Integration of a bio-agent, *P. lilacinus* with carbofuran at 1 kg *a.i.*/ha was found effective in the management of reniform nematode, *R. reniformis* infecting tomato (Parvatha Reddy and Khan, 1988).

Nursery bed treatment with *P. fluorescens* (with 1×10^9 spores/g) and *P. chlamydosporia* (with 1×10^6 spores/g) each at 20 g/m^2 and field application of 5 tonnes of enriched FYM with the above bioagents each at 5 kg significantly reduced reniform nematodes in tomato by 72% over control. The yield increase was up to 21.7% with cost: benefit ratio of 1 : 4.9.

2.2.3.3. Stubby Root Nematode, *Paratrichodorus minor*

(i) Distribution: From India, it has been reported from Delhi and Uttar Pradesh (Prasad and Dasgupta, 1964; Siddiqi, 1962).

(ii) Hosts: *Paratrichodorus minor* has been reported from the rhizosphere of apple, citrus, sugarcane, potato, vegetables, cereals rose and fruits from India.

(iii) Life Cycle: Males are very rare and reproduction is evidently by parthenogenesis. Life cycle on tomato is completed in 21–22 days at 22°C and in 16–17 days at 30°C with a ten fold or more increase in 60 days (Rhoade and Jenkins, 1957).

(iv) Management

(a) Physical Methods: Soil solarization using clear plastic mulch or photo selective (black) polyethylene film reduce populations of *P. minor* on tomato as effectively as fumigation with a mixture of methyl bromide and chloropicrin in warm climates.

(b) Cultural Methods: *Asparagus officinalis* would not support populations of *P. minor* for more than 40 to 50 days. Tomato, normally a good host of this nematode, supported only a low population when asparagus was growing in the same pots. A glycoside (asparagusic acid) is the active principle involved which is toxic to *P. minor* and several other nematode species.

2.2.3.4. Root-knot Nematode, *Meloidogyne incognita* and Wilt, *Fusarium oxysporum* f. sp. *lycopersici* Disease Complex

Jenkins and Coursen (1957) induced wilting in *Fusarium* wilt-resistant tomato variety 'Chesapeake' only when root-knot nematodes were present along with fungal inoculum.

Furthermore, when *M. hapla* was combined with the fungus, only 60% of the plants wilted, whereas *M. incognita acrita* promoted wilt in 100% of the plants.

Bhagawati and Goswami (2000) reported the interaction of *M. incognita* and *F. oxysporum* f. sp. *lycopersici* on tomato (*Lycopersicon esculentum*). It was found that when either or both pathogens were inoculated simultaneously or nematode was inoculated 10 days prior to inoculation of the fungus, the symptoms were visible by 20 days as against no such symptoms in plants where fungus was inoculated 10 days prior to nematodes. The intensity of wilt after 40–60 days of inoculation was significantly higher when simultaneously inoculated or prior inoculation of nematodes compared to prior inoculation of fungus. The number of galls and egg masses and nematode population in soil was significantly reduced when simultaneously inoculated or prior inoculation of nematodes.

(i) Management

(a) Biological Methods: Application of *T. harzianum* at 50 kg/ha (2×10^8 cfu/g) was effective in controlling the disease complex and improving the plant growth parameters and fruit weight (Haseeb *et al.*, 2006).

(b) Integrated Methods

Bioagents and Botanicals: Tomato roots that received *P. lilacinus* along with *T. harzianum* and neem cake were free from root-knot nematodes (*M. incognita*) and did not wilt due to *F. oxysporum* f. sp. *lycopersici* till harvest. The roots were also free from *Fusarium* infection. The above treatment also reduced the % of wilt (10% compared to 90% in control), root-knot index (1.8 compared to 4.4 in control) and increased root colonization with bioagents (74% compared to 0% in control), parasitization of egg masses (56% compared to 0% in control) and eggs/egg mass (44% compared to 0% in control) (Nagesh *et al.*, 2006).

Bioagents, Cultural Methods and Host Resistance: Deep ploughing and exposing soil to hot sun in summer, removal and burning of crop debris, soil application of *T. viride* and *P. lilacinus*, use of wilt resistant varieties like Utkal Pallavi, Utkal Deepti, Utkal Kumari, Utkal Urbasi, etc. help in controlling the disease complex.

Arbuscular Mycorrhizal Fungi and Botanicals: Integration of mustard cake with *Glomus etunicatum* was effective in reducing the damage caused by *M. incognita* and *F. oxysporum* f. sp. *lycopersici* on tomato (Bhagawati *et al.*, 2000).

2.3. BRINJAL

Brinjal (*Solanum melongena*) is a widely grown vegetable crop in Asian countries. It is being cultivated in 0.553 million hectares producing 9.136 million tonnes of brinjal with an average yield of 16.5 tonnes per hectare (Bijay Kumar, 2007) (Tables 2.18 and 2.19). In India, it is adapted to a wide range of climatic conditions from north to south and east to west. In hilly regions, it is grown only in summer. Brinjal is used in a variety of culinary preparations. Pickles and industrially processed foods are also produced. It is rich in vitamin A and B and good for diabetic patients. The major brinjal producing states include

West Bengal, Orissa, Bihar, Gujarat, Andhra Pradesh, Maharashtra, Karnataka, Chattisgarh, Madhya Pradesh, Assam and Haryana.

Table 2.18. Major brinjal producing countries of the world

Country	Area (in '000 ha)	Production (in '000 MT)	Productivity (in MT/ha)	% Share of world production
China	952	17028	17.9	53.5
India	553	9136	16.5	28.7
Egypt	43	1000	23.3	3.1
Turkey	35	930	26.6	2.9
Iraq	23	439	19.1	1.4
Japan	11	396	34.7	1.2
Italy	12	339	27.8	1.1
Sudan	15	272	18.7	0.9
Indonesia	43	252	5.9	0.8
Philippines	21	188	8.8	0.6
Others	120	1856	15.4	5.8
Total	**1829**	**31836**		

Source: FAO (except Indian data) (Indian data source: Indian Horticulture Database 2006)

Table 2.19. Area, production and productivity of leading brinjal growing states in India (2005–06)

State	Area (in '000 ha)	Production (in '000 MT)	Productivity (MT/ha)
West Bengal	152.9	2757.4	18.0
Orissa	127.9	1853.9	14.5
Bihar	53.7	1031.0	19.2
Gujarat	62.7	982.6	15.7
Andhra Pradesh	26.2	549.4	21.0
Maharashtra	32.0	521.6	16.3
Karnataka	14.7	326.0	22.1
Chattishgarh	19.4	266.4	13.7
Madhya Pradesh	14.8	221.9	15.0
Assam	13.0	171.5	13.2
Haryana	11.5	148.7	12.9
Others	24.5	305.8	12.5
Total	**553.3**	**9136.4**	**16.5**

(*Source*: Indian Horticulture Database 2006)

2.3.1. Pests

Brinjal is a very important vegetable crop grown in our country. The shoot and fruit borer *Leucinodes orbonalis*, leafhopper, *Amrasca biguttula biguttula*; ash weevil, *Myllocerus subfacitus;* epilachna beetle, *Epilachna vigintioctopunctata*; red spider mite, *Tetranychus urticae*; gall midge, *Asphondylia* sp.; and stem borer, *Euzophera perticella* are some of the important pests attacking this crop.

2.3.1.1. Shoot and Fruit Borer, *Leucinodes orbonalis*

This is the most important pest on brinjal not only in India, but also in all South Asian Countries. This pest has developed resistance against all groups of insecticides and management is very difficult. Often the extent of damage due to this pest reaches up to 70–80%. Damage is very severe during rainy season and early winter.

(i) Damage: The adult moths lay tiny white eggs singly on flower buds and other plant parts. Eggs hatch into small light brown larvae. Caterpillars feed inside the tender shoots before flowering and cause wilting of the affected shoots (Fig 2.39). Later, the larvae bore into flowers, flower buds and the grown up larvae migrate and bore fruits contaminating them with excreta (Fig 2.39). When the incidence is high, unopened flower buds swell and harbour the borer. Just before pupation, the grown up larvae come out of the fruits and flower buds to pupate in silky cocoons on plant parts or debris.

Figure 2.39. Shoot and fruit borer damage on brinjal

(ii) Management

(a) Cultural Methods

- Use nylon net barrier for raising nurseries to eliminate pest incidence coming from nursery to main field.

- Weekly monitor wilting shoot tips having borer incidence. Cut and destroy such insect damaged shoot tips during pre-flowering and flowering period.
- Regular destruction of larvae in swollen damaged flower buds and fruits after each harvest is compulsory.
- Grow barrier crops like maize all round the crop coupled with clean cultivation.
- Dusting of ash reduces the incidence of shoot and fruit borer.

Removal and destruction of borer damaged shoots during pre-flowering period has potential for reducing the pest population early in the season. Further, growing a barrier crop such as maize, destruction of borer affected fruits, marginally help in reducing the borer damage.

Application of neem cake four times during the crop growth decreases the incidence of borer to 8% from 40% in control. Farmers could realize an additional return of Rs. 35,498/ha.

(b) Use of Pheromone trap: Use of pheromone trap is found to reduce the borer infestation in many parts of India. Use water traps loaded with pheromone for this purpose (Fig 2.40). About 30 traps per acre is known to bring down the pest incidence. Pheromone lure (Lucinlure) in combination with Portable Water Traps (both of PCI) were effective in mass trapping of *Leucinodes orbonalis* to the tune of 1927/acre in Bangalore, 1022/acre in Belgaum and 590/acre in Jogipet in Andhra Pradesh.

Figure 2.40. Sex pheromone and funnel pheromone trap to attract shoot and fruit borer of brinjal

(c) Use of Botanicals

- Apply neem or pongamia cake at 250–500 kg/ha to ridges at flowering. Repeat 2 more times at 30–45 days interval
- Spray NSPE 4% or neem oil 3% at 10 days interval (Table 2.20).
- Mix enodsulfan 35 EC at 2ml/l or cypermethrin 25 EC at 0.75 ml/l with neem soap at 7.5 g/l and spray.

Table 2.20. Effect of neem products for the control of fruit borer in brinjal

Treatment	Mean % of infestation (number basis)	Mean % of infestation (weight basis)
Neem cake extract 10%	55.74	54.57
Neem seed kernel extract 5%	50.89	49.01
Neem oil 3%	50.85	52.48
Carbosulfan 0.04%	39.77	41.77
Untreated check	61.56	59.35

(d) Biological Methods: Inundative release of *Trichogramma chilonis* at 50,000/ha, 5 times at weekly intervals starting from flower initiation stage is also effective. Spray *Bacillus thuringiensis* formulation (1%) at weekly interval followed by release of *Trichogramma chilonis* at 2,50,000/ha (50,000/release-5 times at weekly intervals, starting from flowering) was found to reduce the borer incidence..

Application of *B. thuringiensis* at weekly intervals controlled the fruit borer. In plots treated with *B. thuringiensis* (Delfin), there was 30% reduction in larval population and 48.3% yield increase in brinjal. Application of 5 rounds of *B. thuringiensis* var. *kurstaki* (Biobit) at 1 kg/ha at weekly intervals, right from flower initiation was effective in checking the pest. The per cent fruit damage was low in Biobit (13.44%) as compared to control (19.25%). The yield was also significantly high in Biobit (28.5 kg/plot) as compared to control (14.74 kg/plot).

Eriborus sinicus gave up to 30% parasitism of shoot and fruit borer at Taiwan (AVRDC, 1996).

(e) Integrated Methods: A universally applicable IPM for this pest is yet to be developed. However, following IPM package can be adopted to minimize the fruit borer damage. In areas where brinjal is grown in large areas at a stretch at any time, all such growers have to adopt the IPM to get effective control. Adoption of IPM by a few farmers in isolation is not going to bring down the pest population.

For the management of brinjal shoot and fruit borer, combination of clipping of shoots affected by the borer at weekly interval followed by neem cake application (250 kg/ha) at 30 days after planting and 4% pulverized NSPE/1% neem soap/1% pongamia soap sprays at 60, 75, 90 and 105 days after planting was most effective with 22% borer damage as compared to 46% in control (Krishna Murthy and Krishna Kumar, 2002) (Table 2.21).

2.3.1.2. Leaf-hopper, *Amrasca biguttula biguttula*

This is another important pest of brinjal, particularly during summer. This pest also attacks cotton and okra. If adequate control measures are not taken, 50–60% yield loss can be observed in a span of few weeks.

(i) Damage: Both adults and nymphs suck the sap from leaves. The eggs are inserted into the midrib or veins. On hatching, the nymphs move along the veins and nymphs suck the leaf sap. On the older leaves, the damage is seen as yellowish-green mosaic symptom followed by brown necrosis and curling along the leaf margin (Fig 2.41).

Table 2.21. Incidence of shoot and fruit borer of brinjal under different treatments

Treatment/Dose	Cumulative % of borer infestation	Yield (kgs)
Soil application of neem cake at 250 kg/ha at 30, 60 and 90 DAT.	30.05	5.96
Shoot clipping at weekly interval + neem cake at 30 DAT + NSKE (4%) spray at 60, 75, 90 and 105 DAT.	22.04	6.72
Barrier crop of maize sown 10 days before brinjal planting (2 rows at 30 cm from brinjal and maize at 5 cm form plant to plant) + shoot clipping at weekly interval + NSKE (4%) at 45, 60, 75 and 90 DAT.	23.51	6.53
Barrier crop as above + shoot clipping + endosulfan 700 g *a.i.*/ha alternated with cypermethrin at 50 g *a.i.*/ha at 45, 60, 75 and 90 DAT.	29.44	9.01
Cypermethrin at 50 g *a.i.*/ha at 45, 60, 75 and 90 DAT.	36.62	9.59
NSKE (4%) at 45, 60, 75 and 90 DAT.	28.54	14.33
Control.	45.90	5.33
CD (P = 0.05).	**2.71**	**3.40**

Figure 2.41. Leaf hopper infestation on brinjal leaf

(ii) Management

- Soil application of neem cake 250 kg/ha followed by sprays of NSPE 4% or neem soap 1% at 10 days interval brings down the hopper population.
- Sprays of systemic insecticides like dimethoate 30 EC at 2 ml/l or imidacloprid 200 SL at 0.3 ml/l or acephate 75 SP at 1 g/l at pre-flowering stage brings down the pest population.

2.3.1.3. Ash Weevil, *Myllocerus subfasciatus*

(i) Damage: Ash weevil is another important, endemic pest of brinjal causing plant mortality. Adults feed on the foliage and cause saw teeth like damage along the leaf margin. The adults lay eggs in and around base of the plants in soil and the hatching grubs feed on roots. Affected plants wilt and dry up (Fig 2.42). Wilting is first noticed in patches in a plot. One can notice very few roots in such wilting plants when uprooted. Both the roots and plant base show scraped damage. Grubs pupate in the soil encrustation.

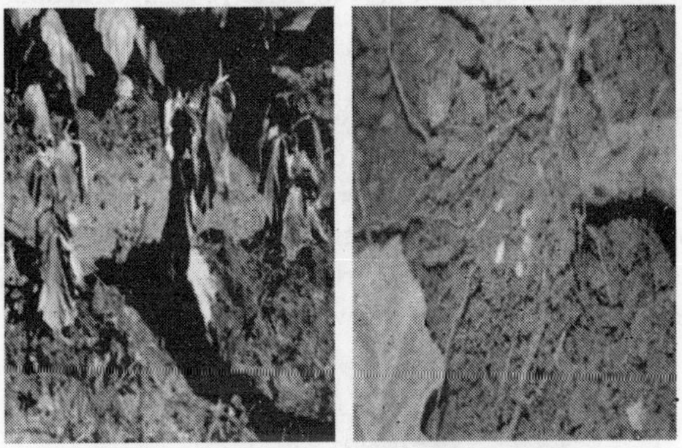

Figure 2.42. Ash weevil damage on brinjal

(ii) Management

- Soil application of neem cake/pongamia cake at the time of transplanting at 250 kg/ha.
- Repeat the neem cake application for 1 or 2 times at 30 and 60 DAP.

2.3.1.4. Epilachna Beetle, *Epilachna vigintioctopunctata*

(i) Damage: Both the grubs and the adult beetles scrape the leaves in semi-circular or half moon shaped fashion (Fig 2.43). Females lay torpedo-shaped, 15–50 yellowish eggs on the ventral surface of leaves. Heavy infestation results in leaf skeletonization. Pupation takes place on the plant itself.

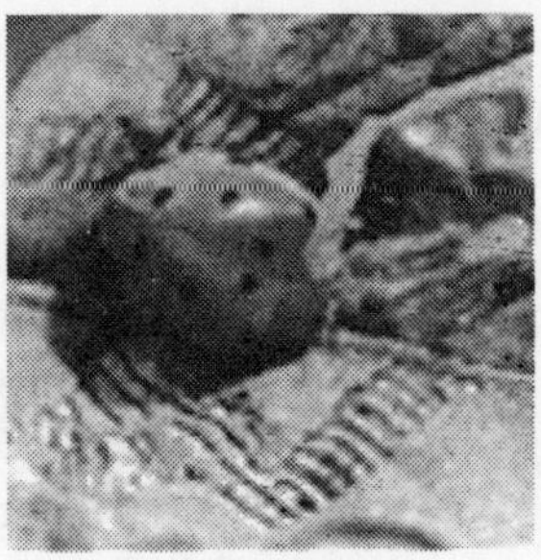

Figure 2.43. Epilachna beetle on brinjal leaf

(ii) Management

- Collection and destruction of adult beetles
- Spray any contact insecticide like carbaryl 50 WP at 3g/l or quinalphos 40 EC at 1.5 ml/l, if required.
- It was effectively kept under check by *Pediobus foveolatus* producing 60–77% parasitism.
- 'Thuricide' dust (3×10^6 spores/mg) recorded total elimination of the grubs of epilachna beetle on brinjal plants within 10 days (Sundarababu, 1970).

2.3.1.5. Red Spider Mite, *Tetranychus urticae*

(i) Damage: Brinjal is often severely affected by mites especially under dry conditions. Larvae are pink in colour while nymphs are greenish red in colour. Spider mites form small web on undersurface of the leaves where they colonize and multiply (Fig 2.44). Tiny mites (larvae, nymphs and adults) feed mainly on the lower surface of the leaves by scrapping the epidermis causing curling, shriveling and yellowing of leaves followed by foliage drying. The affected plants bear few fruits.

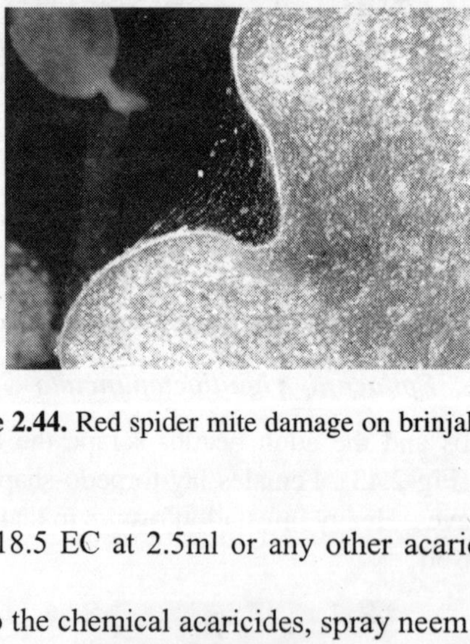

Figure 2.44. Red spider mite damage on brinjal leaves

(ii) Management

- Sprays of dicofol 18.5 EC at 2.5ml or any other acaricide like wettable sulphur 75 WP at 3g/l.
- As an alternative to the chemical acaricides, spray neem soap/pongamia soap 1%.
- Thorough spray on the under surface of the leaves is to be assured.

2.3.1.6. Gall Midge, *Asphondylia capparis*

(i) Damage: In parts of south India, gall midge, *Asphondylia capparis* is a regular pest on flowers and fruits of brinjal and sweet pepper (Tewari *et al.*, 1987; Krishna Kumar *et al.*, 1997). Maggots of this pest feed on developing ovary/fruits. When young flowers are attacked, stamens and pistils degenerate and fruits fail to form. Following the emergence of the fly, the fruit cracks along the inner side of the curve and is rendered unfit for marketing.

(ii) Management

Spraying of 4% NSKE is effective in controlling this pest.

2.3.1.7. Stem Borer *Euzophera perticella*, Leaf Roller, *Eublemma olivacea*, Lace Wing Bug, *Urentius echinus* and Thrips, *Sericothrips solanifolii*

(i) Damage: In certain parts of India, the stem borer, *Euzophera perticella* is serious. Caterpillars bore downwards killing the affected plants (Fig 2.45).

Figure 2.45. Stem borer on brinjal

Brinjal thrips, *Sericothrips solanifolii* damage leaves and fruit. Both nymphs and adults suck the sap from leaves and developing buds. Damaged buds later manifest as scarred fruits and lose market value.

Brinjal is attacked by the leaf roller, *Eublemma olivacea* which causes rolling of leaves.

Lace wing bugs, *Urentius echinus* are found in large number on abandoned or less cared plants especially in summer. Both the adults and the nymphs suck the sap and the affected plants turn yellow.

(ii) Management

- As far as possible raise the nurseries with the nylon net barrier.
- Most of these pests can be managed by application of neem cake at 250 kg/ha.
- Spray 4% NSKE at the time of first flush (Krishna Kumar *et al.*, 1997).
- Repeat the same treatment 2–3 times at 30 days interval.
- Soil application of neem cake/pongamia cake at 250 kg/ha also reduces ash weevil, gall midge and thrips damage.
- The gall midge was found parasitized by *Eurytoma* sp. to the tune of 8–15% (Tewari and Krishnamoorthy, 1986).

2.3.1.8. Mealybug, *Coccidohystrix insolita*

(i) Damage: The mealy bugs attack isolated old plants which become covered with white mass. Insects suck sap from plants making them pale and weak.

(ii) Management

Spraying 0.15% malathion or 0.4% monocrotophos gives effective control of mealybugs. It was also effectively controlled by release of the predator, *Cryptolaemus motrouzieri* (Krishnamoorthy and Mani, 1996).

2.3.1.9. Whitefly, *Bemisia tabaci*

(i) Damage: Numerous white winged fly like insects feed on the plant. They move at slight disturbance. The characteristic symptoms include yellowing of leaves (Fig 2.46). Sooty mould develops on the honeydew excreted by this insect. The pest is serious during summer.

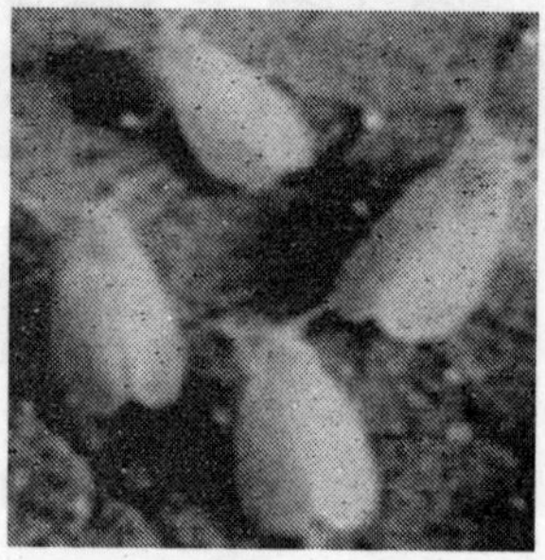

Figure 2.46. Whiteflies on brinjal leaf

(ii) Management

Spraying of neem or pongamia soap at 1% gave effective control of whiteflies on brinjal.

2.3.1.10. Aphid, *Aphis gossypii*

See Fig 2.47.

Figure 2.47. Aphids on brinjal

(i) Management

The seedling bare root dip in 3% neem oil followed by 5% NSKE spray significantly reduced the aphid infestation in brinjal (Table 2.22). Release of *Chrysoperla carnea* at 2 grubs/plant was also found effective.

Table 2.22. Effect of seedling bare root dip in neem products on brinjal aphid

Treatment	% Reduction in root galling	% Increase in yield
Neem cake extract 10%	1.36	24.86
Neem seed kernel extract 5%	1.25	30.93
Neem oil 3%	1.18	34.81
Monocrotophos 0.1%	0.67	62.98
Untreated check	1.81	

2.3.1.11. White Grub, *Holotrichia consanguinea*

(i) Management

Collection of beetles from plants during night hours and killing them in kerosene water reduced the infestation. Fungal pathogens, *Metarrhizium anisopiale* and *Bacillus popillae* var. *holotrichiae* are also effective in reducing the infestation.

Pesticide Residues

The persistence of terminal pesticide residues and waiting periods for brinjal are presented in Table 2.23 (Awasthi, 2001).

Table 2.23. Terminal pesticide residues and their waiting periods for brinjal

Pesticide schedule	Dosage/ Concentration (%)	Terminal residues (ppm/days)	Waiting period (days)
Quinalphos spray at fruiting stage	0.05	0.32/15	21
Monocrotophos spray at fruiting stage	0.05	0.90/3	10
Cypermethrin spray at fruiting stage	0.01	0.003/10	2
Fenvalerate spray at fruiting stage	0.01	0.004/10	6
Carbofuran soil application at planting	0.5 kg *a.i.*/ha	0.11/50	Safe at 1st harvest
Phorate soil application at planting	1.0 kg *a.i.*/ha	1.50/85	Safe at 2nd harvest

2.3.2. Diseases

2.3.2.1. Damping-off, *Pythium aphanidermatum*

(i) Symptoms: The pathogen causes death of plants in the nursery beds.

(ii) Management

Soil solarization using polythene sheet for 30 days during peak summer was most effective and resulted in least occurrence of damping-off (1.5%) when compared to control (30.64%). Seed treatment with *Trichoderma viride* resulted in lowest incidence of damping off (4.7%).

2.3.2.2. Root Rot, *Rhizoctonia solani*

(i) Symptoms: Generally early transplanted brinjal crop is much affected during the months of August-September. Lesions start on stem near collar region at or below the soil level and expand downward into the roots. Initially the bark become wet, soft with macerated tissue. Later on, drooping and wilting of the plant is observed. The disease is more common in poorly drained soil and the fields having prolonged excessive moisture.

(ii) Management

Field sanitation and crop rotation are most important for this soil-borne disease. All infected plants should be uprooted and burnt. Summer ploughing and fallowing should be carried out to reduce the soil inoculum. Long crop rotation, particularly with sugarcane should be followed. Proper drainage should be maintained in the field.

Seed treatment with 0.25% captan greatly reduced the inoculum from the contaminated seeds. When initial symptoms are noticed in the field, drenching the collar region with 0.1% carbendazim is recommended to reduce the disease incidence.

2.3.2.3. Collar Rot, *Sclerotinia sclorotiorum*

(i) Symptoms: Infection takes place on leaves, twigs, flowers and fruits. Water-soaked lesions develop and the infected tissue is macerated by the pathogen. White mass of mycelium grow on the surface and later embedded sclerotia are formed. These sclerotia become black after drying. The pathogen is soil-borne.

(ii) Management

Deep summer ploughing and crop rotation with flooded paddy is beneficial for reducing the soil inoculum. Crop debris and infected plant parts should be collected from the field and burnt without shattering and shedding of sclerotia in the field. Dense planting should be avoided to reduce the shade and humidity in plant canopy.

Spraying of 0.1% carbendazim should be initiated just prior to flowering stage. Alternate spray of 0.25% mancozeb can also be given.

B. subtilis, *T. viride and T. virens* were effective.

2.3.2.4. Phomopsis Blight and Fruit Rot, *Phomopsis vexans*

(i) Symptoms: In the seed beds, the disease appears as damping-off. After transplanting, the leaves show clearly defined, circular, grey to brown spots with light coloured centre. The affected leaves turn yellow and die. The lesions on stem are dark brown becoming grey in the centre. Mostly the stem base is attacked and is characterized by constriction of the base or a grey dry rot. Pale, sunken spots develop on the fruit and may progress to cover the entire fruit surface. The internal portion of the fruit rots. If the fungus enters through the calyx, the whole fruit is mummified due to dry rot.

(ii) Epidemiology: The fungus is seed-borne and also survives between crop seasons on infected plant debris in the field. The fungus requires wet weather and high temperatures for rapid development.

(iii) Management

The following practices may be adopted to manage the disease:

- Use disease-free seeds for planting.
- Collect seeds from only healthy plants and blemish-free fruits.
- Deep ploughing in summer should be practiced.
- Treat seeds in hot water at 50°C for 30 min.
- Spray foltaf or copper oxychloride both at 0.3% concentration.
- Since the pathogen attacks only brinjal, 3 years crop rotation shall be effective for the management of the disease.
- The infected crop debris should be collected and burnt in the field.

2.3.2.5. Leaf Spot, *Cercospora melonginac*

(i) Symptoms: Leaf spots are circular to irregular in shape, usually large and brown to grayish brown in colour (Fig 2.48). Infected leaves are shed.

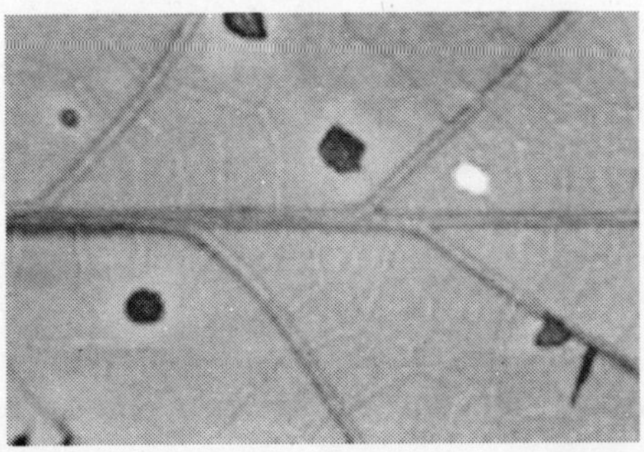

Figure 2.48. *Cercospora* leaf spot on brinjal leaf

(ii) Epidemiology: The pathogen survives in infected plant debris. Moderate temperature, cool and humid weather conditions favour the disease development.

(iii) Management

Spraying 0.1% carbendazim is effective for the management of the disease.

2.3.2.6. Bacterial Wilt, *Ralstonia solanacearum*

(i) Symptoms: Characteristic symptoms of bacterial wilt are rapid and complete wilting of normal grown up plants (Fig 2.49). Pathogen is mostly confined to vascular region.

In advanced cases, it may invade the cortex and pith and cause yellow-brown discolouraion of tissues. Infected plant parts when cut and immersed in clear water, a white streak of bacterial ooze coming out from cut ends is visible.

Figure 2.49. Bacterial wilt affected plants of brinjal

(ii) Management

Crop rotation with cowpea-maize-cabbage, okra-cowpea-maize, maize-cowpea-maize and finger millet-brinjal (Pusa Purple Cluster)-French bean are reported effective in reducing bacterial wilt. Grafting brinjal on *Solanum torvum* rootstock was found effective against wilt.

Bacterial wilt of brinjal was successfully controlled by incorporation of *Pseudomonas fluorescens*.

Brinjal *cvs*. Pusa Purple Cluster, Dingra multiple purple, Arka Keshav, Arka Nidhi, Arka Neelkanth, Gulla, Vijay hybrid and Banaras Giant Green are found resistant.

2.3.2.7. Little Leaf (Mycoplasma)

Brinjal little leaf disease was first reported by Thomas and Krishnaswamy (1939) from Coimbatore, India. The disease is caused by phytoplasma-like organism.

(i) Symptoms: The characteristic symptoms include drastic reduction in leaf size, aggregation of numerous leaves in each branch giving bushy appearance and shortening of internodes (Fig 2.50). Auxiliary buds enlarge but petioles and leaves remain shortened, narrow, pale green, soft and smooth and fruit size is reduced. Purple coloured fruits turn white after infection. At later stage, infection leads to complete sterility of the plant.

(ii) Transmission: The disease is transmitted by leaf hopper, *Hishimonas phycitis*. It is also graft transmissible.

Figure 2.50. Little leaf affected plant of brinjal

(iii) Management

Early sown crop is severely affected by the disease. Hence avoid early transplanting of the crop. Uproot the severely infected plants and burn before spraying with insecticides (4–5 sprays of dimethoate or monocrotophos each at 0.1% at 10 days interval). Removal of weed hosts like *Datura fistulosa* and *Catharanthus roseus* and diseased plants from the field and nearby areas and use of disease resistant/tolerant varieties like Pusa Purple Cluster and Arka Sheel shall keep the disease under check. Periodical spraying of systemic insecticides is recommended to control the vector. During the initial stage of infection, spray 100 ppm streptocycline to check the growth of phytoplasma. The spray should be repeated twice in early sown crop (Anjaneyulu and Ramakrishnan, 1972a). Maximum disease control could be achieved by seedling dip in tetracycline or monocrotophos followed by soil application of phorate (Maheshwari *et al.*, 1988).

Integrated Pest and Disease Management

Nursery

- Deep off-season/summer ploughing to expose resting stages of pests to sunlight.
- Soil solarization for 2–3 weeks during summer using 400 gauge thick polythene sheet.
- Soil application of *Trichoderma harzianum* mixed with FYM.
- Seed treatment with 0.25% captan or carbendazim.
- Seed treatment with *Trichoderma viride* at 4 g/kg of seeds to prevent seed and soil borne infection of fungi and nematodes.
- Raising nursery on raised beds (10 cm above ground level).
- Use of nylon net (40 gauge) to avoid entry of sucking pests.
- Use of bacterial wilt resistant varieties (Arka Nidhi, Arka Keshav, Arka Neelkant, Swarna Shree, Swarna Mani, Swarna Shyamali, Surya, Pant Rituraj, Pusa Purple Cluster).

Main field

- Avoid transplanting before last week of August for prevention of *Phomopsis* stem blight disease.

- Soil application of fresh neem cake at 250 kg/ha 30 DAT and repeat after 35–40 days to reduce shoot and fruit borer incidence.
- Erection of bird perches at 25/ha.
- Release of egg parasitoid, *T. chilonis* at 1 lakh/ha 6 times starting from shoot formation stage at weekly intervals to bring down shoot and fruit borer damage.
- Setting up of yellow sticky traps.
- Spraying of 5% NSKE to control sucking pests in early stages.
- Spraying 0.1% carbendazim after flowering.
- Installing pheromone traps at 13/ha to trap fruit and shoot borer adults.
- Collection and destruction of borer infested fruits from time to time.
- Destruction of root/collar rot, little leaf and bacterial wilt affected plants from time to time.
- Crop rotation with non-host crops like sorghum, wheat, marigold or onion to reduce nematode population in soil.

2.3.3. Nematodes

Root-knot and reniform nematodes are the major nematode problems in successful cultivation of brinjal crop.

2.3.3.1. Root-knot Nematodes, *Meloidogyne* spp.

M. incognita was responsible for 27.30 to 48.55% loss in fruit yield of brinjal (Bhatti and Jain, 1977; Parvatha Reddy and Singh, 1981; Darekar and Mahse, 1988) (Fig 2.51).

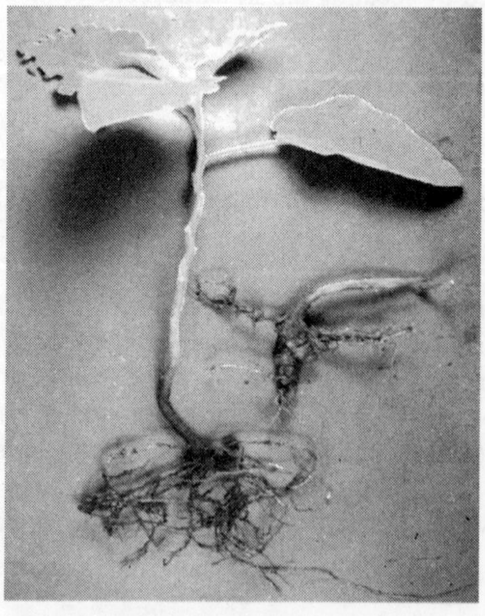

Figure 2.51. Root-knot nematode on brinjal. *Left*–Healthy; *Right*–Infected

(i) Interaction with Other Pathogens: *M. incognita* and *Ozonium texanum* var. *parasiticum* together acted synergistically by reducing the germination of brinjal to 28 per cent (Nath *et al.*, 1976). Combined effect of *M. incognita* and mycoplasma-like organisms (little leaf) on growth of brinjal was more than their individual effects (Dhawan and Sethi, 1977).

(ii) Management

(a) Physical Methods: Burning of paddy husk or saw dust on infested nursery proved to be reasonably effective.

(b) Cultural Methods: Several cultural methods like selection of nematode-free nursery sites, destruction of infested roots after crop harvest, crop rotation with non-hosts (maize, wheat, sorghum) or antagonistic crops (marigold, mustard, sesame), flooding, fallowing, deep summer ploughing and harrowing either alone or in combination proved to be reasonably effective and economical to check multiplication of root-knot nematodes on brinjal.

Decrease in *M. incognita* population on brinjal occurs when the field was left fallow or following crop rotation with marigold, spinach and bottle gourd (Khan *et al.*, 1975).

Summer Ploughing: Ploughing not only leads to disturbance and instability in nematode community but also causes their mortality by exposing them to solar heat and desiccation. Generally, 2 to 3 summer ploughings each at 10 days interval during April-May (40–46°C) have been recorded to reduce 96.5% *M. javanica* population while fallowing itself during the same period registered 44.5% reduction. Additional use of plastic sheets for covering soil either in nursery beds or in field further enhanced nematode reduction. Such an approach also helps in reducing the intensity of weeds, fungi and bacteria in the soil.

The maximum reduction in *M. javanica* population and increase in fruit yield was recorded in brinjal by soil ploughing + covering with polythene sheet, followed by soil ploughing + exposure to sun (Anon., 1989a).

Crop Rotation: Rotation of brinjal with sweet potato (*cv.* Sree Bhadra) reduced the root-knot nematode population by 47% and increased the fruit yield by 22%. Since the sweet potato *cv.* Sree Bhadra is a high yielding one, the farmer will get a good crop within 3 months. Thus it becomes an additional income generating practice (Sheela *et al.*, 2002). Crop rotation with sorghum, wheat and chilli reduced the root-knot nematode population. The population of *M. incognita* larvae decrease in combination of brinjal-wheat, brinjal-chilli-wheat and red gourd-brinjal-mustard. The root-knot larvae were completely eliminated in marigold-marigold-fallow and chilli-cauliflower-cauliflower. Garlic following brinjal also brought about considerable reduction in larvae of root-knot nematode. Intercropping with marigold, onion and garlic is also recommended. The population levels of *M. incognita* on highly infested land dropped to zero after 18 months of continuous cultivation of *Panicum maximum*. Growing of brinjal after *P. maximum* was completely free from root-knot nematodes (Netscher, 1983).

Intercropping: Intercropping brinjal with marigold, garlic or resistant tomato (SL-12) improved plant growth and reduced number of galls and final soil population of nematode (Jain *et al.*, 1990). Marigold (*cv.* Calcutta Yellow)-brinjal cropping sequence has been recommended to the farming community for management of *M. javanica* affecting brinjal (Singh, 1991).

Trap Cropping: Growing of trap crop like knol-khol with brinjal but removed and destroyed before new generation of nematode could emerge helps in reducing the root-knot nematode population (Ayyar, 1926).

Influence of Fertilizers: Row application of fly ash at 0.6 kg/sq. m. enhanced the fruit yield by 27.7% and inhibited the reproduction of *M. incognita* (Khan and Ghadipur, 2004).

Botanicals: Nursery bed treatment with neem cake at 200 g/sq.m. gave maximum reduction in nematode population giving an yield of 20.75 q/ha (Sheela and Nisha, 2004).

Incorporation of neem leaves at 500 g/sq.m. or 250 ml of 5% neem leaf extract/5% neem cake extract/sq.m. in nursery beds of brinjal gave effective control of root-knot nematodes.

Spot application of mustard cake at 25 g/spot and furrow application of saw dust at 210 g/furrow of 3m length were found most effective in reducing the nematode infestation and increasing the yield of brinjal. Similarly, Hazarika (1990) found that spot application of castor/mustard/neem cake at 15 g/spot and furrow application of the above cakes at 100 g/3 m furrow reduced the infestation of *M. incognita* and increased the yield of brinjal. The yield was maximum in spot application of neem cake along with maximum reduction of *M. incognita* population in soil.

(c) Chemical Methods

Nursery Bed Treatment: Treatment of brinjal nursery beds with carbofuran at 3g *a.i.*/m^2 led to 48% increase in yield and at 6g *a.i.*/m^2 gave 60.8% higher yield.

Main Field Treatment: Aldicarb, carbofuran, ethoprophos and phenamiphos each at 1 to 2 kg *a.i.* per ha were found effective in reducing the root-knot nematode population and in increasing fruit yields of brinjal (119 to 145 per cent) (Singh *et al.*, 1978).

Alam *et al.* (1973) reported that oxamyl was effective for the control of root-knot nematodes infecting brinjal. Dichlofenthion was found effective against *M. incognita* in brinjal (Alam *et al.*, 1973).

(d) Biological Methods

Antagonistic Bacteria: Application of *Pasteuria penetrans* in nursery soil of brinjal resulted in 31 and 36% suppression in root galling when seedlings were transplanted to sterilized and nematode-infested soils, respectively. The results were also indicative of easy dissemination of *P. penetrans* to the main field through infected seedlings (Walia *et al.*, 1992).

Nursery bed treatment with *Bacillus macerans* at 25 g/sq.m. + soil drench (2% solution) ten days after sowing gave maximum reduction in nematode population and maximum yield of 38.25 q/ha (Sheela and Nisha, 2004).

Avermectins effectively reduced root-knot nematode infection in brinjal under field conditions (Parvatha Reddy and Nagesh, 2002).

Antagonistic Fungi: Application of 'Royal 350' (*Arthrobotrys irregularis* cultured on oat seed medium) at 140 g/m^2 a month before transplantation of brinjal resulted in good protection against root-knot nematodes. Among the egg parasites, efficacy of *Paecilomyces* *lilacinus* (commercially formulated as 'Biocon' in Philippines) has been found to be comparatively higher in suppressing the population of *Meloidogyne* spp. and *R. reniformis* on brinjal. Soil application of *P. lilacinus* at 20 g/sq. m. to brinjal nursery reduced infection by *M. incognita* by 50% and improved growth of seedlings by 30%.

Application of *Pochonia chlamydosporia* at 50 g/m^2 in the nursery beds significantly increased the plant growth parameters (42 and 50% increase in height and weight of seedlings, respectively over control) and root colonization by the bioagent on transplants. When the above seedlings were transplanted in field, there was reduction in root galling index (5.2 compared to 7.8 in control), number of egg masses in 5 g root (20 compared to 49 in control), number of juveniles in 5 g root (62 compared to 162 in control) and number of juveniles in 100 cc of soil (67 compared to 152 in control) and increased plant growth and yield in eggplant at harvest (Naik, 2004).

(e) Host Resistance: Brinjal *cvs*. Vijaya and Banaras Giant were reported to be moderately resistant to *M. incognita*.

(f) Integrated Methods

Using Bio-agents and Botanicals: Integration of *P. chlamydosporia* (at 100 ml/seed pan containing 1.2 × 10^4 spores/ml) in castor cake (at 40 g/seed pan) amended soil was effective in increasing the seedling weight and colonization of roots with the bioagent. Brinjal seedlings raised in the above treatment transplanted in pots gave maximum increase in plant growth, root colonization and parasitization of eggs of *M. incognita* by *P. chlamydosporia*. The above treatment also gave least root galling and final nematode population both in soil and roots (Rao and Parvatha Reddy, 1993b).

Borkakaty (1993) observed that inoculation of *P. lilacinus* at 4 g/kg of soil in combination with mustard oil cake at 0.5 and 1.0 t/ha increased plant growth with corresponding decrease in number of galls, egg masses and eggs/egg mass of *M. incognita* on brinjal.

Application of 10% neem cake extract (at 20 ml/pot) mixed with spores of *P. lilacinus* (at 1 × 10^6 spores/ml) was effective in increasing plant growth and reducing root galling and final nematode population both in soil and roots. The above treatment also gave maximum parasitization of egg masses of *M. incognita* and spore density of *P. lilacinus* in soil (Rao and Parvatha Reddy, 1994).

Significant reduction in root galling and final nematode population of *M. incognita* were observed in brinjal seedlings which were given bare root-dip treatment in 10% neem leaf suspension mixed with spores of *P. lilacinus* (at 4 × 10^5 spores/ml) for 30 min. Significant increases were also observed in root colonization, parasitization of eggs and spore density of *P. lilacinus* in soil (Rao *et al.*, 1997b).

Root-dip treatment of brinjal seedlings in neem cake extract based formulation of *P. lilacinus* (at 5 × 10^6 spores/ml) for 20 min and planted in pots gave significant increase in plant growth, root colonization , propagule density in soil and parasitization of eggs of *M. incognita* by *P. lilacinus* and drastic reduction in root galling, fecundity and final nematode population in soil and roots (Rao *et al.*, 1998a).

Application of castor cake extract based formulation of *T. harzianum* (at 500 ml/m^2 containing 9.9 × 10^3 spores/ml) to nursery beds of brinjal was effective in producing vigorous seedlings (with maximum seedling weight) with least root galling. The above treatment also increased root colonization and parasitization of *M. incognita* females by *T. harzianum* (Rao *et al.*, 1998c).

Using Arbuscular Mycorrhizal Fungi and Botanicals: Brinjal seedlings raised in nursery treated with *G. fasciculatum* (at 250 g/m^2 containing 16 chlamydospores/g) planted in pots amended with castor cake (applied at 10 g/kg soil 15 days before transplanting) resulted in maximum plant growth, root colonization with *G. fasciculatum* and chlamydospore density in soil and least root galling, fecundity and final nematode population in soil and roots (Rao *et al.*, 1998).

Integration of *G. fasciculatum* and neem cake at 0.5 t/ha was found to be effective in increasing plant growth parameters and yield (30 t/ha compared to 17 t/ha in control) and in reducing the root-knot index (2.6 compared to 4.8 in control) and final nematode population (173.3/250 ml soil compared to 510.8/250 ml soil in control) (Borah and Phukan, 2004).

Using Bio-agents and Arbuscular Mycorrhizal Fungi: Brinjal seedlings raised in seed pans treated with *G. mosseae* (at 100 g/seed pan containing 28-32 chlamydospores/g) dipped in *P. lilacinus* spore suspension (containing 4 × 10^5 spores/ml) for 5 min. and transplanted in pots gave maximum increase in plant growth, root colonization, propagule density in soil of both *G. mosseae* and *P. lilacinus* and parasitization of eggs of *M. incognita*. The above treatment also gave least root galling, fecundity and final nematode population in soil and roots (Rao *et al.*, 1998).

Using Bio-agents, Arbuscular Mycorrhizal Fungi and Botanicals: Strategies to integrate botanicals, bio-agents and endomycorrhizae components for the management of root-knot nematodes infecting brinjal have been developed. These strategies include inoculation of *G. fasciculatum* in the castor cake amended nursery beds followed by the root-dip treatment of mycorrhizal seedlings of brinjal in spore suspension of *P. lilacinus* for the management of *M. incognita* (Rao et al., 1993b).

In nursery, soil application of neem cake at 400 g/m^2 along with *G. mosseae* (at 500 g/m^2 containing 26–32 chlamydospores/g) and *P. lilacinus* (at 2 litres/m^2 containing 6 × 10^5 spores/ml) resulted in production of healthy and vigorous brinjal seedlings colonized with *G. mosseae* and *P. lilacinus* and with least root galling. In field, transplanting of brinjal seedlings raised in the above treatment gave maximum reduction in root galling, egg mass production, fecundity and final nematode population in soil and roots. The above treatment also gave highest fruit yield, root colonization with *G. mosseae* and *P. lilacinus* and egg parasitization with *P. lilacinus* (Rao and Parvatha Reddy, 2001).

Using Two Bioagents: Combined soil application of *P. lilacinus* and *Aspergillus niger* at the time of transplanting brinjal is very effective in reducing root-knot nematodes.

Integration of highly toxic fungus, *Aspergillus niger* (kills most of the infective second stage juveniles) and an egg parasite, *Cladosporium oxysporum* (invades and kills the eggs in egg sac) both at half the doses significantly reduced *M. incognita* population and exhibited better plant growth than when either of the fungal bioagents in brinjal (Goswami and Singh, 2001).

In egg-plant, the nursery seedling stand index was good with seed treatment with *P. fluorescens* (50 g/kg) + soil application (10 g/m^2 seed bed) followed by soil application of neem based *T. harzianum* + *P. fluorescens* and neem based *P. fluorescens*, where the crop stand index was 4.6 and 4.0, respectively.

Using Bio-agents and Chemicals: Split application of *T. harzianum* at 50 kg/ha (10^8 cfu/g) (before transplanting and 45 days after transplanting) + carbofuran at 16.5 kg/ha was effective in increasing the brinjal fruit yield (Haseeb *et al.*, 2004).

Using Cultural and Chemical Methods: The planting of marigold combined with application of carbofuran at 1 kg *a.i.*/ha controls *M. javanica* infestation on brinjal (Singh, 1991).

Nursery treatment with carbofuran at 0.3g *a.i.*/m^2 along with main field treatment with ploughing + exposing the field and ploughing + covering with polythene sheets for 15 days improved the plant growth and yield of brinjal by 20–32% (Sheela *et al.*, 2002). (Table 2.24).

Table 2.24. Integrated management of root-knot nematodes by summer ploughing/fallowing and nursery bed treatment in brinjal

Treatment	Reduction in nematode population (%)		Yield in kg/sq.m.	
	Hisar	Vellayani	Hisar	Vellayani
Ploughing + Fallowing for 15 days	87.00	66.20	28.98	—
Ploughing + Covering with polythene sheet for 15 days	94.90	77.00	18.84	78.00
No ploughing + No covering	77.60	—	—	—

Using Botanicals and Chemical Methods: Application of aldicarb at 1.0 kg *a.i.*/ha in the nursery beds along with neem cake at 400 kg/ha increased yield and reduced root galling (Singh and Gill, 1998) (Table 2.25).

Table 2.25. Integrated management of root-knot nematodes in brinjal with neem cake and nematicides

Treatment	% Reduction in gall index	% Increase in yield
Aldicarb (1.0 kg *a.i.*/ha) + Neem cake (400 kg/ha)	92.90	58.70
Carbofuran (1.0 kg *a.i.*/ha) + Neem cake (400 kg/ha)	68.10	68.10

Using Cultural and Physical Methods: Integrating summer ploughing with soil solarization with polythene mulching effectively reduced *M. javanica* population in brinjal (Jain and Gupta, 1991). Soil solarization of nursery beds (using 100 gauge LDPE clear film for 15 days) and application of neem cake at 200 g/sq.m. proved most effective in reducing root-knot nematode infestation and increasing yield in brinjal. Solarization of nursery beds for 15 days in summer and application of poultry manure at 200 kg/ha gave maximum yield.

2.3.3.2. Reniform Nematode, *Rotylenchulus reniformis*

(i) Management

(a) Chemical Methods: Bare root dip treatment with aldicarb/carbofuran/turbofos at 500 or 1,000 ppm for 15 to 30 min in brinjal economized the reniform nematode management by reducing the quantity of nematicide required (Prasad and Krishnappa, 1981).

(b) Integrated Methods: Integration of a bio-agent, *P. lilacinus* with carbofuran at 1 kg *a.i.*/ha was found effective in the management of reniform nematode, *R. reniformis* infecting brinjal (Parvatha Reddy and Khan, 1989).

2.3.3.3. Root-knot Nematode, *Meloidogyne incognita* and Bacterial Wilt, *Ralstonia solanacearum* Disease Complex

Eggplant is prone to many soil borne diseases among which the bacterial wilt (*R. solanacearum*) in combination with root-knot nematode (*M. incognita*) takes heavy toll every year all over the world (Naik, 2004).

(i) Symptoms: The root-knot nematode, *M. incognita*, present along with *R. solanacearum*, greatly increases the incidence of bacterial wilt of brinjal (Fig 2.52). The root-knot nematode is responsible for breaking bacterial wilt resistance in "Pusa Purple Cluster" cultivar of brinjal (Parvatha Reddy *et al.*, 1979). *M. incognita* interacts with *R. solanacearum* in increasing the speed of development and severity of wilt disease of eggplant. Maximum disease occurred when the bacterium and nematode were inoculated simultaneously. Inoculation with *R. solanacearum* alone resulted in less severe disease. It was concluded that root-knot nematodes modify the plant tissues facilitating bacterial colonization (Nayar *et al.*, 1988).

Figure 2.52. Root-knot nematode and bacterial wilt disease complex in brinjal

(ii) Management

(a) Integrated Methods: Combined application of *T. harzianum* and *P. chlamydosporia* gave effective management of the disease complex and gave least root gall index (RGI), least wilt index (WI) and maximum fruit yield (1.8 RGI, 22.18% WI and 2.024 kg/3 sq.m. yield) followed by *P. fluorescens* + *P. chlamydosporia* (1.81 RGI, 25.13 WI and 1.782 kg/3 sq.m. yield) and *T. harzianum* + *P. fluorescens* (1.93 RGI, 27.28 WI and 1.680 kg/3 sq.m. yield) (Naik, 2004).

2.4. CHILLI

Chilli (*Capsicum annuum*) is valued for its diverse commercial uses. India is a major producer, exporter and consumer of chilli. Indian chillies reach over 90 countries in the world. Bangladesh, Bahrain, Canada, Sri Lanka, Saudi Arabia, USA and UAE are the leading importing countries. In India, chilli is grown in almost all states. It is being cultivated in 0.834 million hectares producing 0.847 million tonnes of chillies with an average yield of 1.0 tonne per hectare. Andhra Pradesh has been the largest chilli-growing state followed by Karnataka and Maharashtra. Productivity of chilli is highest in Andhra Pradesh, followed by Arunachal Pradesh and Punjab. The productivity of chilli in Andhra Pradesh and Punjab is higher since the crop is raised under irrigated conditions compared with Maharashtra and Karnataka. Chilli is rich in vitamins A and C.

2.4.1. Pests

Chilli thrips (*Scirtothrips dorsalis*), yellow mite (*Polyphagotarsonemus latus*) and fruit borers, *Helicoverpa armigera* and *Spodoptera litura* are the major pests of these crops. The thrips and mite damage often occur at the same leaf and is called churda murda complex.

2.4.1.1. Thrips, *Scirtothrips dorsalis*

(i) Damage: Both adults and nymphs suck the sap from young leaves. Affected leaves curl upwards along the margin, and get crinkled and reduced in size (Fig 2.53). When the incidence is severe, leaves drop and cause heavy reduction in yield. The pest is serious during dry monsoon periods and summer months.

(ii) Management

- Spray Acephate 75 SP at 1.5 g/l or fipronil at 1 ml/l or lamda cyhalothrin 5 EC at 0.75 ml/l or imidacloprid 200 SL at 0.3 ml/l alternately at fortnightly interval. Never repeat the same chemical again and again.
- Spray of difenthiuron at 225 g *a.i.*/ha and monocrotophos at 150 g *a.i.*/ha effectively checked the population of thrips on chilli with higher yield.
- Mix acephate 0.5g, 2 ml of pongamia oil and 1 ml sticker and make an emulsion (add a little water and shake thoroughly in a bottle) and make the volume to 1 litre and spray.
- Spraying of Vertimec at 0.3 to 0.4 ml/l was found effective against thrips and mites for 45 days.

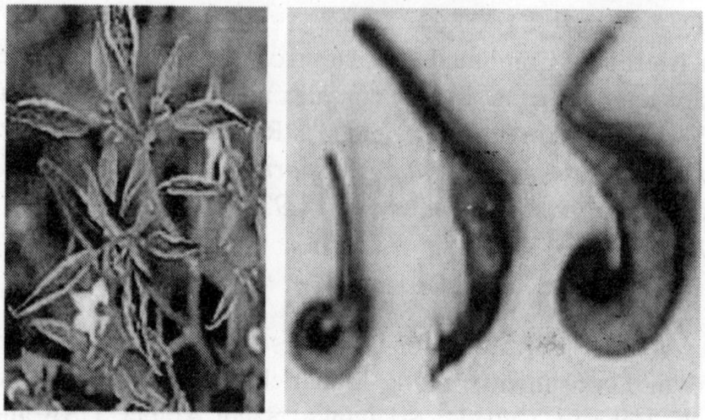

Fig. 2.53. Thrips on leaves and fruits of chilli

- Soil application of aldicarb at 2.8 kg *a.i.*/ha 15 days after transplanting effectively checked thrips and mites and resulted in minimum leaf curl incidence with higher yield of marketable green chilli fruits (Patel *et al.*, 1983).
- Soil application of carbofuran at 0.5 kg *a.i.*/ha 15 days after transplanting followed by spray of 0.1% triazophos commencing from 40 days after transplanting at 10 days interval effectively controlled thrips and increased marketable green chilli fruits (Patel *et al.*, 1997).
- Chilli *cvs.* NS 1101, NS 1701(Namdhari seeds), NP 46A, Pusa Jwala, Arka Lohit and K 2 (Kovilpatti) are tolerant to thrips.
- *Chrysoperla carnea* can be released at 2 grubs/plant.

Pongamia oil 2 ml when mixed with 0.5 g acephate + 2 ml sticker gave excellent control of thrips, *Scirtothrips dorsalis* in chillies, even during summer (Krishna Murthy and Krishna Kumar, 2002). The thrips damage rating was reduced to 20.67 in the above treatment as compared to 45.67 in control. Similarly, the yield was also increased in treated plots over control (Table 2.26).

Table 2.26. Thrips damage and yield under different management programmes in chilli (Summer, 2002)

Treatment	Dose	Rating of thrips damage/5 leaves	Yield kg/plot
Acepahte +Pongamia oil	0.5 g + 2 ml	20.67	1.69
Dimethoate +neem oil	2 ml + 2ml	30.00	1.26
Pongamia oil	20 ml	45.67	1.05
Neem oil	20 ml	40.67	3.50
Dimetoate	2 ml	25.00	0.90
Neem seed powder	4%	40.00	0.05
Acephate	0.75 g	10.67	1.35
Control	—	45.67	0.02
CD 5%	—	**7.33**	**0.10**

2.4.1.2. Pod Borer, *Helicoverpa armigera*

(i) Damage: Tomato fruit borer, *H. armigera* is also a serious pest in chilli and capsicum. It affects the marketable value of the chilli crop to a great extent, if proper care is not taken (Fig 2.54).

Figure 2.54. Pod borer, *Helicoverpa armigera* damage on chilli fruit

(ii) Management

- Two rounds of sprays of *Ha* NPV at 250 LE/ha (6×10^{12} PIB/ha) at 10 days interval were found effective in reducing the loss caused by borers.
- Inundative release of *Trichogramma* sp.-I6 releases of *Trichogramma brasiliensis, T. pretiosum* or *T. chelonis* at 50,000/ha at weekly intervals.
- Use marigold as trap crop (one row of marigold for every 18 rows of chilli) for managing *H. armigera*.
- Installation of pheromone traps at 5/ha.
- Hand collection of larvae on main and trap crops.
- Spraying with 5% NSKE or *Bt* at 500 g/ha.
- Spraying 0.07% endosulfan or monocrotophos or 0.15% carbaryl or 0.05% chlorpyriphos or quinalphos is effective for the management of the pest.

2.4.1.3. Aphid, *Myzus persicae*

(i) Damage: The aphid also becomes a serious pest on chilli occasionally. The infestation may be noticed in a few pockets of a chilli crop. Aphids suck plant sap and cause reduction in plant vigour. Small green insects are clobbered underneath the leaf surface (Fig 2.55). Sooty mold appears on leaves due to secretion of honeydew by aphids. Feeding of ants on aphids may be observed.

(ii) Management

Sprays of systemic insecticide or 1% neem or pongamia soap gives good control of this pest.

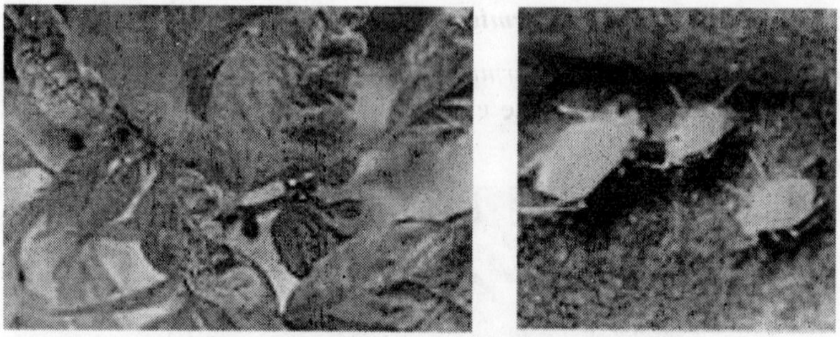

Figure 2.55. Aphid infestation on chilli leaf

Parasitoids, *Aphidius platensis* and *Aphelinus* sp. produced 80–98.8% parasitism in an insecticide-free field (Mani and Krishnamoorthy, 1994). The predators, *Chrysoperla carnea* and *Mallada boninensis*, very effectively controlled aphids on chilli when releases were made at 1: 5 ratio of predator: prey (Mani and Krishnamoorthy, 1997a).

2.4.1.4. Tobacco Worm, *Spodoptera litura*

(i) Damage: Feeding holes on the leaves can be observed due to tobacco worm infestation. Larval entry holes can also be observed near the calyx. The pest infestation causes dropping of young fruits. Mature pods with seed damage, fecal pellets with seed and watery cavity can be seen.

(ii) Management

Collect and destroy egg masses and immature larvae of *S. litura*. Use poison baiting (10 kg rice flour + 1 kg of jaggery + 250 g of methomyl 40 SP) for *S. litura*. Repeat the baiting 2–3 times, if necessary. For *S. exigua,* spray indoxacarb 14.5 SC at 0.75 ml/l or spinosad at 0.75 ml/l or thiodicarb 75 WP at 0.75 g/l. Spraying 0.07% endosulfan or monocrotophos or 0.15% carbaryl or 0.05% chlorpyriphos or quinalphos is effective for the management of the pest.

Two rounds of sprays of *Sl* NPV at 250 LE/ha (6×10^9 PIB/ha) at 10 days interval were found effective in reducing the loss caused by *S. litura*. It can also be controlled by installation of pheromone traps at 5/ha (Lures should be replaced once in 15 days), hand collection of egg masses and early instar larvae, release of *Telenomus remus* at 50,000/ha, installation of bird perches at 50/ha and spraying with 4% NSKE.

2.4.1.5. Leaf Curl Complex

(i) Symptoms: Chilli leaf curl complex is mainly caused by yellow mite and thrips infestation. It is a major problem on the chilli crop. Sometimes, the Gemini virus also causes enation leaf curl symptom. Reduction in size of leaves, shortening of veins, puckering, sometimes mottling and stunting of plants are common symptoms of the disease. Veins are prominently visible on the lower side of the leaves. The infected plants remain stunted and bear no fruits.

(ii) Management

Management of the viruses as well as thrips and mites involves periodical spraying of insecticides and acaricides. Alternate sprays of dicofol at 2.5 ml/litre of water followed by 0.25% wettable sulphur may be taken up to prevent the development of resistance. Two sprays of systemic insecticide are essential for the control of thrips and mite vectors.

2.4.1.6. White Grub, *Holotrichia consanguinea*

(i) Damage: White grubs cause wilting of plants. The plants can be easily pulled out. C-shaped white grubs are found near root region (Fig 2.56) and cause damage to roots.

Figure 2.56. White grubs on capsicum

(ii) Management

Only well rotten farm yard manure should used to prevent root grubs. Collection of beetles from plants during night hours and killing them in kerosene water reduced the infestation. Application of neem cake at 250 kg/ha along with FYM is effective against the pest. In severe cases, application of phorate at 2.5 kg *a.i.*/ha is effective. Fungal and bacterial pathogens, *Metarrhizium anisopiale* and *Bacillus popillae* var. *holotrichiae* are also effective in reducing the infestation.

2.4.1.7. Yellow Mite, *Polyphagotarsonemus latus*

(i) Damage: These infest terminal young leaves like thrips. Mite affected leaves curl downwards and appear twisted (Fig 2.57). Elongation of petioles is another symptom of white mite infestation. Damage is high in hot and humid seasons.

(ii) Management

- Apply wettable sulphur 80 WP at 3g/l or any acaricide (directing the spray on the ventral surface of leaves).
- Spraying of pongamia oil mixed with insecticides given under thrips management will also reduce mites.
- Spray neem seed powder extract 4% at 10 days interval is effective when the pest incidence is low. As and when the pest incidence increases give synthetic acaricide spray.

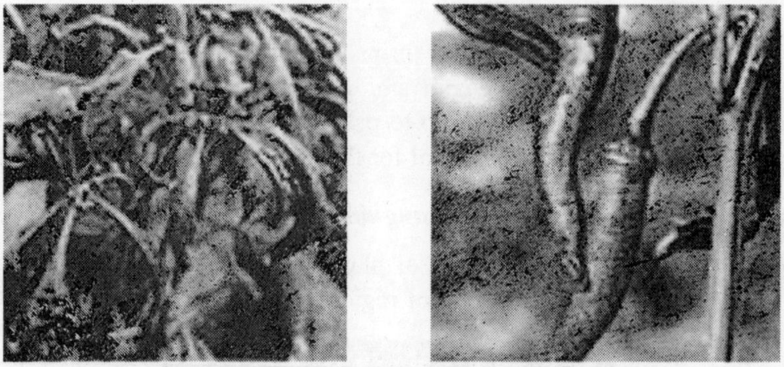

Figure 2.57. Yellow mites on leaves and fruits of chilli

- Spraying of Vertimec at 0.3 to 0.4 ml/l was found effective against thrips and mites for 45 days.
- *Chrysoperla carnea* can be released at 2 grubs/plant.

IPM

The following IPM can be adopted for managing chilli insect pests:

- In nursery, spray a systemic insecticide like imidacloprid (0.3 ml/l) or acephate (0.75 ml/l).
- Dip the roots in a systemic insecticide like imidacloprid (0.3 ml/l) or acephate (0.75 ml/l) for 5 min just before transplant.
- Apply 250 kg/ha neem cake to soil while preparing land.
- Spray acephate 75 SP at 1.5 g/l or fipronil at 1 ml/l or lamda cyhalothrin 5 EC at 0.75 ml/l or imidacloprid 200 SL at 0.3 ml/l alternately at fortnightly interval. Never repeat the same chemical again and again.
- Mix acephate 0.5 g, 2 ml of pongamia oil and 1 ml sticker and make an emulsion (add a little water and shake thoroughly in a bottle) and make the volume to 1 litre and spray.
- In endemic areas where *Helicoverpa* is the major problem, use marigold as trap crop and take 3–4 NPV sprays at flowering.
- In endemic area of *Spodoptera*, give 3–4 sprays of species specific NPV (either *S. litura* or *S. exigua*).
- Monitor for *Spodoptera* eggs and leaf damage. Mechanically collect and destroy eggs and gregarious larvae.
- Give poison baits to grown up *Spodoptera* larvae.

Pesticide Residues

The persistence of terminal pesticide residues and waiting periods for chilli are presented in Table 2.27. (Awasthi, 2001).

Table 2.27. Terminal pesticide residues and their waiting periods for chilli

Pesticide schedule	Dosage/ Concentration (%)	Terminal residues (ppm/days)	Waiting period (days)
Carbaryl spray at fruiting stage	0.10	0.14/10	2
Monocrotophos spray at fruiting stage	0.05	0.21/7	7
Quinalphos spray at fruiting stage	0.05	0.20/15	15
Cypermethrin spray at fruiting stage	0.01	0.06/10	3
Fenvalerate spray at fruiting stage	0.01	0.02/15	5
Carbofuran soil application at planting	1.0 kg *a.i.*/ha	0.05/20	60

2.4.2. Diseases

2.4.2.1. Damping-off, *Pythium aphanidermatum*

(i) Symptoms: The pathogen is responsible for death of seedlings in nursery beds.

(ii) Management

Soil solarization using polythene sheet for 30 days during peak summer was most effective and resulted in least occurrence of damping off (1.28%) when compared to control (30.64%). Seed treatment with *Trichoderma viride* resulted in lowest incidence of damping off (4.7%).

Delivery of *T. harzianum* through solid matrix priming reduced the incidence of damping off in nursery. SMP is a process in which moistened seeds are mixed with an organic carrier and the moisture content is brought to a level just below that required for seed sprouting. SMP can further enhance the efficacy of the antagonist in treated seeds. *Trichoderma* spp. grow on seed surface during priming process and increase in numbers and other microbes may not easily dislodge it.

Integration of soil solarization, application of neem cake and *T. viride* gives effective control of the disease.

2.4.2.2. Phomopsis Blight and Soft Rot, *Phomopsis vexans, Diaporthe vexans*

(i) Symptoms: Infected seeds may cause damping-off in the nursery. Leaves of seedlings show clearly defined circular, light brown spots mostly on lower leaves. Centre of the spots is slightly lighter in colour. Old spots are having numerous black dot-like pycnidia. These spots are papery, often crack and sometimes with shot holes. Almost similar symptoms are observed on the leaves of transplanted crop in the main field. Sometimes irregular big lesions also are observed on leaves, particularly in cool and humid weather. Stem lesions start from basal part of nodal portion as grey dry rot with constriction. This leads to drying of few twigs or partial wilting of the portion of the plant. Pale to light brown sunken spots develop on the old fruits. Individual spots expand and coalesce to cover the entire fruit which leads to fruit rot.

(ii) Management

Deep summer ploughing, seed treatment in hot water at 50°C for 30 min, crop rotation with non-solanaceous crops and collecting and destruction of crop debris shall keep the disease under check. Seeds should be collected from healthy fruits and disease-free field. Seed treatment with 0.25% carbendazim is essential to eliminate any seed inoculum. At least 3-year crop rotation is required with non-host crops to manage the disease. Protective sprays of 0.1% carbendazim at 10 days interval check the disease initiation in the field. Use tolerant varieties like Ramnagar Giant, Pusa Hybrid-5, KS-24 and Green Long.

2.4.2.3. Wilt, *Fusarium oxysporum*

(i) Symptoms: Symptoms include leaf chlorosis, vascular discoloration, and wilting of chilli pepper plants. High temperature and high moisture are conducive to symptom development.

(ii) Management

Reducing saturated conditions in soil is an important step in managing the disease. If economically justified, soil fumigants and solarization may be used to reduce pathogen populations in soil.

Seedling root-dip treatment and soil application of non-pathogenic *Fusarium* (Fo 52) significantly suppressed the wilt incidence. The combination of Fo 52 with *P. fluorescens* gave maximum inhibition of wilt.

2.4.2.4. *Phytophthora* Root Rot, Fruit Rot and Leaf Blight, *Phytophthora capsici*

(i) Symptoms: *Phytophthora* root rot is caused by the fungal organism *Phytophthora capsici*. Infection of the root and lower portion of the stem leads to plant wilting (Fig 2.58), which is the most conspicuous symptom of the disease. In furrow-irrigated fields, there is a row-delimited pattern of wilted plants (Fig 2.58). Over time, wilting is displayed over the entire field. *Phytophthora capsici* produces specialized swimming cells, known as zoospores, which enable spread of the pathogen from plant to plant.

Figure 2. 58. Blackening of chilli plant root, wilting of plant and a field severely affected by *hytophthora* root rot (Courtesy: D. Lindsey)

This is a disease of rainy season, characterized by small, water-soaked spots appearing on fruits leading to complete rot of fruits. On leaves, water-soaked bleached spots appear, resulting in blighting. It causes damage to plants and affect fruit yield.

(ii) Management

Management of *P. capsici* consists of reducing soil water saturation; reducing populations of the pathogen through crop rotation, soil solarization, and application of organic amendments; and restricting pathogen dispersal through mulching and judicious use of fungicides such as metalaxyl and mefenoxam (Ristaino and Johnston, 1999).

Use of disease-free seeds for raising seedlings, collection and destruction of rotten fruits and soil application of neem cake + *Trichoderma viride* reduce the disease incidence.

Treat seeds with chlorothalanil or captan or metalaxyl at 2 g/litre. Spray metalaxyl or mancozeb at 0.2% just before onset of monsoon and 1% Bordeaux mixture or copper oxychloride at 0.3% at 8–10 days interval. Spray of cymoxanil, propineb mancozeb or chlorothalanil at 0.2% is also helpful to fruits.

2.4.2.5. Anthracnose or Dieback, *Colletotrichum dematium*

(i) Symptoms: This is characterized by small grey spots on leaves. It causes necrosis of twigs, lower branches or the entire top of the plant may wither away. Black dots (acervuli) are formed all over the necrotic surface of the affected twig. Later dark brown to black sunken spots having pink spore masses appear on fruits (Fig 2.59). It may cause losses up to 25% under suitable weather conditions. Fruit rot extends to the seed cavity making it internally seed-borne.

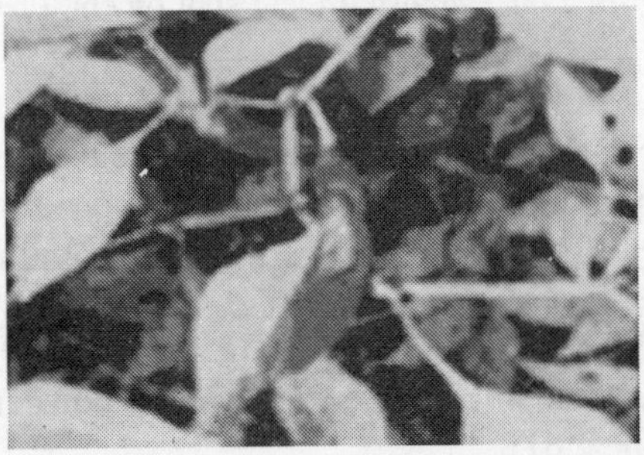

Figure 2.59. Anthracnose on chilli fruit

(ii) Management

Disease-free seeds should be collected from healthy fruits. Avoid apical damage during transplanting and also in the main field particularly at flowering stage. Well drained and porus soil reduces the disease incidence. Pick rotten fruits and destroy.

Treat the seed with 0.25% carbendazim or thiram 75 WP at 2 g/kg seed. Seedlings should be sprayed with 0.1% carbendazim before transplanting. Spray propineb or mancozeb at 0.25% or carbendazim or thiophanate methyl at 0.1% or 0.3% copper oxychloride at fortnightly intervals in the main field. Collect diseased plants and burn. To control post-harvest infection, spraying with benomyl/thiophanate methyl at 1 g/litre helps in reducing the disease incidence.

If it is a seed crop, spraying with metalaxyl mancozeb or cymoxanil mancozeb or Al-Fosetyl at 0.2% at fortnightly intervals gives very good control.

Cv. Arka Lohit is tolerant to anthracnose disease.

2.4.2.6. Leaf Blight, *Alternaria alternata, Cercospora capsici*

(i) Symptoms: Symptoms of *Cercospora* leaf blight are observed as circular spots with light grey to white grey centres with dark brown margins (Fig 2.60). In severe cases, small spots coalesce and lead to defoliation. *Alternaria* leaf spots are covered with a dark grayish brown to black spores. In moist condition, sporulation of both the pathogens was observed on the spots. Disease leads to severe defoliation and affects flowering and fruiting in infected plants.

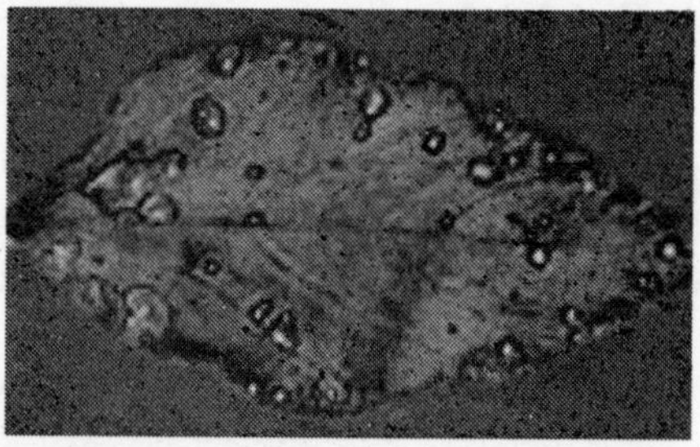

Figure 2.60. *Cercospora* leaf spot on chilli leaf

(ii) Management

Selection of disease-free seed is important to check the primary infection. Plant vigour should be maintained by spraying 0.2% tricel after fourth picking of green chillies. Foliar spray of 0.2% chlorothalanil is a must to protect the crop from the disease.

2.4.2.7. Powdery Mildew, *Levillula taurica*

(i) Symptoms: Powdery mildew infection causes whitish powdery patches both on upper and lower leaf surface (Fig 2.61).

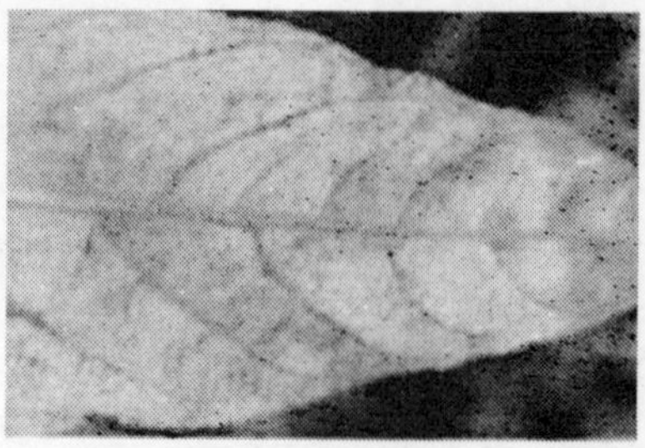

Figure 2.61. Powdery mildew on chilli leaf

(ii) Management

Baycor at 0.2%, Devisulphur at 0.3% and Bayleton at 0.05% have been reported to be very effective for the management of this disease (Keshwal and Choubay, 1983).

2.4.2.8. *Verticillium* Wilt, *Verticillium dahliae* and *V. albo-atrum*

(i) Symptoms: *Verticillium* wilt of chilli is caused by *Verticillium dahliae* and *V. albo-atrum*. Visual symptoms of the disease include stunting, defoliation, and wilting, with discoloration of the vascular system. In the field, the disease first appears in a single plant or a small group of wilting plants (foci) scattered throughout the field (Fig 2.62), depending on the distribution of the pathogen in soil. The disease gradually spreads from these foci, and this expansion may cover the entire area of plantings. Infected plants may exhibit physiological changes such as reduction in photosynthesis, increased plant transpiration, and respiration.

(ii) Management

Application of chloropicrin and metham-sodium fumigants, and use of organic amendments (Lazarovits *et al.*, 2000), may decrease *Verticillium* wilt through reduction of pathogen populations in soil.

Figure 2.62. Early stage of *Verticillium* wilt, and scattered plants of chilli affected by *Verticillium* wilt (Courtesy: D. Lindsey)

2.4.2.9. *Rhizoctonia* Root Rot, *Rhizoctonia solani*

(i) Symptoms: *Rhizoctonia* root rot caused by *Rhizoctonia solani* generally affects seedlings. But, *R. solani* can also infect mature plants and induce root rot (Fig 2.63), which leads to wilting and death of chilli plants. *Rhizoctonia solani* persists in soils and organic debris.

(ii) Management

Management of this fungal pathogen relies on using high-quality seed treated with fungicides such as thiram or captan, and avoiding saturated soil conditions. In fields with high pathogen populations, control measures may include the use of soil fumigants and soil solarisation.

Figure 2.63. Chilli plant infected by *Rhizoctonia solani*, with a reddish-brown discoloration at the soil line level of the stem and root (Courtesy: P. Bosland)

2.4.2.10. Wilt/Collar Rot, *Sclerotium rolfsii*

(i) Management

Soil application of Bioderma (*Trichoderma viride*), avoiding excess irrigation and maintaining wider spacing shall reduce the disease incidence.

2.4.2.11. Bacterial Wilt, *Ralstonia solanacearum*

(i) Management

Cvs. Kandhari, Pungent Pride, Cherry Red, Vattal, Dark Purple, Long Red and Pant C-1 were reported resistant. Manjari and Ujwala developed by Kerala Agricultural University, Vellanikkara, are also highly resistant to bacterial wilt.

2.4.2.12. Bacterial Leaf Spot, *Xanthomonas campestris* pv. *vesicatoria*

Bacterial leaf spot, also called as 'Scab spot' was first observed in the USA on chillies in 1912. At present, it is prevalent in many countries (Patel, 1972) including India where it is present in almost all the chilli growing areas. Loss in chilli yield due to disease has been reported up to 16.6% (Shekhawat and Chakravarti, 1977a).

(i) Symptoms: Small brown spots appear on leaves, petiole, stem and fruit of chillies are distinctly raised (Fig 2.64). These spots turn into greasy or black spots.

(ii) Epidemiology: Pathogen is seed-borne (Kore and Dhaliwal, 1981) and also survives on stray chilli plants (Shekhawat and Chakravarti, 1978b). The bacterium survives for 14 and 10 months in seed at 8 and 10–38°C, respectively. It survives for 150 days in plant debris. Disease spreads to healthy plants through wind splashed rains and to new area through infected seed and planting material.

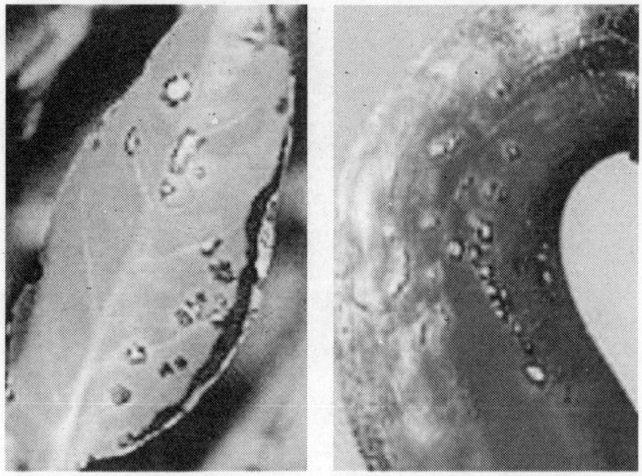

Figure 2.64. Bacterial leaf spot on chilli leaf and fruit

Maximum disease development has been found during July-October (Shekhawat and Chakravarti, 1976) and chilli plants at the age of 42–49 days are found most susceptible (Kore and Dhaliwal, 1981).

(iii) Management

Hot water treatment of seeds at 50°C for 25 min is reported to be effective. Seed treatment with mercuric chloride (1: 1000) is also recommended for the control of the disease. Spraying of 3: 3: 50 Bordeaux mixture or 0.2% zineb or 20 ppm streptocycline or 100 ppm agrimycin + copper oxychloride or copper oxychloride + streptocycline or 200 ppm paushamycin + copper oxychloride is beneficial to check the disease (Raju and Rao, 1984). Chilli *cv.* Jwala is reported to be field resistant.

2.4.2.13. Chilli/Capsicum Mosaic Virus

(i) Symptoms: The important symptoms of this disease are mosaic, mottling, distortion and shortening of leaves (Fig 2.65). The infected plants show stunted growth and in case of early infection, they do not bear fruits. However, in late infection, few fruits are formed but they are small, deformed and unfit for marketing. Sometimes, yellow or green patches or streaks are produced on the fruits.

The mosaic disease is caused by many viruses in chilli/capsicum. The major ones are tobacco mosaic virus (TMV), potato virus Y (PVY), potato virus X (PVX), cucumber mosaic virus (CMV), tobacco etch virus (TEV) and pepper veinal mottle virus (PVMV).

(ii) Transmission: All the viruses are mechanically sap transmitted. TMV and PVX have no insect vectors, but other viruses viz., CMV, PVMV and PVY are transmitted by different species of aphids, *Aphis craccivora, A. gossypii* and *Myzus persicae*. These aphids transmit the viruses in non-persistent manner.

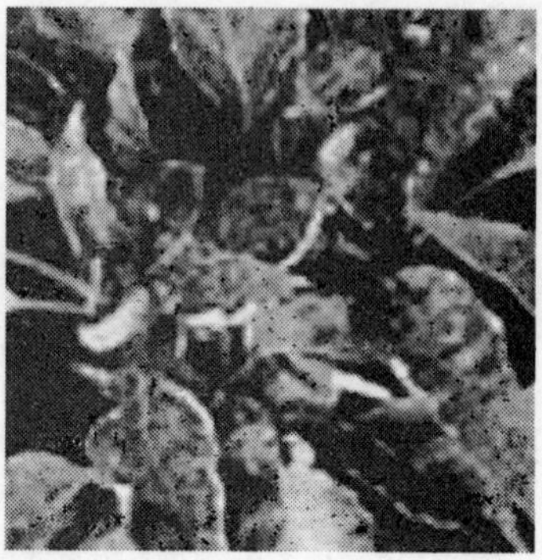

Figure 2.65. TMV on chilli leaves

(iii) Management

- Raise nursery under nylon net.
- Remove all the infected plants from the field.
- Soil application of carbofuran at 1.5 kg *a.i.*/ha at the time of sowing in the nursery followed by 1–2 foliar sprays with 0.05% nuvacron.
- Spraying of 0.05% nuvacron or 0.02% metasystox at 10 days interval to control aphid vectors.
- Spraying of 1.5 to 2.8% mineral oil at 5 to 6 days interval has been found effective in reducing the disease incidence.
- Use tolerant *cvs*. NS 1101 and 1701.

2.4.2.14. Leaf Curl Virus

The disease causes heavy losses in yield and quality of fruits. If the plants are infected within 20–25 days after transplanting, the loss in yield goes up to 80–90%. However, in case of later infections, the loss is comparatively less (Singh *et al.*, 1979b).

(i) Symptoms: Symptoms include abaxial and adaxial curling of the leaves accompanied by puckering and blistering of interveinal areas and thickening and swelling of the veins. In advanced stages, axillary buds are stimulated to produce clusters of leaves which are reduced in size (Fig 2.66). The whole plant assumes bushy appearance with stunted growth. Flowers and fruits are few and small, being curled at the end. The virus has a wide host range. However, important hosts are tobacco, bell pepper and several weed hosts.

(ii) Transmission: The virus is transmitted by the whitefly, *Bemisia tabaci*.

(iii) Epidemiology: The disease spread is more during summer months which favour whitefly multiplication and disease spread.

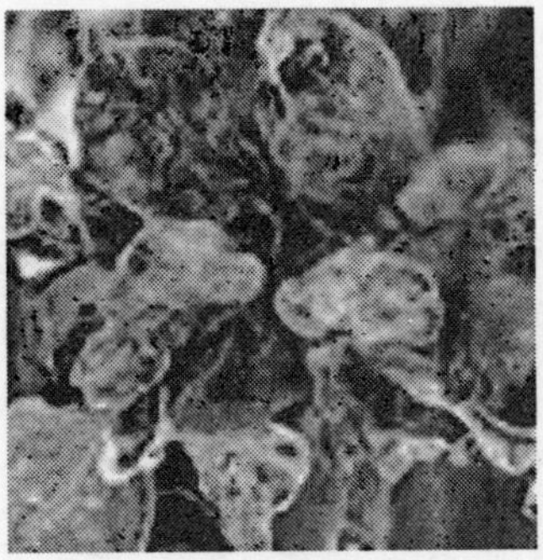

Figure 2.66. Leaf curl virus on chilli leaves

(iv) Management

- Rouging of infected plants from the field.
- Removal of all the weeds from and around the field.
- Soil application of carbofuran at 1.5 kg *a.i.*/ha in the nursery and at the time of transplanting followed by 2–3 sprays of dimethoate or monocrotophos both at 0.05% at 10 days interval prevents whitefly multiplication and spread of the disease (Singh *et al.*, 1979b).

Integrated Management of Insect-borne Viruses

Use of nylon net cover in nursery beds, soil application of carbofuran at 1 kg *a.i.*/ha in combination with foliar application of 0.1% imidachloprid at 10 days interval effectively reduced the vector population as well as the disease incidence.

At Coimbatore, Soil application of carbofuran at 1 kg *a.i.*/ha + covering the nursery bed with nylon net of 400 mesh followed by 2 sprays of 2% neem oil at 15 and 35 DAP recorded lesser virus incidence and highest yield with cost: benefit ratio of 1: 3.3.

At Sabour, Use of nylon net and soil application of carbofuran at 1 kg *a.i.*/ha in seed bed and application of carbofuran at 1.5 kg *a.i.*/ha in the main field a week after transplanting + 3 sprays of nuvacron at 1 ml/litre was most effective in managing the disease with cost: benefit ratio of 1: 5.64.

IDM

- Crop rotation with French beans reduces bacterial wilt.
- Crop rotation with cereals, sesame and mustard, so also intercropping with marigold, onion, garlic reduce nematode infestation.
- Soil solarization of nursery beds for 2–3 weeks using transparent polythene sheets (100–160 gauge) kills soil borne pathogens and nematodes.
- Raising nursery in raised seed beds (10–15 cm) provides good drainage and thereby avoid damping-off.
- Seed treatment with *Trichoderma virde*/*T. harzianum*/*Pseudomonas fluorescens* at 20 g/kg of seeds to prevent damping-off and nematodes.

- Deep summer ploughing helps in exposing pathogens and their resting stages to sun's heat.
- Collection and destruction of diseased plants/plant parts from the field from time to time.
- Soil application of neem cake at 250 kg/ha as basal dose to control nematodes.
- Soil application of FYM enriched with *Trichoderma viride* at 1 MT/ha.

Table 2.28. Pest and disease management in chilli and capsicum

Chemical	Time of application (days after planting)
Monocrotophos 1.5 ml + wettable sulphur 3 g + teepol 0.5 ml/litre of water	15
Monocrotophos 1.5 ml + wettable sulphur 3 g + teepol 0.5 ml/litre of water	25
Cypermethrin 0.5 ml + wettable sulphur 3 g + teepol 0.5 ml/litre of water	40
Endosulfan 2 ml + carbendazim 1 g + teepol 0.5 ml/litre of water	70
Endosulfan 2 ml + metalaxyl and mancozeb (ridomil) 2 g + teepol 0.5 ml/litre of water	80
Monocrotophos 1.5ml + metalaxyl and mancozeb ridomil) 2 g + teepol 0.5 ml/litre of water	90
Monocrotophos 1.5ml + metalaxyl and mancozeb (ridomil) 2 g + teepol 0.5 ml/litre of water	110

Repeat sprays at 130 days after planting, if necessary.

2.4.3. Nematodes

The root-knot nematode is the major limiting factor in successful production of chilli crop.

2.4.3.1. Root-knot Nematodes, *Meloidogyne* spp.

Root-knot nematodes induce much smaller galls on the roots of chillies. *M. incognita* was responsible for 24.54 to 28.00% loss in fruit yield of chillies (Singh *et al.*, 2003).

(i) Symptoms: Root galls on chilli are frequently small.

(ii) Management

(a) Cultural Methods: Decrease in *M. incognita* population on chillies occurs when the field was left fallow or following crop rotation with marigold, spinach and bottle gourd (Khan *et al.*, 1975). Growing of chilli along with marigold gave highest degree of reduction in root-knot nematode population on chilli roots, followed by onion, garlic and asparagus (Trivedi and Tiagi, 1984).

Row application of fly ash at 0.6 kg/sq.m. enhanced the fruit yield by 21.3% and inhibited the reproduction of *M. incognita* (Khan and Ghadipur, 2004).

Botanicals: Soil application of decaffeinated tea leaves effectively reduced root galling on chilli plant (Trivedi and Tiagi, 1984).

(b) Chemical Methods: Aldicarb and carbofuran at 3 per cent (seed treatment) were effective against root-knot nematodes infecting chilli (Kandasamy and Sivakumar, 1981).

Dichlofenthion was found effective against *M. incognita* in chillies (Alam *et al.,* 1973, Saxena *et al.,* 1974).

Aldicarb, carbofuran, ethoprophos and phenamiphos each at 1 to 2 kg *a.i.* per ha were found effective in reducing the root-knot nematode population and in increasing fruit yields of chillies (Handa and Mathur, 1981).

(c) Biological Methods: Pandey and Trivedi (1992) reported a significant reduction in number of galls, eggs/egg mass, hatching and final soil population of *M. incognita* infecting chillies in the presence of *P. lilacinus*.

Avermectins effectively reduced root-knot nematode infection in chilli under field conditions (Parvatha Reddy and Nagesh, 2002).

(d) Host Resistance: Chilli *cvs.* Pusa Jwala, Wonder Hot, Teja, Utkal Abha, Utkal Rashmi were found resistant to root-knot nematodes (Patnaik *et al.*, 2004).

2.4.3.2. Root-knot Nematode, *Meloidogyne javanica* and *Fusarium oxysporum* Wilt Disease Complex

(i) Management

Pseudomonas aeruginosa and *P. lilacinus* when used together significantly reduced infection of the disease complex on chilli (Perveen *et al.*, 1998).

2.5. BELL PEPPER

Bell pepper (*Capsicum annuum*), also known as capsicum or sweet pepper, is a popular vegetable in India. It is relatively a new entrant into our country. It is mainly cultivated in Himachal Pradesh, Uttar Pradesh, parts of Gujarat, Maharashtra, Karnataka, Ranchi region of Bihar and hilly regions of Tamil Nadu. It grows well in summer season in hills and cooler season in the plains.

2.5.1. Pests

2.5.1.1. Gall Midge, *Asphondylia capparis*

(i) Damage: Gall midge is an important pest on sweet peppers. Midge damaged sweet pepper lose their market value. Midge infestation may induce shedding of flowers (Fig 2.67) (Krishna Kumar *et al.*, 1997). Further, pollen germination, seed yield, germination and viability are affected as a result of midge infestation.

(ii) Management

Spraying of 4% NSKE at the time of first flush was found most effective in controlling gall midge infestation (Krishna Kumar *et al.*, 1997).

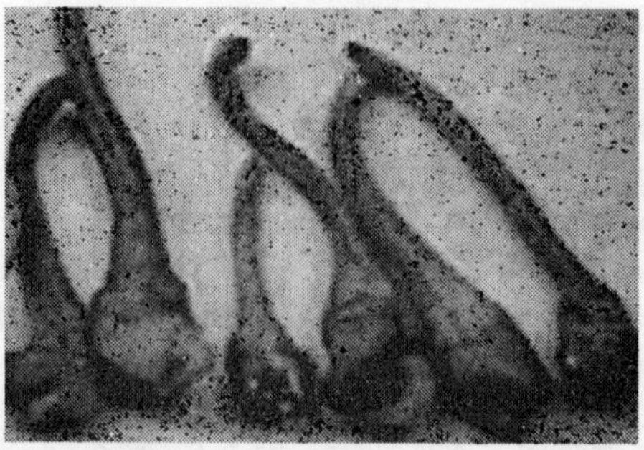

Figure 2.67. Blossom midge infestation on capsicum

2.5.1.2. Thrips, *Scirtothrips dorsalis*

(i) Damage: Both adults and nymphs suck the sap from young leaves. Affected leaves curl upwards along the margin, and get crinkled and reduced in size (Fig 2.68). When the incidence is severe, leaves drop and cause heavy reduction in yield. The pest is serious during dry monsoon periods and summer months.

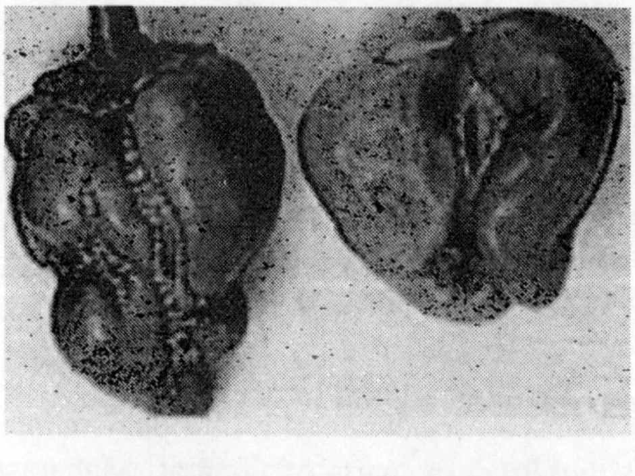

Figure 2.68. Thrips on fruits of bell pepper

(ii) Management

- Spray acephate 75 SP at 1.5 g/l or fipronil at 1 ml/l or lamda cyhalothrin 5 EC at 0.75 ml/l or imidacloprid 200 SL at 0.3 ml/l alternately at fortnightly interval. Never repeat the same chemical again and again.
- Spray of difenthiuron at 225 g *a.i.*/ha and monocrotophos at 150 g *a.i.*/ha effectively checked the population of thrips on chilli with higher yield.
- Mix acephate 0.5 g, 2 ml of pongamia oil and 1 ml sticker and make an emulsion (add a little water and shake thoroughly in a bottle) and make the volume to 1 litre and spray.
- Spraying of Vertimec at 0.3 to 0.4 ml/l was found effective against thrips and mites for 45 days.

- Soil application of aldicarb at 2.8 kg *a.i.*/ha 15 days after transplanting effectively checked thrips and mites and resulted in minimum leaf curl incidence with higher yield of marketable green chilli fruits (Patel *et al.*, 1983).
- Soil application of carbofuran at 0.5 kg *a.i.*/ha 15 days after transplanting followed by spray of 0.1% triazophos commencing from 40 days after transplanting at 10 days interval effectively controlled thrips and increased marketable green chilli fruits (Patel *et al.*, 1997).
- Chilli *cvs*. NS 1101, NS 1701 (Namdhari seeds), NP 46A, Pusa Jwala, Arka Lohit and K 2 (Kovilpatti) are tolerant to thrips.
- *Chrysoperla carnea* can be released at 2 grubs/plant.

2.5.1.3. Aphids, *Myzus persicae*

(i) Damage: Aphid is a serious pest of pepper in parts of South India. Infestation is more, especially on hybrids/improved varieties which are grown under high fertigation and insecticides. Excretion of honey dew by aphids induces sooty mold reducing photosynthesis and fruit quality.

(ii) Management

Though, the coccinellid predators generally regulate aphid population, often spray intervention may be necessary to realize economic yield in sweet peppers. However, indiscriminate use of insecticides would result in aphid resurgence.

2.5.2. Diseases

2.5.2.1. Crown and Fruit Rot, *Phytophthora capsici*

(i) Symptoms: This is a disease of rainy season, characterized by small, water-soaked spots appearing on fruits leading to complete rot of fruits. On leaves, water-soaked bleached spots appear, resulting in blighting. It causes damage to plants and affect fruit yield.

(ii) Management

Minimum plant mortality (16.7%) was recorded when the plants were subjected to *T. viride* (PDBC TV 31) and *T. harzianum* (PDBC TH 10) followed by *T. harzianum* (PDBC TH 18) and *T. viride* (PDBC TV 23) where the mortality was 30%. In fungicidal treatment, 33.3% plant mortality was recorded while in pathogen check, it was highest (46.7%).

2.5.3. Nematodes

The root-knot nematode is an important limiting factor in successful cultivation of capsicum crop.

2.5.3.1. Root-knot Nematodes, *Meloidogyne* spp.

(i) Symptoms: Root galls caused by root-knot nematodes on sweet pepper are frequently small.

(ii) Management

(a) Physical Methods: Burning of paddy husk or saw dust on infested nursery proved to be reasonably effective.

(b) Cultural Methods: Several cultural methods like selection of nematode-free nursery sites, destruction of infested roots after crop harvest, crop rotation with non-hosts (maize, wheat, sorghum) or antagonistic crops (marigold, mustard, sesame) or intercropping with marigold, onion and garlic, flooding, fallowing, deep summer ploughing and harrowing either alone or in combination proved to be reasonably effective and economical to check multiplication of root-knot nematodes on capsicum. Ploughing not only leads to disturbance and instability in nematode community but also causes their mortality by exposing them to solar heat and desiccation. Generally, 2 to 3 summer ploughings each at 10 days interval during April-May (40–46ºC) have been recorded to reduce 96% *M. javanica* population while fallowing itself during the same period registered 45.5% reduction. Additional use of plastic sheets for covering soil either in nursery beds or in field further enhance nematode reduction. Such an approach also helps in reducing the intensity of weeds, fungi and bacteria in the soil.

(c) Chemical Methods: Bare-root dip treatment of capsicum seedlings in 0.1% carbosulfan/monocrotophos for 6 hr eliminates root-knot infection.

(d) Biological Methods

Antagonistic Bacteria: Field application of a nematicidal *Bacillus thuringiensis* strain to capsicum in Puerto Rico reduced galling in roots due to *M. incognita* and increased yield significantly (Zuckermann *et al.*, 1993).

Antagonistic Fungi: Application of 'Royal 350' (*Arthrobotrys irregularis* cultured on oat seed medium) at 140 g/m^2 a month before transplantation of capsicum resulted in good protection against root-knot nematodes. Among the egg parasites, efficacy of *P. lilacinus* (commercially formulated as 'Biocon' in Philippines) has been found to be comparatively higher in suppressing the population of *Meloidogyne* spp. on capsicum.

Bio-agent *P. chlamydosporia* and *Pseudomonas fluorescens* were evaluated under field conditions for their efficacy against root-knot nematode *M. incognita* infecting capsicum. Treatment of the nursery bed with the formulation of *Pochonia chlamydosporia* at 50g/m^2 was significantly effective in reducing the galling index, number of nematodes in roots and soil, increasing the percent parasitization of eggs by bioagent and also yields. Seed treatment with *P. fluorescens* alone and the nursery bed treatment with *P. chlamydosporia* alone were effective. The main purpose of these studies was to produce capsicum seedlings that are colonized by *P. chlamydosporia* before transplanting so that they could carry the bio-agent to the field. During this process the field soil would be enriched with the propagules of eco-friendly component of management (here it is a bio-agent) in two or three seasons (Naik, 2004).

(e) Host Resistance: Capsicum *cvs*. Mississipi-68, Santanka, Anaheim Chile and Italian Pickling are reported to be resistant to root-knot nematodes (Hare, 1951). Black Indica, Naharia and Pant C1 were reported to be resistant to *M. incognita*, while All Big as moderately resistant. California Wonder and Naharia were resistant to *M. javanica*, while

Early California Wonder as moderately resistant. Naharia was reported to be resistant to *M. arenaria*.

(f) Integrated Methods: Integration of *P. fluorescens* and *P. chlamydosporia* in the nursery bed has proved significantly effective in reducing the root-galling index (*M. incognita*), number of nematodes in the roots and soil and increasing the yield of the capsicum crop under field conditions (Naik, 2004).

In capsicum, the seedling stand was good where the combination of neem based *P. fluorescens* and *T. harzianum* was used (4.4), followed by seed treatment with *P. fluorescens* + soil application of *T. harzianum* (Naik, 2004).

2.5.3.2. Root-knot Nematode, *Meloidogyne incognita* and Bacterial Wilt, *Ralstonia solanacearum* Disease Complex

Capsicum is prone to many soil borne diseases among which the bacterial wilt (*R. solanacearum*) in combination with root-knot nematode (*M. incognita*) takes heavy toll every year all over the world (Naik, 2004).

(i) Management

Integrated Methods: Combined application of neem based formulations of *P. fluorescens* and *P. chlamydosporia/T. harzianum* at 40 g/m^2 in nursery beds and transplanting these seedlings in the main field resulted in significant reduction in disease index and root-knot index in capsicum to the tune of 70% and increased the crop yield by 37% (Rao *et al.*, 2002) (Table 2.29). Combinations of the bioagents did not affect the colonization of the individual bioagents on the roots and hence the transplants carried the bioagents to the main field. This has resulted in the effective management of the pathogens involved in the disease complex.

Table 2.29. Effect of integration of neem-based bioagents on the growth of transplants and management of disease complex and yield of capsicum

Treatment	Seedling weight (g)	Root-knot index (1–10)	Disease index (1–9)	Yield in kg/4 m^2
Seed treat. with *P. fluorescens*	421	5.6	6.4	4.3
Seed treat. with neem-based *P. fluorescens*	428	5.2	6.7	4.7
Nursery treat. with *P. fluorescens*	435	4.6	5.4	4.8
Nursery treat. with *P. chlamydosporia*	364	4.4	7.3	3.2
Nursery treat. with *T. harzianum*	374	4.8	7.0	3.4
Nursery treat. with *P. fluorescens* + *P. chlamydosporia*	463	4.1	5.2	4.0
Nursery treat. with *P. fluorescens* + *T. harzianum*	493	3.8	3.5	5.1
Control	340	8.7	8.2	2.6
CD at 5%	**27.20**	**0.49**	**0.38**	**0.25**

Naik (2004) reported that combination of *T. harzianum* along with *P. fluorescens* increased the yield (3.320 kg/3m^2 plot) followed by combination of *T. harzianum* and *P. chlamydosporia* (2.960 kg/3m^2). The least gall index was present where the combination of *T. harzianum* and *P. fluorescens* was used (1.5) followed by *T. harzianum* + *P. chlamydosporia* (1.7) and *P. fluorescens* + *P. chlamydosporia* (1.7). But among combination treatment of bio-agents, all the treatments were on par (*P. fluorescens* + *T. harzianum*, *P. fluorescens* + *P. chlamydosporia*, *T. harzianum* + *P. chlamydosporia*). Results related to the percent bacterial wilt disease (*R. solanacearum)* incidence also showed similar trend in which *T. harzianum* + *P. fluorescens* performed very well and the percent incidence was 16.90 followed by *T. harzianum* + *P. chlamydosporia* (17.20).

Table 2.30. Management of insect pests in solanaceous vegetable crops

Crop	Pest	Symptoms	Pesticide/Bioagent	Remarks
Potato	Cutworm, *Agrotis* spp.		*Steinernema carpocapse* soil application of 5 billion infective juveniles through irrigation water on detection of cutworm larvae.	Entomophilic nematodes are highly susceptible to desiccation, so apply through irrigation water.
Tomato	Fruit borer	Damaged leaves	Endosulfan-2 ml/l during flowering period and bored fruits after fruiting	After flowering on 28, & 35 DAP (See IPM given separately)
			NPV-200LE/ha	Spray on 25, 30, 35 & 40 days after planting
			Trichogramma brasiliensis (Egg parasitoid) at 250,000 parasites/ha	5 releases at weekly intervals after 25 DAP
			Trichogramma brasiliensis/ T. chilonis/T. pretiosum, 50,000 at weekly intervals/ 6 times from 25th day after transplanting	First release to be initiated when first moth is caught in pheromone traps or eggs are noticed.
	Serpentine leaf miner	Mines on the leaves	NSKE 4% (40 gm/l); Neem formulations-5–7.5 ml/l; Triazophos-1.5 ml/l	Once in 20 days
Brinjal	Shoot & Fruit borer *Leucinodes* orbonalis	Drooping of young shoots and holes plugged with	Apply neem cake/ pongamia cake at 250 kg/ha at flowering and spray cypermethrin-	Prune the drooping shoots and burn before flowering and repeat cake and

Table. 2.30. (Contd.)

Crop	Pest	Symptoms	Pesticide/Bioagent	Remarks
		excreta before flowering and bored fruits with holes plugged with excreta in fruiting stage	0.5 ml/l	insecticide when required
	Ash weevil *Myllocerus subfaciatus*	Saw like damage in the margins of leaves. If drooping plants are uprooted, very few roots with feeding damage and few grubs can be soon	Neem cake/pongamia cake at 250 kg/ha	Apply around the base of plants while planting and repeat after 60 DAP
	Aphids	Colony of aphids on tender parts	Phosphamidon-0.5 ml/l	As and when observed
	Epilachna beetle	Scrapped leaf lamina	Carbaryl-4 g/l; endosulfan-2 ml/l	As and when observed
	Midge	Dried or malformed flowers and malformed fruits	NSKE 4%; phosphamidon-2 ml/l	Once in 2–3 weeks after flowering
	Coccidohystrx insulata		*Cryptolaemus montrouzieri*, 2 beetles/plant. Once, as soon as infestation is noticed.	Release only in endemic areas.
Chilli/ Capsicum	Aphids	Clustering of aphids in young leaves	Phosphamidon-0.5 ml/l; Monocrotophos-1.25 ml/l	As and when observed
	Mites	Leaves curl downwards	Dimethoate-2 ml/l; Wettable sulphur-3 gm/l; Dicofol-2 ml/l	As and when observed
	Thrips	Leaves curl inside	Phosphamidon-0.5 ml/l; Monocrotophos-1.25 ml/l	As and when observed
	Midge	Drying of flowers and distorted fruits	Neem cake/pongamia cake-250 kg/ha; Phosphamidon-0.5 ml/l; Monocrotophos-1.25 ml/l	Apply cakes after flowering. Give sprays if required at 15 days interval
	Helicoverpa armigera		*Ha* NPV-250 LE/ha	

Table 2.31. Solanaceous vegetable crop varieties resistant to biotic stresses

Vegetable crop	Pest/Disease	Resistant varieties
Tomato	Fusarium wilt	HS-110, Sel-28
	Bacterial wilt	Arka Abha, Arka Alok, Arka Shreshta, Arka Abhijit, Megha, Shakthi, Sun 7610, Sun 7611, BT-1, BT-10, LE 79-5.
	Root-knot nematode	Pusa -120, Pusa Hybrid-2, Arka Vardan.
	Leaf curl virus	Avinash-2, Hisar Anmol, H-44, H-36, H-86, H-88.
	Powdery mildew	Arka Asish.
Brinjal	Bacterial wilt	Arka Nidhi, Arka Keshav, Arka Neelkant, Swarna Shree, Swarna Shyamali, Surya, SM 6-6, SM 6-7, BB-7, BB-44.
	Little leaf	Pusa Purple Cluster (Field resistant).
	Phomopsis blight	Pusa Bhairav.
Chilli	TMV, CMV, Leaf curl	Pusa Sada Bahar, Punjab Lal, Pusa Jwala.
	Mosaic, leaf curl	Pant C-1.
	Powdery mildew	Arka Suphala (T).
	Powdery mildew, Viruses	Arka Harita, Arka Meghana.

3 BULBOUS VEGETABLE CROPS

3.1. ONION AND GARLIC

Onion (*Allium cepa*) is one of the most important commercial vegetables. The bulbs are consumed both in mature and tender stages. Even the leaves and stalks of inflorescence are used as vegetables. It is being cultivated in 0.695 million hectares producing 9.248 million tonnes of onions with an average yield of 13.3 tonnes per hectare (Bijay Kumar, 2007) (Tables 3.1 and 3.2). It is grown in western, northern as well as in southern India. Maharashtra, Gujarat, Bihar, Karnataka, Andhra Pradesh, Madhya Pradesh, Rajasthan, Haryana, Uttar Pradesh, Orissa and Tamil Nadu are the major onion-growing states of India.

Table 3.1. Major onion producing countries of the world

Country	Area (in '000 ha)	Production (in '000 MT)	Productivity (in MT/ha)	% Share of world production
China	901	19040	21.1	30.2
India	695	9248	13.3	14.7
USA	65	3346	51.2	5.3
Turkey	77	2070	26.8	3.3
Pakistan	128	1765	13.8	2.8
Russia	123	1759	14.3	2.8
Iran	50	1685	33.7	2.7
Egypt	43	1302	30.6	2.1
Brazil	58	1138	19.5	1.8
Mexico	40	1115	28.1	1.8
Others	1179	20500	17.4	32.6
Total	**3359**	**62968**		

Source: FAO (except Indian data) (Indian data source: Indian Horticulture Database 2006)

Garlic (*Allium sativum*) is mainly used for flavouring vegetables and meat dishes. It is carminative, stimulant and cures digestive disorders. It has been found to be very effective in curing diabetes and high blood pressure. It is being cultivated in 0.107 million hectares producing 0.436 million tonnes of garlic with an average yield of 4.1 tonnes per hectare. Madhya Pradesh, Gujarat, Orissa, Maharashtra, Uttar Pradesh and Rajasthan are major garlic-growing states. More than 50% production comes from Madhya Pradesh and Gujarat only.

Table 3.2. Area, production and productivity of leading onion growing states in India (2005–06)

State	Area (in '000 ha)	Production (in '000 MT)	Productivity (MT/ha)
Maharashtra	154.0	1878.8	12.2
Gujarat	84.9	1876.6	22.1
Bihar	49.8	1011.7	20.3
Karnataka	139.0	1008.5	7.3
Andhra Pradesh	39.4	670.0	17.0
Madhya Pradesh	40.0	639.8	16.0
Rajasthan	44.5	427.2	9.6
Haryana	15.5	317.7	20.5
Uttar Pradesh	21.2	283.1	13.3
Orissa	26.8	243.9	9.1
Tamil Nadu	29.2	233.8	8.0
Others	50.8	657.3	12.9
Total	**695.1**	**9248.4**	**13.3**

(*Source*: Indian Horticulture Database 2006)

3.1.1. Pests

3.1.1.1. Thrips, *Thrips tabaci*

(i) Damage: Thrips are the most important insect pests on onions. Adults are yellowish to yellowish brown with narrow wings. Eggs are laid in clusters in leaf tissues. Numerous nymphs and adults are observed between leaf sheaths and stems lacerating the epidermis of leaves and sucking the exuding cell sap. The affected leaves show silvery blotches which later turn into brownish colour (Fig 3.1). The leaves get distorted from tips downwards and the plant ultimately wilt and dry away. Seedlings are killed due to heavy infestation and the plant growth is retarded. The shape and size of bulbs as well as yield is affected by thrips.

(ii) Management

- Soil application of neem cake at 250 kg/ha at 30 and 60 days after planting.
- Soil application should be followed by 2 to 3 sprays of neem seed powder extract 4% or neem soap 1% at 10 days interval.
- Soil application of phorate or carbofuran granules (1 kg *a.i.*/ha) is also recommended. Alternate spraying of 0.07% endosulfan (2 ml/litre of water), or 0.07% nuvacron

(2 ml/litre of water) or 0.01% cypermethrin (1 ml/litre of water) along with a sticker (Triton or Sandovit) at fortnightly intervals is effective.

Figure 3.1. Thrips damage on onion leaves

- Spray fipronil 5 SC at 1ml/l or lamda cyhalothrin 5 EC at 0.5 ml/l or acephate 75 SP at 1g/l alternated with spray of azadirachtin formulation (10,000 ppm) at 2 ml/l.
- Thrips are weak fliers and carried to long distances by wind. Blocking adult thrips can reduce the initial and subsequent pest load on onion. Barrier crops (2 rows of maize all-round onion crop) can be employed for this purpose.
- Coccinellid predators were higher in plots where maize was used as a barrier crop. *Scymnus nubilis*, *Chrysoperla* spp. and *Latus* spp. are predaceous on thrips.
- Onion *cvs*. Arka Lalima, White Persian, Grano, Sweet Spanish, Crystal Wax, Spanish White, Bombay White, Pusa Red, Udaipur 103, N 780, N 53, and White Poona Red are reported to be resistant/tolerant to thrips.
- Sprinkling of water through jet nozzles prevents thrips multiplication. Spraying with 5% NSKE during vegetative stage is also effective against thrips.
- Use yellow sticky traps at 10/ha to prevent thrips damage.

Barrier Cropping for Management of Thrips

Thrips are weak fliers and carried to long distances by wind. Blocking the adult thrips can reduce the initial and subsequent pest load on onion. Barrier crops can be employed for this purpose.

Lowest number of thrips was recorded in 2 rows of maize planted in zig zag manner (10.49/plant) and in one inner row of wheat and one outer row of maize (11.19/plant) compared to 26.54 thrips/plant where no barrier was present. Blocking of adult thrips was highest in one inner row of wheat and one outer row of maize (86.66%) and in 2 rows of maize planted in zig zag manner (66.00%) at 50 cm height. At 100 cm height, blocking was highest with 2 rows of maize planted in zig zag manner (76.44%) followed by one inner row of wheat and one outer row of maize (67.41%). On overall basis, the % blocking was highest in 2 rows of maize planted in zig zag manner (80.45%) followed by one inner row of wheat and one outer row of maize (79.17%). Coccinellid predators were higher in plots where maize was used as a barrier crop.

3.1.1.2. Leaf Eating Caterpillar, *Spodoptera litura, S. exigua*

(i) Damage: It is a polyphagous pest. It is active in nursery stage. Larvae damage leaves of onion and garlic.

(ii) Management

Fields should be properly ploughed to expose and kill the pupae in the soil. Flood irrigation may be done to drown the hibernating caterpillars. Digging trenches around the infested field prevents migration of larvae. Release of *Trichogramma brasiliensis* at 50,000/ha at weekly interval should be done 4–5 times. Anthocorid bugs are potent predators of leaf eating caterpillars.

3.1.1.3. Head Borer, *Helicoverpa armigera*

(i) Damage: The head borer is a serious pest of onion seed crop in northern India. The larva of this insect cuts the pedicel of the flower and feeds on the stalk. Single larva damages many flower stalks. The fully-grown larva is greenish with dark brown grey lines along the side of body and measures about 35–45 mm in length.

(ii) Management

Spraying endosulfan (2–3 ml/litre of water) along with sticker (Triton/Sandovit) is recommended to control the insect.

3.1.1.4. Onion Maggot, *Hylemia antiqua*

(i) Damage: The adults of this insect appear like housefly. The flies lay their eggs on old leaves or on soil and larva enters into soil and damage disc portion of onion bulb. Infested plants turn yellowish brown and finally dry up. The affected bulbs rot in storage.

(ii) Management

Regular crop rotation should be followed and phorate should be applied in the soil before transplanting.

3.1.1.5. Cut Worm, *Agrotis ipsilon*

(i) Damage: The larvae of this insect are seen in nursery beds and newly transplanted onion fields. The tender plants are found damped at ground level during the night. Young larvae feed gregariously on foliage but later segregate and enter into soil. Young larvae are yellowish grey and later on become brown, greasy to touch and coil when disturbed. They cut the seedlings at ground level during night and hide during day.

(ii) Management

Soil application of carbofuran (1 kg *a.i.*/ha) at the time of planting is recommended. Chlorpyriphos (5 ml/litre of water) also gives good control of this pest.

3.1.1.6. Mites, *Aceria tulipae*

(i) Damage: These are small white mites, which suck the sap from the leaves making them curl, discolour, and get stunted. They also damage the dormant bulbs in the storage. The mites feed upon the epidermal cells of bulb scales, causing the cells to collapse.

(ii) Management

- Soil application of neem cake at 30 and 60 days after planting followed by spray of 4% neem seed powder extract or 1% neem soap at 10 days interval.
- Spray acaricides like wettable sulphur 80 WP at 3 g/l.

Pesticide Residues

The persistence of terminal pesticide residues and waiting periods for onion are presented in Table 3.3. (Awasthi, 2001).

Table 3.3. Terminal pesticide residues and their waiting periods for onion

Pesticide schedule	Dosage/ Concentration (%)	Terminal residues (ppm/days)	Waiting period (days)
Carbaryl spray at fruiting stage	0.10	1.86/25	20
Malathion spray at fruiting stage	0.05	1.46/5	6
Cypermethrin spray at fruiting stage	0.01	ND*/Bulbs	Nil
Fenvalerate spray at fruiting stage	0.01	ND*/Bulbs	Nil
Carbofuran soil application at planting	1.0 kg *a.i.*/ha	0.42/75	90
Phorate soil application at planting	1.0 kg *a.i.*/ha	0.25/75	60

* ND – Not detectable

3.1.2. Diseases

3.1.2.1. Damping-off, *Pythium* sp., *Rhizoctonia solani, Fusarium* sp.

(i) Symptoms: Cold, wet soils often encourage the development of damping-off symptoms very early in the seedling's growth. Seedlings may fall over and die as a result of breakdown of plant tissues at the soil line. Sometimes damping-off occurs before the seedling even emerges. The disease is usually caused by *Pythium*, *Rhizoctonia* or *Fusarium* fungi, either alone or in combination. Damping-off can occur in the field, or in the nursery if conditions are too wet.

(ii) Management

Use disease-free seeds for sowing. Seed treatment with *T. harzianum* at 5 g/kg and sowing in solarized soil helps in reducing the disease incidence. Overcrowding of seedlings and water stagnation in the nursery beds should be avoided. Drenching the soil with 1% Bordeaux mixture or 0.2% copper oxychloride reduces the disease incidence.

3.1.2.2. Basal Rot, *Fusarium oxysporum* f. sp. *cepae*

Basal rot is caused by the soil-borne fungus *Fusarium oxysporum* f. sp. *cepae* and generally occurs when soil temperatures are very warm (optimum 29°C).

(i) Symptoms: The early symptoms in the field are yellowing of leaves and tip dieback. As the disease progresses, the whole plant may collapse and, if the plant is pulled, it often comes out without any roots attached since they have decayed. The basal plate of the onion becomes pinkish-brown and secondary bacterial rots may develop in the affected area (Fig. 3.2). If infection occurs late in the season, the symptoms may not show up until the onions are in storage.

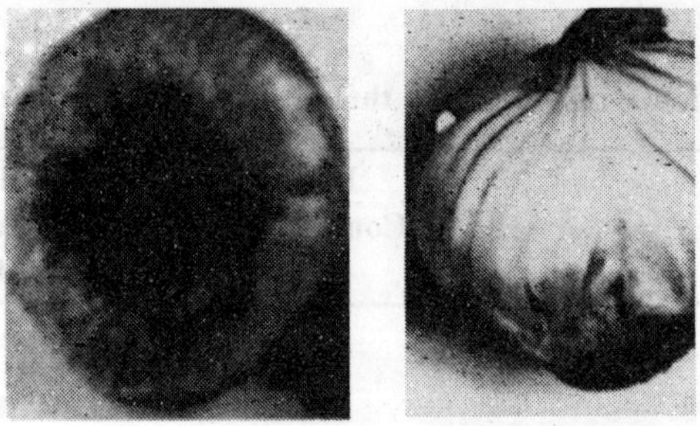

Figure 3.2. Basal rot on onion bulbs

(ii) Management

Mixed cropping and crop rotation reduces the incidence of the disease. Losses can be reduced with a 4-year rotation out of onions. Seedling dip in antagonists like *Pseudomonas cepacis* and *T. viride* gives protection to seedlings. Soil solarization using polythene sheet of 750 gauge in summer season for 20–30 days reduces the infectious propagules, which in turn reduces the disease incidence. *T. viride* at 1250 g + 50 kg FYM if applied in soil before planting gives good control of basal rot in seed crop. Onion *cvs*. Arka Lalima, Arka Niketan, Arka Pragati are reported to be resistant/tolerant to basal rot.

3.1.2.3. Purple Blotch, *Alternaria porri*

(i) Symptoms: The disease is severe in Maharashtra, Gujarat, Tamil Nadu, Rajasthan, Uttar Pradesh and Punjab causing damage up to 50% (Thind and Jhooty, 1984). The disease is more damaging to the seed crop. The disease appears on leaves as purplish spots with dark or red purplish margin and a yellow halo that extends upwards and downwards to some extent (Fig 3.3). The seeds do not develop or remain shriveled on the infected seed stalks and heavily infected seed stalks are prone to breaking. The spots on leaves rapidly enlarge and the leaves are gradually shed.

Figure 3.3. Purple blotch on onion leaf and inflorescence

(ii) Epidemiology: The temperature range of 21–30°C and high humidity favoured disease development. The fungus survives from one season to another as mycelium and spores in leaves and stalks left as plant debris, as chlamydospores and is also seed borne in nature.

(iii) Management

Heat treatment of onion bulbs at 35°C for 8 hr before planting gives effective control of the disease. Summer sown crop escapes the disease infection. Spraying of 0.2% mancozeb gives effective control. Onion *cvs*. Arka Kalyan, Arka Kirtiman, Arka Lalima and Agrofound Dark Red are reported to be tolerant to purple blotch. Crop rotation with coarse cereals also reduces blotch incidence.

3.1.2.4. Blight, *Stemphyllium vesicarium*

(i) Symptoms: The disease causes blight in northern parts of India and mainly damages radical leaves of the seedling at 3–4 leaf stage. But it has also been recorded on the inflorescence. The disease is serious in other parts of the world. The perfect stage of the pathogen is *Pleospora allii*.

Lesions are initially light yellow to brown and water-soaked. They elongate, often reaching the leaf tips, and become dark brown to black. The disease can become serious following periods of more than 24 hr of rainy weather.

(ii) Management

Summer ploughing of infested field reduces inoculum potential of blight. Field sanitation, collection and burning of crop residues minimizes infection rate.

The disease can be controlled by 4 sprays on bulb crop and 5 sprays in seed crop with 0.25% mancozeb along with sticker Triton-X.

3.1.2.5. Downy Mildew, *Peronospora destructor*

The pathogen always attacks the oldest leaves. This disease can rapidly destroy the foliage and thus prevents the formation, ripening and storage of onion bulbs. The disease is very damaging to onion seed crop when the flower stem is destroyed before the seeds ripen.

(i) Symptoms: The first sign of downy mildew is a purple-grey, velvety growth on otherwise green leaves (most easily seen in the early morning) (Fig 3.4). The disease often starts in patches and is favoured by cool (less than 22°C) and humid weather. Diseased leaves quickly turn pale-green, then yellow, and collapse and die. The pale-green and yellow phase is characterized by oval-shaped lesions that often become infected with other diseases such as purple blotch or bacterial infections. Several cycles of sporulation and infection can occur.

Under favourable conditions, the diseased areas become covered with white to purplish fruiting bodies. Often, the leaves fold over at the affected areas and the leaf tip withers. Bulbs of such plants do not attain full growth and are often soft and immature. The fungus overwinters in infected crop residues and also persists from one year to another as mycelium in infected bulbs in storage.

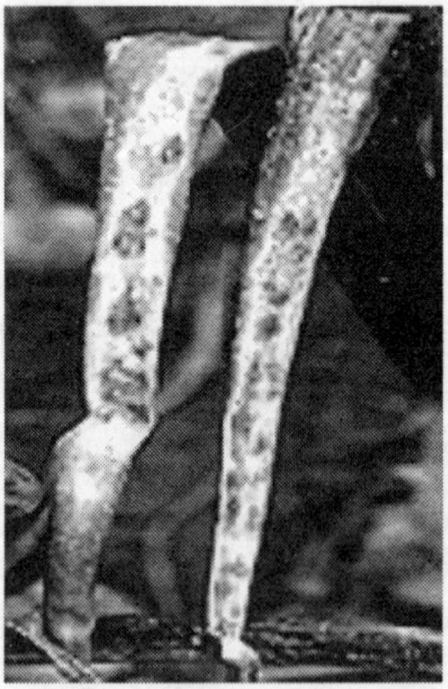

Figure 3.4. Downy mildew infection on onion leaves

(ii) Management

Regular fungicidal sprays with difolatan or cumin or ridomil all at 0.1% are recommended for the control of downy mildew disease (Mirakhur *et al.*, 1977; Mir, 1987).

3.1.2.6. Smut, *Uromyces cepulae*

(i) Symptoms: The fungus causes considerable losses at the seedling stage. The first symptom of the disease appears on the cotyledons and young leaves in the form of longitudinal blisters which are black and contain the fruiting bodies of the fungus. The seedlings often die before emergence. If the plant survives, the disease becomes systemic and they remain at a vegetative stage for the whole growing season. The bulbs also become covered with blackish lesions (Fig 3.5).

Figure 3.5. Onion smut

The pathogen is soil-borne and the disease is more prevalent in temperate region. The onion is susceptible to this disease for a very short time *i.e.*, between germination and seedling emergence. A cool, wet spring increases the incidence of smut infection because the onion seedlings grow slowly and the flag leaf is in the soil for a longer period. Similarly, planting onion seeds too deep will also make them more likely to be infected. Smut spores survive in the soil for many years, and even long crop rotations may not reduce disease incidence. The disease is spread when contaminated soil or set onions are transferred to smut-free areas.

(ii) Management

Seed treatment with thiram or captan at 3 g/kg of seed checks soil-borne infection. Thiram and captan both at 0.2% may be applied to the soil before sowing the seeds. The soil may be treated with formalin to eliminate the pathogen in the soil. Growing onions from transplants avoids the disease.

3.1.2.7. White Rot, *Sclerotium cepivorum*

White rot, caused by the soil-borne fungus *Sclerotium cepivorum*, is a very destructive disease that begins in the field and can carry over into storage.

(i) Symptoms: The first above-ground symptoms are yellowing and dieback of the leaf tips, followed by a collapse of the affected leaves; however, these symptoms alone could easily be confused with other types of damage (*i.e.*, onion maggot). When the bulbs and roots are examined, a white, fluffy mold and soft rot will be observed. Masses of tiny black sclerotia can also be seen within this mold (Fig 3.6). These sclerotia remain in the soil for many years. Infected bulbs can rot in storage boxes and stain other bulbs. White rot typically develops in patches in the field and is less of a problem when soils are warm (higher than 24°C) and dry.

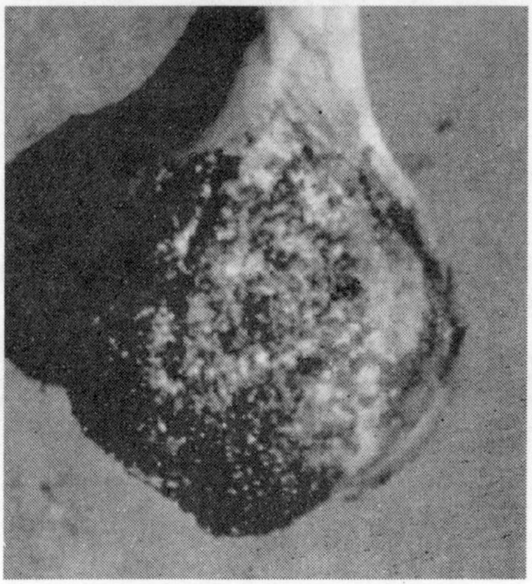

Figure 3.6. White rot of onion

(ii) Management

Removal and destruction of infected plants reduces the inoculum potential in the soil. Early planting of onion and garlic is an important management strategy to avoid white rot in areas with infected soil. Soil solarization for 8–11 weeks during summer using transparent sheet resulted in reduction of *S. cepivorum* to undetectable levels in upper 20 cm of soil. Application of *T. harzianum, T. viride, T. virens* and *Bacillus subtilis* to the solarized soil effectively controlled the disease.

3.1.2.8. Bacterial Stalk Rot, *Pseudomonas gladioli* pv. *allilicola*

The disease was observed for the first time on onion at Kanpur, Uttar Pradesh during 1971 (Swarup *et al.*, 1973). The disease incidence varies from 12.8 to 35.6% in seed crop and 8.5 to 30.6% in bulb crop (Kishun and Swarup, 1979b).

(i) Symptoms: The disease is characterized by internal brown discolouration and dry rotting in the basal portion of the flower stalk resulting in partial to complete withering of inflorescence prior to seed formation. Such rotting may extend internally up to the seed bulb. Outwardly, the plants appear normal till either inflorescence gets dried up or stalks topple down. Yellowing of outer leaves from tip downwards followed by drying is seen in bulb crop. During storage, the inner scales of affected bulbs become water-soaked followed by browning and drying which results in the hollowing of the bulbs.

Besides onion, the bacterium is pathogenic to garlic, tuberose, rice, bajra, sorghum, cowpea and other plants (Swarup, 1983).

(ii) Epidemiology: The pathogen survives in diseased plant debris in soil. Onion seedlings get infected in the field, and then it invades the bulbs and survives along with them, causing dry rot in the storage. Partially rotten bulbs while sprouting produce diseased plants in the next season (Kishun and Swarup, 1979b). The pathogen also survives in diseased plant debris at 6–8°C up to 12 months and at room temperature (25–30°C) for 6 months (Kishun and Swarup, 1979b). The bacterium enters the host through wounds and spread in storage by contact of diseased and healthy bulbs.

(iii) Management

Soaking of onion bulbs in ceresan wet, sandoz seed dressing-6334, 2000 ppm aretan, agrimycin-100, 200 ppm streptocycline for 2 hr reduced the disease incidence in seed crop (Kishun and Swarup, 1982).

Onion lines 5001-12, 5308 (bulb crop), 5223-2, 5404-3, 6106, 6621 and VL-14 (seed crop) remain free from infection under natural conditions (Kishun and Swarup, 1982).

3.1.2.9. Storage Soft Rot, *Erwinia carotovora* sub sp. *carotovora*

(i) Symptoms: Soft rot is one of the more prevalent causes of loss in storage onions (Fig 3.7). The soft rot bacterium can enter the neck tissues as plants approach maturity. In the field, plants wilt and die. As the rot progresses, invaded scales become soft and foul-smelling. Onions with mechanical injuries, sunscald, or bruises are particularly susceptible to bacterial soft rot, especially if they have been held under warm, humid conditions.

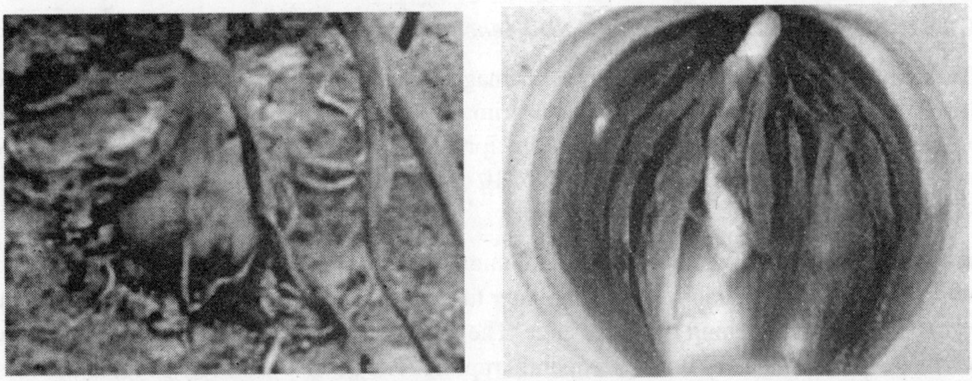

Figure 3.7. Soft rot of onion

(ii) Management

Storage of bulbs in well aerated and ventilated storage structures and eliminating the rotted bulbs by periodical turning of the storage heaps are recommended for minimizing the losses during storage.

3.1.2.10. Onion Yellow Dwarf Virus

(i) Symptoms: The first sign of infection in plants grown from naturally infected onion bulbs is a series of short yellow streaks which appear at the base of the first leaf emerging through the neck of the bulb. In general, all the leaves emerging after the appearance of first symptoms show signs of disease and those previously developed remain apparently healthy. Under favourable conditions, the leaves showing short yellow streaks at the base become yellow throughout and also crinkled, somewhat flat and fall off. Flower stalks of infected plants producing seed show yellow streaks extending upwards from the base. Later, the streaks coalesce, the stalks become yellow throughout, twists and curls in a characteristic manner. The infected plants are stunted in growth. The flower clusters of infected mother plants are smaller and have fewer flowers than normal plants. Besides onion, the virus also infects garlic, narcissus, Chinese naked lily and jonquil.

(ii) Transmission: The virus is mechanically transmitted. The most important aphid vectors include *Aphis rumicis, A. maidis* and *Rhophalosiphum prunifoliae.*

(iii) Management

- Obtain virus-free planting material by indexing.
- Virus-free seed stocks of bulbs can be produced in disease-free areas.
- Rouge virus infected plants in the field from time to time.

3.1.3. Nematodes

Stem and bulb, root-knot and reniform nematodes have been recognized as the limiting factors in successful production of onion and garlic.

3.1.3.1. Stem and Bulb Nematode, *Ditylenchus dipsaci*

The stem and bulb nematode cause great damage to the onion and garlic crops in temperate areas of many countries. The nematode is primarily observed in heavy soils. The nematode problem becomes serious with abundant precipitation or a high soil humidity. The nematode causes serious damage at inoculum level of 10 or more nematodes/500 g of soil. *D. dipsaci* was responsible 'bloat' disease of onion.

(i) Symptoms: The leaves of the infected onion plants become abnormally thickened and twisted; they become flabby, droop, and later turn yellow. In a more severe infestation, the sprouts die before reaching two-leaf stage. The inner scales of the bulbs are infested more often than the outer scales which frequently rupture. The bulbs eventually become soft and begin to rot.

The leaves of garlic plants turn yellow and die but no thickening or twisting is evident (Fig 3.8). The infected garlic bulbs become loose (the cloves separate from the base) and cannot be used for seed even if they tolerate winter storage well because the plants which form often die during the vegetative period.

Figure 3.8. The stem and bulb nematode (*Ditylenchus dipsaci*) on garlic. A – Garlic crop in field infected with *D. dipsaci*. B – *Left*, healthy; *Right*, infected. (Courtesy: Sasser, 1971)

(ii) Life Cycle: The fourth stage juveniles penetrate the stem and leaf tissue through the stomata. Egg laying begins at temperatures of 1–5°C with the optimum at 13–18°C. *D. dipsaci* completes one generation in 19–23 days at 15°C. Nematode activity stops at 36°C. the nematode prefers the cool moist climatic conditions existing in the upland tropics and wet winter seasons in the subtropics. *D. dipsaci* can parasitize plants in both heavy and light soils, although a higher incidence of infestation seems to occur on heavy soils.

(iii) Spread and Survival: The nematode can survive in the soil without a host plant for more than one year and the fourth juvenile stage can survive in anabiosis for many years. The nematode can be spread by transportation in infested bulbs, plant residue and adhering soil. Seed-borne infections also are responsible for long distance spread in onion. Other hosts and weeds are responsible for maintaining infestations between onion and garlic.

(iv) Management

(a) Physical Methods: The nematodes can be exterminated by soaking the plants in water containing 0.5% formalin at 24°C, followed by a hot water treatment (with addition of 0.5% formalin at 49°C for 20 min).

Soaking of onion sets or garlic cloves in cold water for 3 to 4 days will also kill the nematodes.

(b) Cultural Methods: The diseased plants and their remnants should be gathered and destroyed.

Bulbous plants, pea, carrot and beans should be excluded from the crop rotation system in order to control the onion race. Onion and garlic may be rotated with wheat, barley, maize, rape, white clover, red clover, Lucerne or lettuce to overcome the stem and bulb nematode damage.

(c) Chemical Methods: Seed stocks should be disinfected with methyl bromide (30 g of methyl bromide/m^3 for 20 hr). Garlic cloves may also be disinfected by fumigation with methyl bromide (30g of methyl bromide/m^3 for 8 hr) before autumn sowing.

Row application of thionazin at 4.5 g/metre row also gives effective control of the stem and bulb nematodes.

3.1.3.2. Root-knot Nematodes, *Meloidogyne* spp.

Vadivelu and Rajendran (1986) reported *M. arenaria* on onion from Palladam district of Tamil Nadu. The infected plants showed patchiness and unthrifty growth with 'neck twisting' symptoms. The root system was very much reduced and bulbs were sleek and elongated.

(i) Symptoms: Symptoms of root-knot on onion are very discrete, the main symptom being the presence of the protruding egg masses (Fig 3.9).

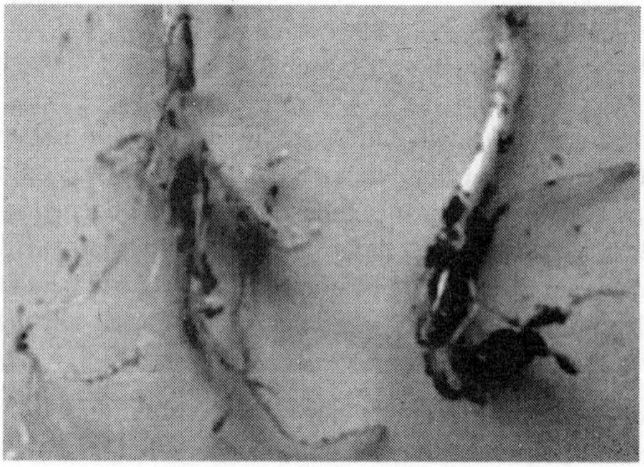

Figure 3.9. Root-knot nematode on onion

(ii) Management

(a) Physical Methods: Hot water treatment of onion bulbs at 43.5C for 120 min eliminates the nematode infection.

(b) Cultural Methods: The lowest populations of plant parasitic nematodes were recorded in onion-cowpea-fallow cropping sequence (Vetrivelkalai and Subramanian, 2006).

(c) Integrated Methods: Root dip treatment of onion seedlings in *P. lilacinus* + neem cake suspension for 1 hr before transplanting proved effective.

3.1.3.3. Reniform Nematode, *Rotylenchulus reniformis*

(i) Management

(a) Host Resistance: Onion *cv.* Evergreen is reported to be resistant to the reniform nematode.

Table 3.4. Management of insect pests in bulbous vegetable crops

Crop	Pest	Symptoms	Pesticide/Bioagent Dose/litre	Remarks Remarks
Onion	Thrips	Yellowing and crinkling of leaves	Monocrotophos-1.25 ml	As and when observed
	Capsule borer	Presence of borer	Phosphamidon-0.5 ml	As and when observed
	Capsule borer in seed crop	Presence of borer	Fenvalerate-0.5 ml	As and when observed

Table 3.5. Bulbous vegetable crop varieties resistant to biotic stresses

Vegetable crop	Pest/Disease	Resistant varieties
Onion	Purple blotch, Basal rot, Thrips	Arka Pitamber, Arka Kirtiman, Arka Lalima.
	Purple blotch, *Alternaria pori*	Arka Kalyan.
Garlic	Purple blotch, *Stemphylium* disease	Agri-found White (G-41), G-1, G-50, G-51

4 MALVACEOUS VEGETABLE CROPS

4.1. OKRA

Okra (*Abelmoschus esculentus*) is an annual vegetable crop grown in tropical and sub-tropical regions. It is being cultivated in 0.376 million hectares producing 3.684 million tonnes of okra with an average yield of 9.8 tonnes per hectare (Bijay Kumar, 2007) (Table 4.1). West Bengal, Bihar, Orissa, Gujarat, Andhra Pradesh, Chattishgarh, Maharashtra, Assam, Uttar Pradesh, Haryana and Karnataka are the major okra-growing states. Tender green fruits are cooked in curry and soup. The root and stem are used for clearing cane juice in preparation of 'gur'. High iodine content of fruits helps to control goitre while leaves are used in inflammation and dysentery. It is a good source of vitamins A, B and C. The fruits also help in cases of renal colic, leucorrhoea and general weakness. The dry seed contains 13–22% good edible oil and 20–24% protein. The oil is used in soap, cosmetic industry and as vanaspati while protein is used for fortified preparations. The crushed seed is fed to cattle for more milk production and the fibre is utilized in jute, textile and paper industry.

Table 4.1. Area, production and productivity of leading okra growing states in India (2005–06)

State	Area (in '000 ha)	Production (in '000 MT)	Productivity (MT/ha)
West Bengal	65.8	718.9	10.9
Bihar	56.2	714.1	12.7
Orissa	71.4	618.8	8.7
Gujarat	41.0	350.3	8.5
Andhra Pradesh	25.6	332.7	13.0
Chattishgarh	18.1	184.7	10.2
Maharashtra	29.0	182.7	6.3
Assam	9.6	119.7	12.5
Uttar Pradesh	8.8	110.2	12.5
Haryana	14.0	93.7	6.7
Karnataka	7.7	63.2	8.2
Others	28.9	194.9	6.7
Total	**376.1**	**3684.0**	**9.8**

(*Source*: Indian Horticulture Database 2006)

4.1.1. Pests

Hoppers, shoot and fruit borer and petiole maggot are the main pests of okra.

4.1.1.1. Leaf Hopper, *Amrasca biguttula biguttula*

(i) Damage: These are polyphagous and destructive pests of malvaceous crops that lay pear shaped, elongated and yellowish white eggs in the veins on the under surface of leaves. Both the adults and nymphs suck the cell sap (Fig 4.1). Nymphs move diagonally when disturbed. There are 10 to 12 overlapping generations in a year. Nymphs and adults suck cell sap and inject toxic saliva into the plant tissues. Plants lose their vitality and affected leaves turn yellow and curl upwards. When the infestation is high, typical hopper burn symptoms are produced. The leaves turn brick red and show large necrotic spots (this is often confused with fungal infection). The pest infestation is more severe during summer. Potentially they can cause 50–100% yield loss if adequate control measures are not taken on time.

Figure 4.1. Leaf hopper on okra

(ii) Management

- Apply neem cake at 250 kg/ha to soil immediately after germination and repeat it after 30 days.
- In the initial stages of crop before flowering, spray systemic insecticide like acephate 75 SP (1 ml/l) or imidacloprid (0.3 ml/l). Once the fruit harvest starts, avoid systemic insecticides as harvesting is done once in 3–4 days.
- Spray neem or pongamia soaps at 0.5% or pulverized neem seed powder extract (NSPE) 4%. Give the lower surface of the leaves thorough spray, as the pest is generally found there.

Sprays of 1% neem/pongamia soap were highly effective. Spraying 5% NSKE acts as oviposition deterrent, repellent and growth inhibitory effect on leaf hoppers and resulted in lower percentage of normal adult emergence.

Soil application of neem and pongamia cakes immediately after germination and repeated at 30 days interval twice at 250 kg per ha. reduced the hopper (*Amrasca biguttula biguttula*) incidence to less than 3/plant (Krishna Murthy and Krishna Kumar, 2002).

Two alternate sprays of deltamethrin at 10 g *a.i.*/ha or fenvalerate, cypermethrin or permethrin all at 50 g *a.i.*/ha proved most effective and recommended for the management of leaf hoppers (Krishna Kumar *et al.*, 1989).

Lower population of leaf hoppers was noticed with sprays of poultry manure + herbal leaf extract and neem cake + herbal leaf extract.

Okra *cvs*. Janardan, EMS-8, Ludhiana Selection-2 (a sister line of Punjab Padmini) and Punjab Padmini had a lower level of jassid population.

4.1.1.2. Shoot and Fruit Borer, *Earias vittella, E. insulana*

(i) Damage: *Earias vittella*, commonly called as spotted bollworm, is very destructive to okra. It is active through out the year reaching peaks during March-May and August-October. Full grown caterpillars are 18 to 24 mm long, stout, spiny and brownish coloured. Adults are 13 to 15 mm long having pale white fore-wings with a broad wedge-shaped horizontal green patch in the middle. Hind wings are silvery-creamy in colour. Wing expanse is about 30 to 34 mm. Eggs are light brown in colour and are laid singly on buds, shoots and fruits. Eggs hatch in about 7 days. Young larvae are spindle shaped with yellow spots on tanned skin colour. The pest has about 8 to 12 overlapping generations in a year. *E. insulana* is found in drier regions. Its fore-wings are completely green. Its moths are smaller in size and larvae almost similar except in size.

Female lays green colored eggs with longitudinal ridges on buds, flowers and fruits. When the crop is young, larvae bore into tender shoots and tunnel downwards, which wither, drop down and growing points are killed (Fig 4.2). With the formation of fruits, the caterpillars bore inside these and feed on inner tissues which become deformed in shape with no market value (Fig 4.2). Like brinjal shoot and fruit borer, the infestation is seen on shoots before flowering and after flowering they feed exclusively on fruits.

Figure 4.2. Shoot and fruit borer damage on shoots and fruits of okra

(ii) Management

- Collect and destroy the affected fruits and stems.
- Apply neem cake at 250 kg/ha after germination and repeat two more times at 30 days interval and spray neem soap 1% or NSPE 4% at 10 days interval. However, please note that botanicals get washed away due to rains. Under such conditions spray effective contact insecticides like indoxacarb 14.5 SC at 0.75 ml/l. Care must be taken for maintaining a waiting period of at least one week to avoid residues in the produce.
- Spray any contact insecticide. Safe waiting periods are to be strictly followed.
- Avoid staggered planting.

Soil application of neem and pongamia cakes at 250 kg per ha immediately after germination and repeated twice at 30 days interval reduced the borer incidence to less than 7%. Similarly significant reduction in hopper (*Amrasca biguttula biguttula*) (< 3/plant) and petiole maggot (*Melanagromyza hibisci*) (< 5%) were also noticed. Cakes were better than seed powders in reducing pest incidence and neem cake was the best treatment. This treatment was also found to reduce yellow vein mosaic virus and powdery mildew diseases (Krishna Moorthy and Krishna Kumar, 2002) (Table 4.2).

Table 4.2. Efficacy of neem and pongamia cakes on major pests and YVMV disease in okra

Treatment	Incidence of petiole maggot (%)	Mean no. of Hoppers/ 3 leaves	Incidence of fruit borer %	Incidence of YVMV
Neem seed powder	13.79	3.63	11.21	0.66
Pongamia seed powder	15.73	3.97	16.54	1.33
Neem cake	4.76	2.77	5.22	0.00
Pongamia cake	2.44	3.04	6.22	5.00
Untreated control	16.67	7.24	24.24	80.64

Spraying of 5% NSKE also gives effective control of shoot and fruit borer. Spraying of 5% NSKE and 3% neem oil effectively checked the fruit borer damage in okra (Table 4.3).

Table 4.3. Effect of neem products in the control of fruit borer in okra

Treatment	Fruit borer damage (%)	
	Weight basis	Number basis
Neem cake extract 10%	7.72	3.26
NSKE 5%	1.87	0.36
Neem oil 3%	1.44	0.23
Carbosulfan 0.044%	2.77	0.53
Untreated check	37.71	33.77

Two alternate sprays of deltamethrin at 10 g *a.i.*/ha or fenvalerate, cypermethrin or permethrin all at 50 g *a.i.*/ha proved most effective and recommended for the management of fruit borer (Krishna Kumar and Srinivasan, 1984).

Releases of *T. chilonis* brought about 49.22% reduction in fruit borer (*E. vitella*) in Tamil Nadu. Three applications of *B. thuringiensis* (Dipel) at 0.5 kg/ha at weekly intervals reduced damage by *E. vitella.*

Okra *cvs*. Selection-2 (Gujarat Agricultural University, Anand), Parbhani Kranti, EMS-8 and Ludhiana Selection-2 had lower borer infestation on shoots and fruits.

4.1.1.3. Aphids, *Aphis gossypii*

(i) Damage: This is a polyphagous pest, feeding in colonies and completely covers the shoot tips, buds and lower surface of leaves (Fig 4.3). The aphids are very small, soft bodied insects which are whitish-green in colour. Aphids multiply parthenogenetically in large numbers in a very short period. Both nymphs and adults suck the sap due to which plants loose their vitality. Leaves curl downwards and there is retardation of growing shoots. They also excrete honeydew on which sooty mould develops which hampers the photosynthetic activity of the plant.

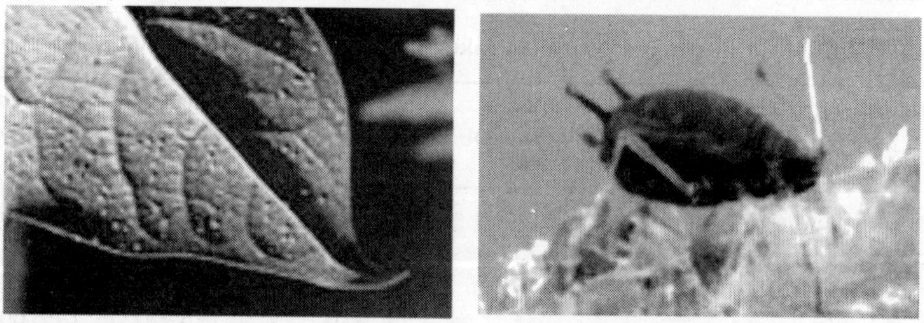

Figure 4.3. Aphid infestation on okra leaves

(ii) Management

- Clipping and destruction of infested shoots and plant parts.
- Thorough spraying of neem or pongamia soap (1%) or pulverized neem seed powder extract (NSPE) 4%.
- During pre-flowering period, spraying a systemic insecticide like dimethoate 30 EC (2 ml/l) or acephate 75 SP or acetamiprid.

4.1.1.4. Stem Fly/Petiole Maggot, *Melanagromyza hibisci*

(i) Damage: This is a dipteran fly and attacks the crop throughout the growth period. Generally drying of leaves is noticed. If petioles of such leaves are opened, yellow maggot and pupae can be noticed. Pest attack is noticed immediately after germination which may either cause mortality or swelling of the particular area of attack. Some times cracking of main stem is also observed.

(ii) Management

- Apply neem cake at 250 kg/ha immediately after germination and repeat after 30 days. This should be followed by sprays of NSPE 4% or neem soap 1% at 10 days interval after flowering.
- Stem fly is heavily parasitized in the field to an extent of 36.7% by *Eurytoma* sp. (Krishna Kumar and Srinivasan, 1988).

Soil application of neem and pongamia cakes immediately after germination and repeated at 30 days interval twice at 250 kg per ha. reduced the petiole maggot (*Melanagromyza hibisci*) incidence to less than 5% (Krishna Murthy and Krishna Kumar, 2002).

4.1.1.5. White Fly, *Bemesia tabaci*

(i) Damage: It is a polyphagous vector transmitting yellow vein mosaic virus disease (YVMV). Adults are tiny insects with white wings and pale body. Eggs are attached by thin, small whitish stalks. It lays eggs on the lower surface of leaves. Both the adult and the nymph suck leaf sap and exude honey dew (Fig 4.4). As a result of sucking, the plants lose their vitality and become pale yellow. Photosynthetic activity is affected due to sooty mold development. At present many varieties and hybrids resistant to YVMV are available in market and hence, disease management is very easy.

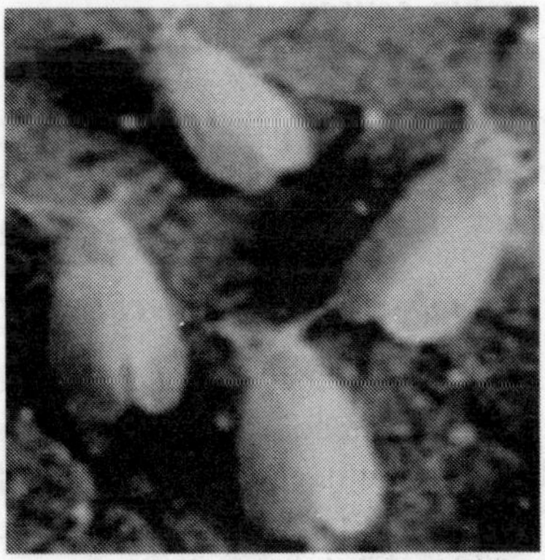

Figure 4.4. Whitefly infestation on okra

(ii) Management

- Grow YVMV resistant varieties/hybrids.
- Apply neem cake at 250 kg/ha at germination and again at 30 DAP followed by sprays of pulverized neem seed powder extract 4% or neem oil 1%.
- Spray imidacloprid 200 SL at 0.3 ml/l or thiomethoxam at 0.3 gm/l (should not to be sprayed after flowering stage).

Spraying of 5% NSKE effectively controlled the whitefly in okra resulting in over 80% reduction in whiteflies followed by 3% neem oil spray with 78% reduction over untreated control (Table 4.4).

Table 4.4. Effect of foliar application of neem products for the control of whiteflies in okra

Treatment	Mean population of whiteflies/3 leaves	% Reduction over untreated check
Neem cake extract 10%	9.2	67.9
NSKE 5%	5.7	80.1
Neem oil 3%	6.3	78.0
Carbosulfan 0.044%	23.9	16.5
Untreated check	28.6	—

4.1.1.6. Pod Borer, *Helicoverpa armigera*

(i) Damage: This pest is occasionally serious during monsoon season on okra. Damage is restricted to the apical end only. Since okra fruits are harvested once in 2–3 days, damage by *H. armigera* is not very high.

(ii) Management

Application of 5 rounds of *Ha* NPV at 500 LE/ha at weekly intervals right from flowering gave effective control of the pest.

4.1.1.7. White Grub, *Holotrichia consanguinea*

(i) Management

Collection of beetles from plants during night hours and killing them in kerosene water reduced the infestation. Fungal and bacterial pathogens, *Metarrhizium anisopiale* and *Bacillus popillae* var. *holotrichiae* are also effective in reducing the white grub infestation.

4.1.1.8. Mites, *Tetranychus cinnabarinus, Oligonychus coffeae*

(i) Damage: The infestation of mites is mostly observed during the warm and dry periods of the summer season. Larvae are pink in colour while nymphs are greenish red in colour. Adults are oval in shape, reddish brown in colour. Colonies of larvae, nymphs and adults feed on the under surface of the leaves (Fig 4.5). The affected leaves gradually start curling and finally get wrinkled, crumpled and fall down. This in turn affects the growth and fruit-forming capacity of the plants.

Figure 4.5. Spider mite infestation on okra leaf

(ii) Management

- Spray dicofol 18.5 EC at 2.5 ml/l or any other acaricide like wettable sulphur 50 WP at 3 g/l.
- As an alternative to synthetic acaricides, spray 1% neem soap/pongamia soap.
- Thorough spray on the under surface of the leaves is to be ensured for getting good control.

Pesticide Residues

The persistence of terminal pesticide residues and waiting periods for okra are presented in Table 4.5. (Awasthi, 2001).

Table 4.5. Terminal pesticide residues and their waiting periods for okra

Pesticide schedule	Dosage/ Concentration %	Terminal residues (ppm/days)	Waiting period (days)
Monocrotophos spray at fruiting stage	0.05	0.03/7	3
Quinalphos spray at fruiting stage	0.05	0.22/15	12
Cypermethrin spray at fruiting stage	0.01	0.23/5	3
Fenvalerate spray at fruiting stage	0.01	0.19/7	5
Endosulfan spray at fruiting stage	0.07	0.37/3	2
Carbaryl spray at fruiting stage	0.10	0.25/7	Nil
Carbofuran soil application at planting	0.5 kg *a.i.*/ha	0.20/45	50
Phorate soil application at planting	1.0 kg *a.i.*/ha	0.16/52	60

4.1.2. Diseases

4.1.2.1. Leaf Spot, *Cercospora abelmoschi*

(i) Symptoms: The fungus causes no definite leaf spot but grow as sooty to dark olivaceous ectoparasite covering the entire leaf lamina. The disease is initiated as small, isolated fungal growth on lower surface of leaves, but very soon it spreads to both sides without any definite area (Fig 4.6). The pathogen severely reduces the photosynthetic area. Infected leaves fall down very quickly soon after drooping and rolling. The disease causes severe loss in seed crop as well as late sown crop grown for fruits. Sometimes light brown lesions are also observed after washing out of spores from leaves due to rainfall.

(ii) Management

Sowing of okra should be completed during first week of July. The defoliated leaves should be collected and burnt in the field to reduce the inoculum at source.

Three sprays of 0.05% bitertanol or 0.025% tridimefon or 0.8% Bordeaux mixture at 500 litres/ha starting from the appearance of symptoms at weekly intervals is effective for the management of leaf spot disease.

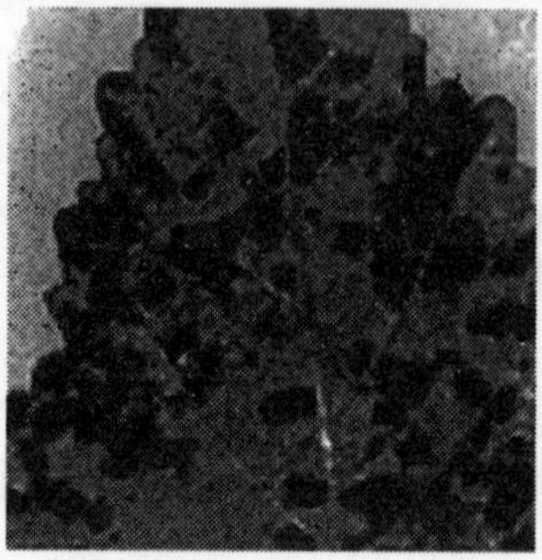

Figure 4.6. *Cercospora* leaf spot on okra

4.1.2.2. Powdery Mildew

(i) Symptoms: The powdery mildew disease usually appears at the end of last harvesting of the fruits. The disease is characterized by the appearance of tiny, white to dirty grey spots. They become powdery as they enlarge. The superficial powdery mass may ultimately cover the entire green surface of the host (Fig 4.7). The effect of severe infection is premature defoliation and death of the plant parts.

Figure 4.7. Powdery mildew on okra leaves

(ii) Management

Spraying of 4% NSKE and 1% neem oil effectively controlled the disease.

4.1.2.3. Yellow Vein Mosaic Virus (YVMV)

This is the most important and devastating virus disease of okra. Sastry and Singh (1974) estimated that if the plants are infected within 20 days after germination, the loss in yield was recorded up to 98%. The plants that infected at 35 and 50 days after germination, the loss in yield was estimated to be 83 and 49%, respectively.

(i) Symptoms: YVMV is caused by Gemini virus and transmitted by whitefly. Characteristic symptoms appear as yellow vein and veinlets leaving green tissue in interveinal area

(Fig 4.8). Severely infected leaves sometimes become completely yellow. Fruits also change in colour to yellow and become hard in early stage of development. The infected plants are stunted and bear very few yellow coloured fruits.

Figure 4.8. YVMV infection on okra leaves

(ii) Transmission: In nature, the virus is transmitted by whitefly, *Bemisia tabaci*.

(iii) Epidemiology: Vector population and virus incidence were more during March to June when the atmospheric temperature remained high and humidity is less which is favourable for whitefly multiplication and spread of the disease.

(iv) Management

Kharif okra should be sown early between last week of May and second week of June under Konkan conditions for obtaining higher yield with less YVMV incidence. Removal of malvaceous hosts, and weeds (especially *Hibiscus tetraphyllus* and *Althea rosea*) around okra field helps in reducing the disease incidence. Rouging of diseased plants as soon as they are noticed in the field helps to reduce the disease incidence. Infested crop should not be used for seed collection. The appearance of YVMV can be delayed by mulching the soil with coloured polythene sheet.

Sowing of 4 to 5 rows of sorghum or pearl millet or maize all round the okra field at least 60 days before sowing okra has been found beneficial for the management of YVMV. Border cropping along with 3 to 4 foliar sprays of either dimethoate or monocrotophos both at 0.1% at 10 days interval has been found more effective (Singh, 1990).

Seed treatment with 0.025% imidachloprid is effective in controlling the vector. Carbofuran at 1.0 kg *a.i.*/ha should be applied to the soil at the time of sowing seeds. The whitefly population and yellow-vein mosaic virus incidence could be minimized significantly by 4 to 5 foliar sprays of either dimethoate or monocrotophos both at 0.1% at 10 days interval followed by 1 or 2 sprays 0.2% mineral oil (Sastry and Singh, 1983a; Singh *et al.*, 1977).

Use resistant okra cultivars like Arka Anamika, Arka Abhay, Pusa Sawani, Harbajan, Parbhani Kranti to manage YVMV.

4.1.2.4. Enation Leaf Curl Virus

The disease was first observed at Bangalore during 1986 by Singh and Dutta. This disease is most destructive virus disease which is severe during kharif season. The disease is caused by Gemini virus and transmitted by whitefly.

(i) Symptoms: The characteristic symptoms include thickening of the veins, warty and rough texture of leaves, leaf curling and twisting of the main stem and lateral branches. Twisting of leaf petiole was found to be prominent. In severe cases, newly emerged leaves also exhibit bold enations, thickening and curling. The virus infected plant is unable to bear its weight. The stem and leaf petiole are bent and zig zag in shape. All flowers do not set the fruits. If the fruit is set, seeds are not fertile and aborted. The characteristic small green out growths are observed on vein and veinlets on the lower side of the leaves (Singh and Dutta, 1986).

(ii) Transmission: The virus is readily transmitted by grafting. The natural transmission occurs through whitefly, *Bemisia tabaci* (Singh and Dutta, 1986).

(iii) Management

- The infected plants should be rouged out as soon as they are noticed in the field.
- The spread of the disease can be reduced by foliar sprays of dimethoate or monocrotophos both at 0.05% at 10 days interval.

Integrated Pest and Disease Management

- Deep off-season/summer ploughing is useful to expose nematodes and resting stages of insects.
- Using YVMV resistant varieties (Arka Abhay, Arka Anamika, Parbhani Kranti, Varsha Uphar) for planting.
- Seed treatment with *Trichoderma viride* at 4 g/kg of seeds.
- Erection of bird perches at 25/ha.
- Intercropping with onion or cowpea is useful.
- Crop rotation with non-host crops such as sesame, marigold, onion or mustard reduces nematode infestation.
- Sowing of sorghum/maize all around okra field to prevent the entry of fruit borer.
- Soil application of fresh neem cake at 250 kg/ha to reduce fruit borer incidence.
- Setting up of yellow sticky traps for monitoring whitefly at 10 traps/ha.
- Installation of pheromone traps for *H. armigera* and *Erias vitella* at 5 traps/ha.
- Release of egg parasitoid, *T. chilonis* at 1 lakh/ha.
- Collection and destruction of borer infested fruits from time to time.
- Destruction of YVMV affected plants from time to time.

Table 4.6. Pest and disease management in okra

Days after sowing	Management	Pests and diseases controlled
15	Imidacloprid 0.5 ml/litre of water*	Leafhoppers, aphids, white fly
25	Imidacloprid 0.5 ml + mancozeb 2 g/l of water*	Leaf hoppers, aphids, white fly and leaf spot diseases
40	Cypermethrin 0.5 ml + wettable sulphur 3 g/l of water	Fruit borer and powdery mildew
60	Cypermethrin 0.5 ml + wettable sulphur 3 g/l of water	Fruit borer and powdery mildew

*Seed treatment with imidacloprid (Gaucho) at 40 g/kg seed can also be taken up instead of foliar sprays.

4.1.3. Nematodes

Root-knot and reniform nematodes have been recognized as the major nematode pests of okra.

4.1.3.1. Root-knot Nematodes, *Meloidogyne* spp.

M. incognita was responsible for 28.08 to 90.90% loss in fruit yield of okra (Bhatti and Jain, 1977; Parvatha Reddy and Singh, 1981), while *M. javanica* caused 20.20 to 41.20% loss in yield (Jain *et al.*, 1986).

(i) Interaction with Other Pathogens: .Root-knot nematodes cause collar rot disease complex in association with *Pythium aphanidermatum* and *Ozonium texanum* var. *parasiticum* in okra.

(ii) Management

(a) Cultural Methods: Decrease in *M. incognita* population in okra occurs when the field was left fallow (Khan *et al.*, 1975). The maximum reduction in *M. javanica* population and increase in fruit yield was recorded in okra by soil ploughing + covering with polythene sheet, followed by soil ploughing + exposure to sun (Anon., 1989).

Crop Rotation: Decrease in root-knot nematode population in okra occurs following marigold, spinach and bottle gourd (Khan *et al.*, 1975). Out of 15 different cropping sequences studied, tomato-onion-resistant tomato *cv.* Hisar Lalit-okra was found to be the best in terms of root-knot nematode reduction and economic returns (Kanwar, 1990). Crop rotations having okra-garlic-cluster bean and okra-coriander-resistant tomato *cv.* Hisar Lalit effectively managed the root-knot nematode population (Anon., 1993b).

Rotation of okra with sweet potato (*cv.* Sree Bhadra) reduced the root-knot nematode population by 21% and increased the fruit yield by 20.57% (Sheela *et al.*, 2002). Crop rotation with cabbage, marigold and wheat significantly reduced the root-knot nematode population. Crop rotation with cereals and marigold helps in reducing the nematode population in soil. In cropping sequence involving cabbage-marigold-wheat-kochia, the population of root-knot nematode larvae reduced to such an insignificant level that growing okra following the above sequence did not bring about increase in number of larvae (Alam *et al.*, 1977).

Intercropping: Atwal and Mangar (1969) showed that root exudates from sesame (*Sesamum orientale*) have nematicidal properties against *M. incognita*. When okra was grown in *M. incognita* infested soil it was only slightly attacked and there were fewer nematodes compared with when okra was grown in the absence of sesame.

Intercropping of okra with sesame decreased root-knot nematode penetration in okra and also delayed the life cycle of *M. incognita* (Tanda and Atwāl, 1988). Similarly, intercropping of okra with marigold, margosa and Persian lilac have been recorded to reduce nematode population and root galling.

Botanicals: Singh and Sitaramaiah (1966) applied finely divided oil cakes to root-knot infested soil and noted a reduction in disease incidence in okra. Soil application of neem cake at 2 t/ha recorded lowest root-knot index (1.6), followed by fresh neem leaves at 5 t/ha (1.6) and silk worm litter at 2.5 t/ha (1.7). Significant highest yield was also obtained with neem oil cake (6.2 t/ha) followed by neem leaves (6.1 t/ha) (Sahoo *et al.*, 2004b).

Best results in respect of root-knot reduction due to *M. javanica* and increase in yield of okra were obtained by amending the soil with saw dust at 2.5 tonnes per ha 3 weeks before planting and then applying N through urea at 120 kg per ha (Singh and Sitaramaiah, 1971b).

Spot application of groundnut cake and sunflower cake each at 0.25 t/ha, saw dust and coal ash each at 5 t/ha significantly reduced the root-knot nematode population (Bhosle *et al.*, 2006).

Application of press mud at 15 t/ha, a week prior to sowing is recommended for effective management of root-knot nematodes in kharif season with cost: benefit ratio of 1: 3.1. Application of fresh neem or karanj cake at 1 t/ha at the time of final land preparation is recommended.

Furrow application of neem cake at 100 g/2 m furrow, spot application of neem cake at 15 g/spot and spot and furrow application of mustard and castor cakes gave 58.88, 48.36, 35.86, 35.53, 34.87 and 31.91% increase in yield, respectively and reduced the nematode infestation (Talukdar, 1993).

Bala and Sukul (1987) observed that Eugenol, an active nematicidal principle from *Ocimum sanctum* showed systemic effect in reducing root-knot infection in okra. Ahuja and Mukhopadhyaya (1984) observed that application of neem leaves and seeds of Persian lilac maximally decreased the root-knot index due to *M. incognita* and significantly increased the yield of okra fruits.

(b) Chemical Methods: Seed treatment with aldicarb and carbofuran both at 6 and 12 per cent was effective against the root-knot and reniform nematodes infecting okra (Sivakumar *et al.,* 1976). Jain and Bhatti (1981) reported effective control of *M. javanica* on okra by seed treatment with phenamiphos at 2, 4 and 6 per cent.

Alam *et al.* (1973) reported that oxamyl was effective for the control of root-knot nematodes infecting okra. Aldicarb, carbofuran, ethoprophos and phenamiphos each at

1 to 2 kg *a.i.* per ha were found effective in reducing the root-knot nematode population and in increasing fruit yields of okra (56 to 87 per cent).

Drill application of carbofuran at 1.5 kg *a.i.*/ha below the seed level reduces the incidence of *M. incognita* in okra.

(c) Biological Methods

Antagonistic Bacteria: Root-knot nematodes in okra can be managed by the application of *Bacillus macerans* at 1.2×10^8 cells/sq.m. before sowing.

Antagonistic Fungi: *P. lilacinus* reduced root-knot nematode egg masses by 91% after 60 days and by 96% after 90 days, while aldicarb reduced egg masses by 90 and 91% after 60 and 90 days after inoculation, respectively in okra. In general, *P. lilacinus* and aldicarb treatments greatly suppressed the numbers of nematode galls and egg masses on okra plants. However, it was evident that *P. lilacinus* was more effective in reducing nematode egg masses than aldicarb (Ibrahim and Rezk, 1988).

The soil incorporation of *P. lilacinus* (in rice hulls and bran soil mix) was effective in reducing the gall index (2.7 as against 5.0 in control), nematode population in soil (77.3% over control) and in increasing the number of okra fruits (121.7 as against 86.0 in control) and fruit weight (3.26 kg as against 2.11 kg in control) (Davide and Zorilla, 1986).

(d) Host Resistance: Okra *cv.* Red Wonder was reported to be resistant, while *cv.* Parbhani Kranthi was found moderately resistant to *M. incognita*. Clemson Spineless was reported to be moderately resistant to *M. arenaria*.

(e) Integrated Methods

Using Bio-agents and Botanicals: Integration of *Paecilomyces lilacinus* with neem cake gave effective control of *M. incognita* on okra.

Application of 5 per cent inoculum of *Arthrobotrys conoides* to the pot soil amended with FYM, effectively reduced the larval penetration of *M. incognita* and root galling was reduced by 30-40 per cent in okra (Srivastava and Swarup, 1986).

Soaking of okra seeds in 10% castor cake suspension mixed with spores of *P. lilacinus* (1.5×10^6 spores/ml) for 30 min and sowing in soil drenched with 10% castor cake suspension at 20 litres/6 sq.m.) was effective in reducing root galling, final nematode population of *M. incognita* and increasing the fruit yield, root colonization, propagule density in soil and parasitization of eggs by *P. lilacinus* (Rao *et al.*, 1997a).

Using Botanicals and Chemicals: An integrated management of *M. incognita* infecting okra using neem or karanj oil cake at 0.5 t/ha along with carbofuran at 1 kg *a.i.*/ha gave maximum reduction in root galling with consequent increase in okra fruit yield (Parvatha Reddy and Khan, 1991).

Application of subabool (*Leucaena leucophila)* leaves at 40 g/kg soil which are allowed to decompose for four weeks before sowing + carbofuran at 1 kg *a.i.*/ha while sowing of okra seeds resulted in minimum galling (35.4 galls/plant) in comparison to control (64 galls/plant) and better plant growth parameters in okra (Paruthi *et al.*, 1987).

Using Bio-agents, Chemicals and Botanicals: The combined treatment with *P. lilacinus* at 4 g/kg soil + carbosulfan 25 EC at 0.2% + poultry manure at 2.5 t/ha + FYM at 2.5 t/ha gave maximum increase in plant growth parameters and yield (9.2 t/ha compared to

2.0 t/ha in control), and decrease in number of galls (83.5 compared to 183.6 in control), egg masses per root system (23.6 compared to 70.5 in control) and final nematode population in soil (200 compared to 585 in control) (Das and Sinha, 2005).

Using Cultural and Chemical Methods: Summer ploughing + seed treatment with carbofuran at 3% *a.i.* w/w + main field treatment with aldicarb at 1 kg *a.i.*/ha led to 76–79% decrease in nematode population and 35.1% higher yield over untreated check. Further, Summer ploughing + mulching transparent polythene sheet + seed treatment with carbosulfon at 3% *a.i.* w/w led to 32.5% higher yield. Summer solarization + treated seeds + use of neem cake at 400 kg/ha was most effective treatment and gave 50% higher okra yield.

Using Botanicals and Physical Methods: Integration of soil solarization for 15 days in summer and application of neem cake at 200 kg/ha is effective in the management of root-knot nematodes and in getting higher yields.

Using Physical and Chemical Methods: Soil solarization with single layer of polyethylene mulch for 20 days during June and application of carbofuran at 0.5 kg *a.i.*/ha gave least root galling (42/plant compared to 245/plant in control) (Sharma *et al.*, 2005).

4.1.3.2. Reniform Nematode, *Rotylenchulus reniformis*

(i) Interaction with Other Pathogens: Okra plants succumbed to wilt at an early stage when the reniform nematode, *R. reniformis* and *R. solani* were present together (Kumar and Sivakumar, 1981).

(ii) Management

(a) Cultural Methods: Application of potash in combination with phosphorus or nitrogen or potash alone checks the reniform nematode multiplication on okra to a great extent (Sivakumar and Meerazainuddin, 1974).

(b) Chemical Methods: Treatment of okra seeds with carbofuran at 1.25g *a.i.* w/w economized the nematode management by reducing the quantity of nematicide required (Sivakumar *et al.*, 1973).

(c) Biological Methods: Seed treatment with avermectin 100% recorded maximum fruit yield and highest reduction in *R. reniformis* population. Avermectin treated plants also recorded significantly increased total phenol content and enzymatic activities such as peroxidase and IAA oxidase (Jayakumar *et al.*, 2004).

(d) Integrated Methods: Application of aldicarb at 1 kg *a.i.*/ha + neem cake at 0.5 t/ha followed by carbofuran at 1 kg *a.i.*/ha + neem cake at 0.5 t/ha proved most effective in reducing the *R. reniformis* population and in increasing the growth of okra plants (Krishna Rao *et al.*, 1987).

Deep ploughing (20 cm) followed by fallowing for one month, fallowing for one month after weeding, integration of aldicarb application at 0.8 kg *a.i.* per ha at sowing after either of the cultural practices or deep ploughing (20 cm) together with carbofuran or aldicarb seed treatment resulted in the control of the reniform nematode (*Rotylenchulus reniformis*) and better yield of okra (Lakshmanan and Sivakumar, 1981).

4.1.3.3. Root-knot Nematode, *Meloidogyne incognita* and Wilt, *Fusarium oxysporum* f. sp. *vasinfectum* Disease Complex

(i) Management

(a) Biological Methods: Among botanicals and biocontrol agents evaluated, *Trichoderma harzianum* was the most effective treatment in improving plant growth followed by neem seed powder, neem cake and *Aspergillus niger*. Highest suppression of root colonization of the fungus (*F. oxysporum* f. sp. *vasinfectum*) was achieved in plants treated with neem seed powder (7.5), neem cake (12.5), *A. niger* (15.0) and *T. harzianum* (17.5). The greatest suppression of root-knot development was achieved by *T. harzianum* followed by neem seed powder, neem cake and *A. niger* (Abuzar and Haseeb, 2006).

4.1.3.4. Root-knot Nematode, *Meloidogyne incognita* and Root Rot, *Rhizoctonia solani* Disease Complex

(i) Management

(a) Biological Methods: *Paecilomyces lilacinus* can be effectively used against *Rhizoctonia solani – M. incognita* complex of okra (Shahzad and Ghaffar, 1989).

Table. 4.7. Management of insect pests in okra

Pest	Symptoms	Pesticide/Bioagent Dose	Remarks
Fruit borer	Bored fruits and larvae inside with excreta	Fenvalerate/cypermethrin at 0.5 ml/l	Once in 15 days after flowering
Petiole maggot	Death of 10–20 days old plants and drying of leaves and maggots in dead plant base and drying leaves	Carbofuran-15 kg/ha; Monocrotophos-1.25 ml/l	Soil application while sowing/spray on 7 and 20 days after sowing
Aphids/ white fly/ Jassid	1. Colony of aphids on young leaves and fruits 2. White small flies observed on under surface of leaves 3. Yellowing of leaves and green young jassids under the leaf surface	Phosphamidon-0.5 ml; Monocrotophos-1.25 ml/l	21 and 35 DAS

Table. 4.8. Okra varieties resistant to biotic stresses

Pest/Disease	Resistant varieties
YVMV	Pusa Sawani, Arka Abhay, Arka Anamika, Hisar Unnat, DVR-1, DVR-2, IIVR-10, Varsha Upkar, P-7, Pusa A-4, Parbhani Kranti (T), Punjab Kesari.
YVMV & fruit borer	Pusa A-4.

5 CRUCIFEROUS VEGETABLE CROPS

5.1. CABBAGE AND CAULIFLOWER

Cabbage (*Brassica oleracea* var. *capitata*) is an important winter vegetable of cole group. It is a rich source of vitamins A, B and C and also contains minerals. It is being cultivated in 3.121 million hectares producing 67.315 million tonnes of cabbages with an average yield of 21.568 tonnes per hectare (Bijay Kumar, 2007) (Table 5.1). It covers about 4% of total area under vegetables. India comes next to China in cabbage production. In India, it is being cultivated in 0.267 million hectares producing 5.922 million tonnes of cabbage with an average yield of 22.1 tonnes per hectare (Bijay Kumar, 2007) (Tables 5.1 and 5.2). It is now grown almost throughout the year. West Bengal, Bihar, Orissa, Assam, Maharashtra, Gujarat, Uttar Pradesh, Haryana, Karnataka, Chattishgarh and Madhya Pradesh are the major cabbage-growing states.

Table 5.1. Major cabbage producing countries of the world

Country	Area (in '000 ha)	Production (in '000 MT)	Productivity (in MT/ha)	% Share of world production
China	1719	33817	19.5	50.2
India	267	5922	22.1	8.8
Russia	166	4051	24.5	6.0
Korea	42	2603	61.8	3.9
Japan	53	2287	42.9	3.4
Ukraine	71	1475	20.9	2.2
Poland	39	1382	35.6	2.1
Indonesia	64	1293	20.1	1.9
USA	30	1100	36.9	1.6
Romania	55	1012	18.5	1.5
Others	615	12375	20.1	18.4
Total	**3121**	**67315**	**21.5**	

Source: FAO (except Indian data) (Indian data source: Indian Horticulture Database 2006)

Cauliflower (*Brassica oleracea* var. *botrytis*) is the most popular winter vegetable among cole crops. It is rich in vitamin C. It is being cultivated in 0.292 million hectares producing 5.260 million tonnes of cauliflowers with an average yield of 18.0 tonnes per hectare (Bijay Kumar, 2007) (Tables 5.3 and 5.4). West Bengal, Bihar, Orissa, Maharashtra, Haryana,

Gujarat, Assam, Madhya Pradesh, Uttar Pradesh, Punjab and Chattishgarh are the major cauliflower-growing states. With the development of new varieties, it is now being grown in non-traditional areas like Andhra Pradesh, Tamil Nadu and Kerala.

Table 5.2. Area, production and productivity of leading cabbage growing states in India (2005–06)

State	Area (in '000 ha)	Production (in '000 MT)	Productivity (MT/ha)
West Bengal	74.7	1982.7	26.5
Bihar	33.7	931.6	27.6
Orissa	36.6	578.3	15.8
Assam	24.3	480.3	19.8
Maharashtra	15.0	381.0	25.4
Gujarat	19.0	329.2	17.3
Uttar Pradesh	7.0	209.9	30.0
Haryana	12.3	172.2	14.0
Karnataka	6.4	134.2	20.8
Chattishgarh	6.8	114.0	16.8
Madhya Pradesh	3.8	75.0	20.0
Others	27.7	533.1	19.2
Total	**267.4**	**5921.6**	**22.1**

(*Source*: Indian Horticulture Database 2006)

Table 5.3. Major cauliflower producing countries of the world

Country	Area (in '000 ha)	Production (in '000 MT)	Productivity (in MT/ha)	% Share of world production
China	363	7369	20.3	40.5
India	292	5260	18.0	28.9
USA	69	1288	18.6	7.1
Spain	25	442	17.4	2.4
Italy	18	431	23.7	2.4
France	27	375	13.8	2.1
Mexico	25	359	14.2	2.0
Poland	14	239	17.0	1.3
UK	19	219	11.7	1.2
Pakistan	11	206	18.2	1.1
Others	130	1992	15.3	11.0
Total	**995**	**18180**		

Source: FAO (except Indian data) (Indian data source: Indian Horticulture Database 2006)

Table 5.4. Area, production and productivity of leading cauliflower growing states in India (2005–06)

State	Area (in '000 ha)	Production (in '000 MT)	Productivity (MT/ha)
West Bengal	65.6	1666.2	25.4
Bihar	59.8	938.5	15.7
Orissa	45.1	638.9	14.2
Maharashtra	13.0	328.9	25.3
Haryana	21.0	302.8	14.4
Gujarat	16.1	277.8	17.2
Assam	16.7	247.8	14.8
Madhya Pradesh	8.6	137.4	16.0
Uttar Pradesh	6.7	134.4	20.0
Punjab	5.6	131.4	23.5
Chattishgarh	9.6	123.5	12.9
Others	24.0	332.7	13.9
Total	**291.8**	**5260.1**	**18.0**

(*Source*: Indian Horticulture Database 2006)

5.1.1. Pests

5.1.1.1. Diamondback Moth (DBM), *Plutella xylostella*

(i) Damage: Diamond back moth is a major pest of almost all the cruciferous crops and is cosmopolitan in distribution. It is a small grayish brown insect bearing whitish triangular spots on posterior margins of the forewings and while resting, the wings come together and form a diamond pattern. This is a major pest of cruciferous crops, particularly cabbage and cauliflower during January-June months and also during drought periods in monsoon. Eggs are yellowish white, and are laid singly on tender leaves. A single female may lay about 40–60 eggs. Incubation period is 3–6 days. The first instar larvae mine the epidermal surface of the leaves. Second instar onwards the larvae feed externally by making holes in the leaves (Fig 5.1). Pupation occurs inside silken cocoon on the plants, which lasts for 4–6 days. The complete life cycle from egg to egg takes place in about 20–25 days.

(ii) Management

(a) Cultural Methods: Sprays of 4% NSKE/4% pulverized NSPE were found to be highly effective against DBM. Field sprayed with NSKE/NSPE had obtained higher yield, incurred 15% less cost and realized 43% higher net returns. The performance of powder formulation of 'Soluneem' was on par with 4% NSKE in controlling DBM. Sprays of neem soap or pongamia soap both at 1% were highly effective in reducing DBM (Table 5.5), aphids (*Brevicoryne brassicae*) and leaf webber (*Crocidolomia binotalis*). Further, the soap sprays significantly increased the yield of cabbage and cauliflower and were relatively safe to the parasitoid, *Cotesia plutellae* (Srinivasan and Krishna Moorthy, 1993; Krishna Moorthy *et al*., 1998; Krishna Murthy and Krishna Kumar, 2002).

Figure 5.1. DBM damage on cabbage

Table 5.5. Efficacy of neem and pongamia soaps on DBM and yield of cabbage

Treatment	DBM incidence/plant	Yield (t/ha)
NSKE	3.00	109.60
Bt	3.00	92.67
Soluneem	3.33	99.60
Neem oil	9.00	66.53
Pongamia oil	7.67	73.33
Neem soap	2.33	111.33
Pongamia soap	4.67	116.67
Control	36.33	43.87

Use of Neem and Pongamia Soaps: Under field conditions, neem and pongamia soap sprays were highly effective in reducing DBM, aphids (*Brevicoryne brassicae*) and leaf webber (*Crocidolomia binotalis*) and were at par with NSKE. They were superior to many commercial synthetic insecticides like dichlorvos, cypermethrin, combination formulation of cypermethrin and profenfos in controlling DBM. Further, the soap sprays significantly increased the yield of cabbage. On cauliflower too, the neem and pongamia soap sprays were effective and increased the yield significantly. The data on DBM, its parasitoid *Cotesia* *plutellae* and yield of cabbage and cauliflower are given in Tables 5.6 to 5.8. The soap sprays were also found to be relatively safe to *C. plutellae* (Krishna Murthy and Krishna Kumar, 2002).

Table 5.6. Effect of different botanicals and chemicals on the incidence of DBM in cabbage (2001–2002)

Treatment	Dose	No. of DBM larvae/10 plants	
		Winter	Summer
Neem soap	0.5%	10.66	5.33
Neem soap	0.75%	1.33	10.67
Neem soap	1%	0.33	15.00
Pongamia soap	0.5%	1.33	2.67
Pongamia soap	0.75%	0.33	7.00
Pongamia soap	1%	0.00	14.33
NSPE	4%	0.00	2.33
Diclorvos	0.1%	0.00	—
Bt	0.15	0.00	12.00
Cypermethrin	0.0125%	4.00	65.33
Control	—	5.67	92.33
C.D. (P = 0.05)	—	**6.15**	**12.15**

Table 5.7. Effect of different botanicals on yield and the incidence of larval parasitoid *Cotesia plutellae* in cabbage (2001–2002)

Treatment	Dose (%)	Yield t/ha		***Cotesia plutellae*** **cocoons/10 plants**
		Summer	Winter	
Neem seed powder	4	88.00	51.00	3.66
Neem soap	0.5	60.92	37.73	1.66
Neem Soap	0.75	85.00	38.73	1.33
Neem soap	1.0	83.89	51.00	2.33
Pongamia soap	0.5	70.00	37.73	2.00
Pongamia soap	0.75	74.11	47.00	0.33
Pongamia soap	1.0	87.78	52.33	1.33
Diclorvos	0.1	48.89	—	0.00
Bt	0.1	79.33	41.46	0.00
Cypermethrin	0.0125	51.89	27.20	0.00
Control	—	33.78	13.53	2.67
CD (P = 0.05)	—	**16.34**	**6.13**	**2.14**

Table 5.8. Effect of different botanicals on diamondback moth population and yield of cauliflower (2001–2002)

Treatment	Dose (%)	DBM/10 plants	Yield (t/ha)
Neem soap	0.5	4.33	6.18
Neem soap	0.75	5.00	5.27
Neem soap	1.0	4.33	7.83
Pongamia soap	0.5	6.67	3.23
Pongamia soap	0.75	3.00	5.85
Pongamia soap	1.0	1.33	7.90
NSPE	4	0.00	7.62
Bt	0.1	5.33	7.53
Cypermethrin	0.0125	13.33	4.82
Control	—	33.00	3.24
CD (P = 0.05)		**3.69**	**2.73**

The use of both neem and pongamia soaps in the cabbage IPM was successfully validated in more than 50 farmer's field during 2001 and 2002. Among them, 15 farmers used only soap sprays while others used 1–2 insecticide applications (which was not really necessary) and managed the pests very well and reduced the cost towards pesticides. These farmers were previously spraying insecticides more than 12 times for controlling insect pests of cabbage.

The soaps lack systemic effect and the residual toxicity is limited. Therefore, the coverage on the crop surface is very important while spraying to get effective control. During summer months if adults of DBM are found in large numbers, then they have to be controlled first using light traps during night. Otherwise, larval population may not get reduced as the eggs laid by the high adult population hatch and develop without any problem due to low residual toxicity.

(b) Biological Methods: Under natural conditions in an undisturbed ecosystem (insecticide-free environment), the parasitoid, *Cotesia plutellae* is capable of excersizing about 70–80% control of DBM in India (Nagarkatti and Jayanth, 1982; Chandramohan, 1994) (Fig 5.2).

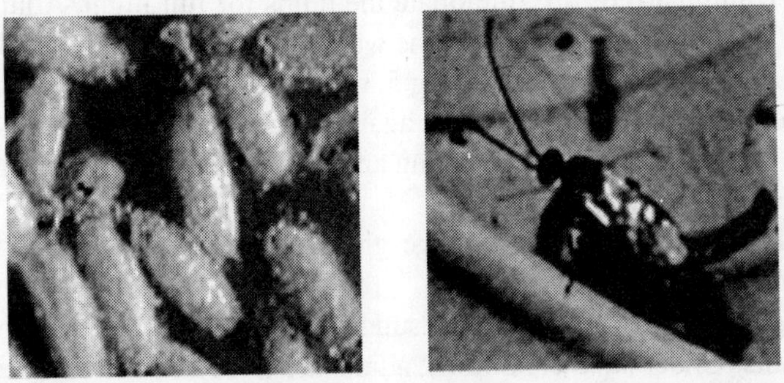

Figure 5.2. Pupae and adult of DBM parasitoid, *Cotesia plutellae*

The parasitoid, *Didegma semiclausum* producing excellent parasitism up to 68% in Nilgiris, showed its full potential in places of cooler temperatures ranging from 15–20°C and high hills (Chandramohan, 1994).

Bacillus thuringiensis formulations have been found effective for the management of DBM. 'Biotrol' WP (25×10^6 spores/mg) was effective in controlling the diamondback moth on cabbage (Rajamohan and Jayaraj, 1978). 'Thuricide' HPSC and 'Dipel' WP (16×10^6 IU/mg) at 1 and 1.5 g/l could effectively check *P. xylostella* on cabbage (Varma and Gill, 1976). Weekly sprays of 'Dipel' at 0.5 kg/ha were recommended for the control of *P. xylostella* and *Crocidolomia binotalis* (Krishnaiah *et al.*, 1981). Application of 5 rounds of *B. thuringiensis* var. *kurstaki* (Halt) at 1 kg/ha at weekly interval gave effective control of larval population (0.61–0.81 larva/plant) (79.9% reduction in larval population over check) as compared to control (3.02 larvae/plant). The yield was also high in Halt (70.2 kg/plot) as compared to control (53.8 kg/plot).

The highest mortality of DBM (52.5% in WP and 70.0% in Emulsified suspension formulation) was recorded in case of *Beauvaria bassiana* isolated from DBM. Emulsified suspension formulation of *B. bassiana* will be an ideal organic input as foliar mycoinsecticide in cabbage ecosystem. Application of 5 rounds of *Paecilomyces farinosus* at weekly intervals significantly brought down the larval population of DBM and increased the marketable yield. Natural epizootic of *Zoopthora radicans* was recorded on cabbage causing 68.6% mortality to the insect larvae (Gopalakrishnan, 1998). Application of 5 rounds of *Paecilomyces farinosus* at 1.7×10^8 spores/ml + Triton X-100 (0.01%) at weekly interval significantly brought down the larval population to 1.07/10 plants (as compared to 5.15/10 plants in control) and increased the fruit yield to 43.69 kg/plot (as compared to 30.88 kg/plot in control) (Gopalakrishnan, 1998).

(c) Integrated Methods

- Sow Indian mustard as a trap crop.
- Spray neem seed powder extract 4 % or neem soap 1% or pongamia soap 1% by giving thorough coverage to the crop canopy.
- Install light traps to control adults if required. For one acre plot use 3–4 light traps (60 or 100 W bulbs). They have to be hung above a bucket half filled with water. Otherwise hung the bulb above a gunny bag (slanting below). Gunny bag has to be smeared with grease or oil. Illuminate the bulbs for full night. Adults of DBM will get attracted to light and trapped in the water/oil. Use the light traps for 3–4 days for effective control of DBM adults (Fig 5.3).
- As this pest has developed resistance against many insecticides, spray of insecticides has to be done carefully after choosing an effective insecticide currently available in the market.
- Follow the detailed of IPM package given at the end of the section of pests of cruciferous vegetables.

Even after following IPM, some times sudden influx of DBM adults may come from neighbouring plots. Under such conditions, adults are to be controlled by using light traps and then only spray of an effective insecticide.

Figure 5.3. Light trap to attract DBM larvae

IPM using Indian Mustard as a Trap Crop: IPM using Indian mustard as a trap crop involves planting of paired rows of mustard after every 25 rows of cabbage/cauliflower and spraying of NSKE 4% at primordial formation. Two more sprays of NSKE 4% may be given at 10–15 days interval after the first spray. The IPM gave 152% more returns than pure cabbage crop. IPM controls diamondback moth, leaf webber (*Crocidolomia binotalis*); stem borer (*Hellula undalis*); aphids (*Brevicorne brassicae, Hyadaphis erysimi*) and bug (*Bagrada cruciferarum*) (Srinivasan and Krishna Moorthy, 1991) (Figs. 5.4 and 5.5).

Figure 5.4. IPM in cabbage using Indian mustard as a trap crop

Figure 5.5. IPM in cauliflower using Indian mustard as a trap crop

5.1.1.2. Leaf Webber, *Crocidolomia binotalis*

(i) Damage: This pest is widely distributed in the Indian subcontinent, South Asia and Australia. It is a serious pest of all cruciferous crops. Full grown caterpillars are pale brown in colour with a paler dorsal stripe and two dorso-lateral, one lateral and one sub-lateral series of black spots and specks. The moths lays eggs in clusters of 40–100 on leaf surface which hatch in 5–7 days. Young larvae feed gregariously and later web the leaves together and feed within (Fig 5.6). These leaves contain excreta of the leaf webber rendering either total loss or giving overall dirty appearance to the heads and hence the less market value for the produce. This results in rotting of cabbage heads and cauliflower curds. The larval period lasts for 25–50 days while pupation occurs in soil, which varies from 15–40 days.

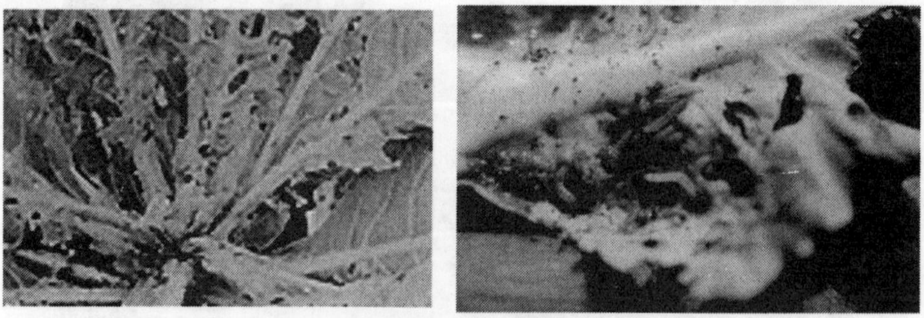

Figure 5.6. Leaf webber damage on cabbage leaves

(ii) Management

- Collection and destruction of egg masses and gregarious larvae.
- Spray any contact insecticide on the foliage when the larvae are observed on the crop.

- Spray neem seed kernel extract (NSKE) 4% or pulverized neem seed powder extract (NSPE) 4%.

5.1.1.3. Stem Borer, *Hellula undalis*

(i) Damage: Though this pest is sporadic in nature, its infestation prevents the head formation (Fig 5.7) and hence is serious. Many times more than 50% of the plants show multiple head formation. Sometimes multiple head formation may also be due to poor quality of the seeds. Caterpillars are pale whitish brown in colour and have 4 to 5 purplish brown longitudinal stripes. The adult moths lays eggs singly on the leaves, the hatched larvae mine the leaves along side veins. The grown up larvae bore into stems preventing head initiation which results in multiple shoots and multiple head formation with low market value. The damage is more severe during March-July months.

(ii) Management

- Spray the seedlings with an effective contact insecticide before planting followed by one more spray at 10 DAP, particularly for summer and kharif crop (between March-July).
- Spray pulverized neem seed powder extract (NSPE) 4%.

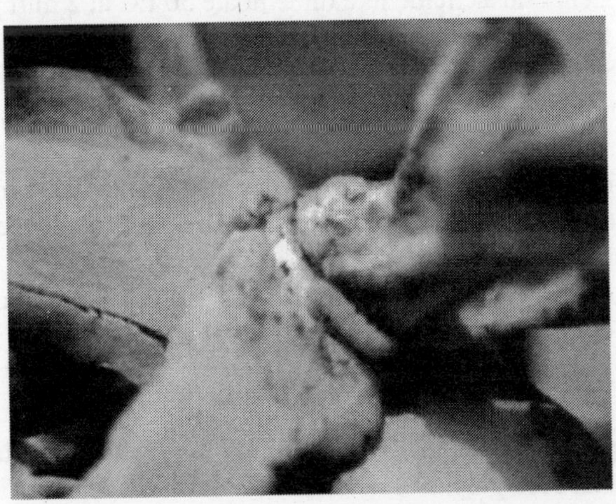

Figure 5.7. Stem borer damage on cabbage

5.1.1.4. Aphids, *Brevicoryne brassicae, Hyadaphis erysimi*

(i) Damage: Aphid is very serious during dry and hot periods of the year. *Brevicoryne brassicae* is more dominant than *Hyadaphis erysimi* and its incidence is high during February-June. *B. brassicae* have white mealy covering while, it is lighter in colour in *H. erysimi*. These are parthenogenetic. Both nymphs and adult aphids suck the sap and devitalise the plants affecting the quality of head or curds formed (Fig 5.8). Aphids also produce copious amount of honey dew which makes the plants sticky favouring the growth of sooty mold as a result of which a black coating is formed on affected plants, hindering the photosynthesis. These aphids become severe when contact insecticides are sprayed regularly.

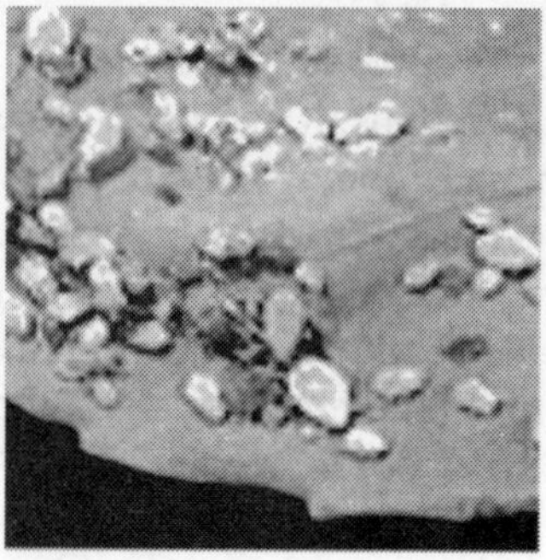

Figure 5.8. Aphid infestation on cabbage leaf

(ii) Management

- Spray any systemic insecticide like dimethoate 30 EC at 2 ml/l when the aphids are observed on the crop.
- Spray pulverized neem seed powder extract (NSPE) 4%.
- Follow mustard trap crop IPM as mustard will attract all aphids after flowering and main crop will be spared of the pest incidence.
- Cauliflower intercropped with non-crucifer host plants like sunflower, tomato and marigold were highly effective in reducing the aphid incidence and enhancing the number of natural enemies resulting in higher yields.
- Natural epizootic of fungus *Erynia neoaphidis* causing 80.6% mortality was reported on cabbage (Gopalakrishnan and Mohan, 1993).

5.1.1.5. Tobacco Caterpillar, *Spodoptera litura*

(i) Damage: It is a polyphagous pest. The moths lay dirty white eggs in groups and young larvae up to third instar feed gregariously and skeletonise the leaves (Fig 5.9). Caterpillars are pale brown in colour with a greenish to violet tinge. Gown up larvae completely devour the leaves and other plant parts in case of severe incidence. When the incidence is high they attain cut worm status, hide during daytime and come out during night and devastate the crop. During severe infestation, the entire crop may be defoliated overnight.

(ii) Management

- Collection and destruction of egg masses and gregarious feeding larvae is the most effective and cheapest method. Regular scouting has to be followed to locate egg masses and location of skelitonised leaves.
- Poison baiting (10 kg rice/wheat flour + 1 kg of jaggery + 500 g of methomyl formulation per acre). Mix the flour with jaggery in the morning. In the evening add

insecticide and mix carefully by wearing gloves and using a stick. Sprinkle the mixture in furrows. The fermenting jaggery attracts the caterpillars and after feeding the bait they get killed. Baiting has to be done 2–3 days until complete control is achieved.

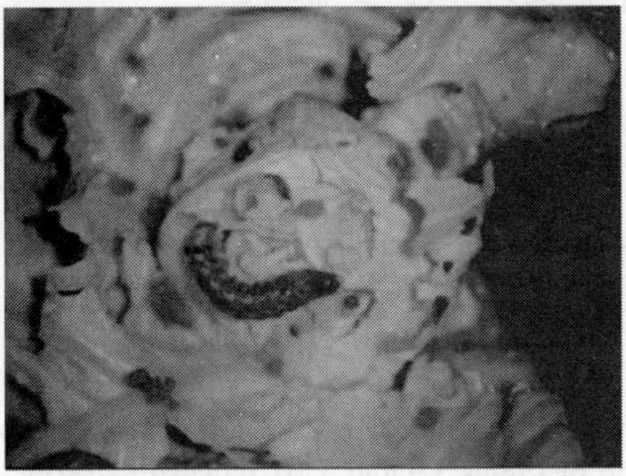

Figure 5.9. Cabbage head damaged by *Spodoptera litura*

- NSKE 4% spray controls early stage larvae. Large larvae have to be controlled by piercing with a sharp thick iron needle in head and killing.
- The exotic egg parasitoid, *Telenomus remus* alone was found to give effective control (Krishnamoorthy and Mani, 1985).

5.1.1.6. Gram Caterpillar, *Helicoverpa armigera*

(i) Damage: This is a minor pest of cruciferous vegetables which feed on the foliage deteriorating the market quality of heads and curds. Solitary larvae with hairs and big holes plugged with excreta are found on heads.

(ii) Management

- Mechanical control will be very effective for gram caterpillar than spraying insecticides as the grown up caterpillars are likely to hide inside the hole and insecticide will not reach it by spraying.
- Young larvae may be reduced by spraying 0.07% endosulfon or 0.1% dichlorvos.

5.1.1.7. Mustard Saw Fly *Athalia lugens proxima*

(i) Damage: This pest is occasionally serious on cabbage and cauliflower. One can see black larvae gregariously feeding on leaves.

(ii) Management

- As and when observed, spray quinalphos or chlorpyrifos both at 0.05% for the management of mustard saw fly.

5.1.1.8. Cabbage Butterfly, *Pieris brassicae*

(i) Damage: This is widely distributed in north America, Europe, China, Burma and India. In India, it is found along Himalayan region and parts of north India. It is not found in south India. Eggs are laid in clusters on undersurface of leaves. A female may lay more than one cluster of eggs. Young larvae are gregarious. They disperse later and skeletonise the leaves or even bore into heads. Pupation is on leaves and stems. Incubation, larval and pupal period are 3, 5 and 7 days, respectively.

(ii) Management

Mechanical removal of egg masses and gregarious larvae is effective. Spraying of *Bacillus thuringiensis* also gives good control of this pest. 'Thuricide' (30×10^6 spores/mg) at 0.4% was found effective for the control of *Pieris brassicae* and *P. xylostella* on cauliflower.

Cabbage butterfly is found heavily parasitized (80-90%) by a gregarious parasitoid, *Cotesia glomerata* and an ichneumonid, *Hypersota abeninus* in Solan region of Himachal Pradesh (Sood *et al.*, 1995).

Pesticide Residues

The persistence of terminal pesticide residues and waiting periods for cabbage and cauliflower are presented in Tables 5.9 and 5.10. (Awasthi, 2001).

Table 5.9. Terminal pesticide residues and their waiting periods for cabbage

Pesticide schedule	Dosage/ Concentration (%)	Terminal residues (ppm/days)	Waiting period (days)
Endosulfan spray at head formation stage	0.07	1.40/7	5
Monocrotophos spray at head formation stage	0.05	0.08/3	2
Phosalone spray at head formation stage	0.05	1.00/10	10
Metalaxyl spray at head formation stage	0.20	0.57/20	21
Cypermethrin spray at head formation stage	0.01	0.17/10	7
Fenvalerate spray at head formation stage	0.01	0.06/15	11
Carbofuran soil application at planting	1.0 kg *a.i.*/ha	0.44/50	60
Phorate soil application at planting	1.0 kg *a.i.*/ha	0.60/50	60

Table 5.10. Terminal pesticide residues and their waiting periods for cauliflower

Pesticide schedule	Dosage/ Concentration (%)	Terminal residues (ppm/days)	Waiting period (days)
Endosulfan spray at head formation stage	0.07	1.40/7	5
Endosulfan spray at curd formation stage	0.07	0.44/10	3
Quinalphos spray at curd formation stage	0.05	0.65/15	20
Phosalone spray at curd formation stage	0.05	0.65/10	6
Cypermethrin spray at head formation stage	0.01	0.11/10	8
Fenvalerate spray at head formation stage	0.01	0.05/15	9
Phorate soil application at planting	1.5 kg *a.i.*/ha	0.42/60	70

5.1.2. Diseases

5.1.2.1. Damping-off, *Pythium aphanidermatum, Rhizoctonia solani*

(i) Symptoms: The fungus attacks in the seedling stage causing damping-off as well as wire stem. The pathogen is responsible for bottom rot, head rot and foot rot in older plants.

Rhizoctonia solani attacks the seedlings at the bottom of the young stem (hypocotyl) or at the soil line. The tissue becomes water-soaked and rapidly collapse causing the seedlings to topple over and die. At later stages, plants are brownish to black just above and below soil line. The stems are usually somewhat smaller than normal but tough and woody.

(ii) Epidemiology: The fungus survives in soil. Secondary spread takes place by irrigation water and during cultural operations.

(iii) Management

Well drained nursery beds, use of well decomposed organic manure, wide spacing between the rows, lesser density of seedlings and light but frequent irrigations helps in reduction of damping-off. Good control of damping-off was obtained through soil incorporation of wheat bran preparation of *T. harzianum* in commercial nursery. Seed treatment with 0.2% carbendazim or benomyl or *Trichoderma* spp. was found to be as effective as soil treatment for managing the disease. Soil application of 0.1% carbendazim or 0.2% rovrol is effective.

Soil drenching with 0.6% Bordeaux mixture commencing 15 DAS and subsequent 2 more drenchings at 10 to 15 days interval gives effective control of the disease.

5.1.2.2. Downy Mildew, *Peronospora parasitica*

The disease occurs generally in nursery and to a lesser extent in the main field.

(i) Symptoms: The pathogen infects stem, leaves and flowers. The disease is characterized by the appearance of purplish brown spots on the undersurface of the leaves. On the upper surface of the leaves, the lesions are tan or yellow. Thin grayish-white downy fungal growth occurs on the under surface of leaves (Fig 5.10). Infected leaves dry up, shrivel and fall off. Flower parts are deformed. The stem show dark brown and sunken lesions or streaks which later develop the downy growth of the fungus. Heads formed are very small in size.

(ii) Epidemiology: The pathogen survives on diseased plant debris. Low temperature, high humidity and high moisture favour the disease development.

(iii) Management

- Seeds should be cleaned properly so that fragments or parts of the pod which may contain oospores are not present.
- Crop rotation with non-cruciferous crops is effective in reducing the disease incidence.
- Spraying with ridomil MZ or aliette both at 0.2% or copper oxychloride at 0.3% gives effective control of the disease.

Figure 5.10. Downy mildew on cabbage leaves

5.1.2.3. White Rot or Stalk Rot, *Sclerotinia sclerotiorum*

(i) Symptoms: Leaves when infected turn light green and later yellow and are shed prematurely. Black soft rot often appears on petioles and midrib touching the soil and are covered with fluffy growth of the fungus. Stem at the ground level shows girdling. Fluffy mycelium of the fungus grows on these infected plant parts with numerous sclerotia embedded. Cool (20–25°C) and humid (95–100% RH) conditions are ideal for disease initiation and spread (Sharma and Sharma, 1985).

(ii) Management

Avoiding excess moisture with less frequent irrigation and reducing the plant populations by following wider spacing will create microclimate less congenial for disease development. Application of FYM or saw dust reduces the disease incidence. Soil amendment with sunflower inflorescence, rape seeds and pine needles reduce the disease incidence (Gupta *et al.*, 1986). Follow cabbage/cauliflower-paddy rotation. Remove diseased leaves and destroy them.

Application of benomyl and MBC both at 0.1% was found effective in reducing the disease incidence (Gangopadhyay, 1984). Soil application of *T. harzianum* (TH 4) and *Aspergillus niger* (Sanjeevini and Kalisena) were highly effective.

5.1.2.4. *Alternaria* Black Spot, *Alternaria brassicae*

(i) Symptoms: On leaves several small, dark brown zonate spots are produced expanding rapidly to form circular lesions up to 1 cm in diameter. The enlargement of lesions may lead to formation of concentric circles which coalesce in the centre (Fig 5.11). In humid weather, the fungus may cause bluish growth in the centre of the spots which hampers the photosynthetic activity of the plant thereby hampering the overall productive potential of the plants. Sometimes cauliflower heads are also infected by the pathogen. The heads show browning, starting at the margin of the individual flowers or flower clusters.

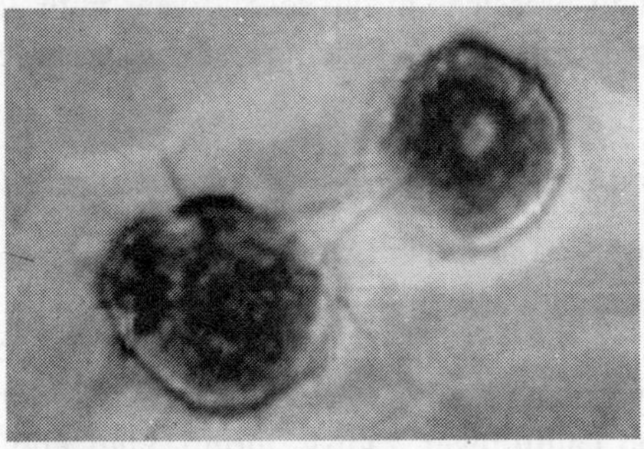

Figure 5.11. *Alternaria* leaf spot on cabbage

(ii) Epidemiology: The fungus is seed-borne and survives in diseased crop debris. The disease is favoured by hot moist weather.

(iii) Management

Seed treatment with hot water maintained at 50°C for 30 min is effective. Giving wider spacing of 60 × 50 cm and removal of basal leaves from time to time help to reduce the disease incidence. Seed treatment with *T. viride* and *P. fluorescens* and soil treatment of nursery beds with neem cake enhanced seed germination and seedling stand. Spraying with captan or copper oxychloride both at 0.3% is effective in controlling the disease.

5.1.2.5. Club Root, *Plasmodiophora brassicae*

(i) Symptoms: Roots enlarge to form spindle-shaped growth (Fig 5.12). The pathogen is responsible for temporary flagging of leaves on bright days leading to their wilting. Secondary invasion of soft rot bacteria follows. This helps in accumulation of toxic materials in plants.

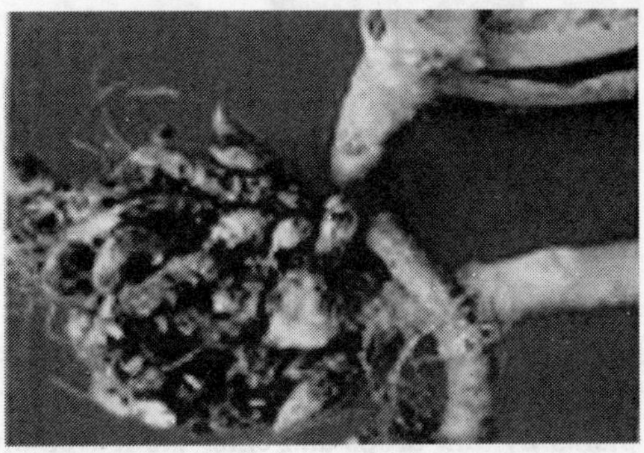

Figure 5.12. Club root affected cabbage roots

(ii) Management

Trap cropping with rayosak among susceptible brassicas in the first year and tori sarson and radish among resistant brassicas in the second year with 30 days optimum trapping period were found most effective in reducing the disease intensity and increasing the yield.

5.1.2.6. Black Rot, *Xanthomonas campestris* pv. *campestris*

The disease was first observed in the USA on cabbage in 1891. In India, the disease was first described in 1928 on cabbage from Pune, Maharashtra. Now the disease is known to be prevalent in most of the crucifer growing areas of the country. Severe losses due to the disease in heavily manured soils near big towns is reported (Patel *et al.*, 1949a). Black rot also occurs on cauliflower, radish, turnip, knol khol and mustard.

(i) Symptoms: The first sign of the disease appears near the leaf margins, characterized by chlorosis, which progresses towards the centre of the leaf blade in 'V' shape (Fig 5.13). In affected portions, veins and veinlets turn brown and finally black. In case of severe infection, the leaves wither. The vascular blackening may extend to the main stem causing systemic spread of the infection. Infected plants become stunted, often one side of cotyledons turn yellow to black, bend down and drop-off prematurely. Many lower leaves are also shed off early. Young plants may be killed in the seed bed itself. In cabbage and cauliflower, if the heads are already formed, soft-rot of the head may set in.

Figure 5.13. Black rot affected plants of cabbage

(ii) Epidemiology: Black rot is seed-borne. The pathogen survives for longer periods in naturally infected seed stored at room temperature or at lower temperature (10°C) (Bandopadhyay and Chattopadhyay, 1986a). The pathogen also survives in infected plant debris in soil (Trivedi and Patel, 1972). The bacterium also survives in resident form on a weed, *Centenella asiaticum* (Kishun and Chand, 1988d). The pathogen spread to the neighbouring healthy plants by rain drops and enter the leaves through natural openings and wounds. Infected seeds and transplants are mainly responsible for wide spread of the disease. Cabbage planted in September-October and January, and cauliflower in October showed higher disease incidence than cabbage planted in November and February, and cauliflower in February (Bandopadhyay and Chattopadhyay, 1986b).

(iii) Management

- Hot water treatment of seeds at 50ºC for 30 min (Shekhawat *et al.*, 1982).
- Seed treatment with 0.1% mercuric chloride for 30 min, 100 ppm plantamycin (Kishun, 1981a, 1984a), 100 ppm agrimycin-100 (Rao and Kishun, 1986), 100 ppm streptocycline (Shah *et al.*, 1985) or 0.01% aureomycin (Trivedi and Patel, 1972).
- Drenching of seed beds with 5% commercial formalin or1% Bordeaux mixture.
- Spraying cauliflower with 100-200 ppm agrimycin-100 (Bera, 1986), 500 ppm carbendazim (Saini and Parashar, 1981); and cabbage with erythromycin (Mondel and Mukherjee, 1978) are found effective in reducing the disease.
- Three years crop rotation with non-host crops. Crop rotation with rice was found to be best in reducing the black rot disease.
- Strict field sanitation will help in checking the disease.
- Rouging of diseased plants helps in reducing the disease incidence.
- *B. subtilis* was found effective as seed treatment, seedling root-dip treatment and soil drenching.
- Cauliflower *cvs.* Pant Shubra (Singh *et al.*, 1987) and Pusa Ice (Sharma *et al.*, 1977) are reported to be resistant.

Integrated Pest and Disease Management

Nursery

- Deep off-season ploughing.
- Soil solarization for 2–3 weeks during summer using 400 gauge thick transparent polythene sheet.
- Green manuring with sunnhemp in July–August.
- Hot water treatment of seeds at 50°C for 30 min or seed soaking in 100 ppm streptocycline solution for 30 min to control black rot disease.
- Raising nursery on raised beds (10 cm above ground level) for good drainage, to avoid damping-off.
- Seed treatment with *T. harzianum/T. viride* at 4 g/kg to prevent infection of soil/ seed borne fungal diseases.
- Apply 1 kg FYM enriched with 50 g of *T. harzianum/T. viride*/m^2 nursery bed.
- Spraying *Bt* formulation (1 ml or 1 g/litre) at 10 days after sowing to control DBM.
- Raise seedlings in portrays in nursery under net house conditions wherever such facility is available.
- Spray indoxacarb 14.5 SC (0.5 ml/l) or novaluron (0.75 g/l) or quinalphos (1.5 ml/l) a day before transplanting seedlings to control stem borer and early infestation of DBM.

Main field

- Growing 2 rows of Indian mustard as a trap crop after every 25 rows of cabbage to trap DBM, leaf webber and aphids. One row of mustard is sown 15 days before cabbage planting and second after 25 days after cabbage planting. However, ensure that first and last row of the plot are mustard.
- Giving wider spacing of 60 × 50 cm to reduce the incidence of *Alternaria* black spot, black rot and downy mildew.
- Installation of light traps at 8/ha to attract DBM adults.
- Removal of basal leaves to reduce the incidence of *Alternaria* black spot, black rot and downy mildew.
- Spray 0.25% mancozeb or 0.2% chlorothalanil to control foliar diseases.
- Spray trap crop (mustard) with quinalphos 40 EC at 1 ml/l or neem soap 10 gm/l at 5 days after germination and repeat 2 or 3 times to protect mustard foliage.
- Spray cabbage crop with *Bt* (1%) or neem or pongamia soap (0.75%) at 10 days after planting.
- Thorough spray with 4% pulverized neem seed powder extract (NSPE) or 1% pongamia soap or neem soap to cabbage at 20 DAP (at primordial stage) and repeat 2–3 times at 10–15 days interval according to pest incidence. These will control diamondback moth, leaf webber, aphids *etc*. NSKE is safe to *Cotesia plutellae*, the dominant parasitoid of DBM.
- If aphids are found, spray systemic insecticide since soap does not control aphids due to improper coverage.
- Avoid spray of NSPE after curd formation in cauliflower.
- If NSPE or soaps are not available, spray *Bt*.

5.1.3. Nematodes

Stunt and root-knot nematodes are the major limiting problems in successful cultivation of cabbage and cauliflower crops.

5.1.3.1. Stunt Nematode, *Tylenchorhynchus brassicae*

Siddiqi *et al*. (1972) encountered *T. brassicae* as the most prevalent species around the roots of cabbage and cauliflower in Uttar Pradesh.

(i) Distribution and Hosts: The populations of *T. brassicae* were exceptionally high in Aligarh, Kanpur, Ghazipur and Rampur districts of U.P. In general, all crucifers are good hosts of *T. brassicae*, but cabbage, cauliflower and tomato proved to be most efficient hosts. The multiplication rate on cabbage (*cv*. Suttons Earliest), cauliflower (*cv*. Patna) and Knol-khol (*cv*. Purple Vienna) was 13, 18 and 29, respectively. Common weeds such as *Chenopodium album, Trianthema monogyna* and *Portulaca oleracea* were found to be hosts of *T. brassicae*.

(ii) Symptoms: *T. brassicae* caused reduction in fresh weight of plants and suppressed root growth, resulting in reduced water absorption.

(iii) Life Cycle: The optimum soil temperature (30°C) and soil moisture (25–30%) for reproduction of *T. brassicae* was same for both cabbage and cauliflower. Further, high temperature (35–40°C) and low moisture are highly unfavourable for survival of *T. brassicae* in the absence of host. However, at moderate temperature and moisture conditions, the nematode is capable of surviving even up to 240 days.

(iv) Interaction with Other Pathogens: *Tylenchorhynchus brassicae* alone did not affect the percentage emergence of cauliflower seedlings. *R. solani* was highly destructive by itself, but the adverse effect on the emergence of seedlings was greater when nematode infestation occurred along with fungal infection (Khan *et al.*, 1971).

(v) Histochemistry: An increase in total soluble phenols, ortho-dihydroxy phenols, free amino acids and proteins, but not in total carbohydrates, was found in the roots of cabbage and cauliflower infected with 10,000 *T. brassicae* (Alam *et al.*, 1976). Total phenols, ortho-dihydroxy phenols and free amino acids increased with increasing seedling age at inoculation (Khan *et al.*, 1980).

(vi) Management

(a) Cultural Methods: There was sudden decline in population of *T. brassicae* when cabbage and cauliflower were rotated with wheat (Siddiqi *et al.*, 1973). Intercropping of cabbage and cauliflower with neem reduced the multiplication rate of *T. brassicae* and improved the plant health (Siddiqi and Saxena, 1987a, b). The mustard (rabi), radish (summer), sesame (kharif) sequence considerably reduced the population of *Tylenchorhynchus* spp. (Haque and Gaur, 1985).

Botanicals: Seed dressing with latex of *Calotropis gigantea, C. procera, Euphorbia milii, E. nerifolia* and *E. tirucalli* significantly inhibited *T. brassicae* multiplication on cabbage and cauliflower and improved plant growth (Siddiqi and Alam, 1989a).

Bare-root dip of cabbage and cauliflower seedlings in leaf extracts of neem and Persian lilac (Siddiqi and Alam, 1989b), interculture of seedlings of marigold, neem and Persian lilac reduced multiplication rate of *T. brassicae* and improved plant growth.

Amendment of soil with oil cakes of neem, groundnut, mustard and castor (equivalent to 100 kg N/ha) was effective in reducing the population of *T. brassicae* around the roots of cabbage and cauliflower (Siddiqi *et al.*, 1976). Neem cake was the best.

(b) Chemical Methods: Aldicarb, carbofuran, ethoprophos and phenamiphos each at 1 to 2 kg *a.i.* per ha were found effective in reducing the root-knot nematode population and in increasing flower yields of cauliflower (Ahuja, 1983).

Aldicarb at 0.5 kg *a.i.* per ha and carbofuran at 1.5 kg *a.i.* per ha were effective in reducing *T. brassicae* and *T. dubius* population and in increasing yields of cabbage (Varma *et al.*, 1978).

(c) Host Resistance: *Brassica rapa cv.* Sarson (yellow mustard) was found to be least susceptible to *T. brassicae*, and could be grown on infected soil in preference to *B. nigra* (black mustard) (Khan *et al.*, 1986).

5.1.3.2. Root-knot Nematode, *Meloidogyne incognita*

(i) Interaction with Other Pathogens: Maximum wilt score due to *F. oxysporum* f. sp. *conglutinans* (4.0) and wilt symptom index (80.0) was observed in cauliflower plants which received *M. incognita* 14 days before the inoculation of wilt fungus (Pathak and Keshari, 2004).

5.1.3.3. Root-knot Nematode, *Meloidogyne incognita* and Club Root, *Plasmodiophora brassicae* Disease Complex

(i) Management

(a) Integrated Methods: PGPR strains (*Pseudomonas fluorescens, Bacillus subtilis*) combined with fungal biocontrol agents (*Trichoderma viride, T. harzianum*) were found to be effective in reducing the nematode-fungal disease complex in cabbage (Loganathan *et al.*, 2001). The bioformulation mixture of *P. fluorescens, T. viride* and chitin effectively reduced the disease complex in cabbage and cauliflower both under greenhouse and field conditions (Samiyappan, 2003) (Table 5.11).

Table 5.11. Efficacy of bioformulation mixtures against club root disease–root-knot nematode complex in cabbage under greenhouse conditions

Treatment	Club root index	Nematode incidence	
		Population	Root-knot index
Trichoderma viride	25.99 (30.65)	129	2.66
Pseudomonas fluorescens	28.20 (32.07)	112	2.33
T. viride + *P. fluorescens*	25.33 (30.22)	114	2.33
T. viride + chitin	25.44 (30.29)	108	2.33
P. fluorescens + chitin	25.66 (30.43)	111	2.00
T. viride + *P. fluorescens* + chitin	22.22 (28.12)	108	2.00
Chitin alone	31.70 (34.26)	139	3.00
Carbendazim	19.90 (26.49)	264	4.66
Carbofuran	39.90 (39.17)	106	1.66
Cabendazim + carbofuran	15.00 (22.79)	103	1.66
Plasmodiophora brassicae alone	48.90 (44.37)	0.033	0.133
Meloidogyne incognita alone	0.03 (0.60)	280	5.00

Table 5.12. Management of insect pests in cruciferous vegetable crops

Crop	Pest	Symptoms	Pesticide/ Bioagent	Dose/litre	Remarks
Cabbage/ Cauliflower	Diamond back moth	Young leaves with holes, mostly on growing part (primordia) and greenish larvae	NSKE 4%	40 g	20, 30–35 & 45–50 DAP
			Bacillus thuringiensis formulations		15, 30, 40 and 50 DAP
			Dipel	1 ml	
			Delphin	1 g	
	Gram caterpillar	Solitary larvae with hairs and big holes plugged with excreta on formed heads	—	—	Hand picking and killing of larvae as and when observed
	Leaf webber/ stem borer	Group of hairy larvae feeding on leaves and damaged heads may rot	Monocrotophos	1.25 ml	As and when observed
			Methyl Parathion	2 ml	
			Cypermethrin	0.5 ml	

Table 5.13. Cruciferous vegetable crop varieties resistant to biotic stresses

Vegetable crop	Pest/Disease	Resistant varieties
Cabbage	Black rot	Pusa Mukta.
	Black leg	Pusa Drum Head.
Cauliflower	Black rot	Pusa Snowball K-1.
	Black rot & curd blight	Pusa Shubhra.
	Downy mildew	Pusa Hybrid-2.
	Curd blight	Pusa Synthetic.

6 LEGUMINOUS VEGETABLES

6.1. PEA

Pea (*Pisum sativam*) is an important vegetable crop widely grown throughout the world. As a cool season crop, it is extensively grown in temperate zone; but restricted to cooler altitudes in the tropics and winter season in the sub-tropics. A rich source of proteins (25%), amino acids and sugars (12%), green peas are an all time favourite vegetables. It is rich in vitamins A, B, C and minerals. It is being cultivated in 0.314 million hectares producing 2.299 million tonnes of peas with an average yield of 7.3 tonnes per hectare (Bijay Kumar, 2007) (Tables 6.1 and 6.2). Garden pea is grown in Uttar Pradesh, Madhya Pradesh, Himachal Pradesh, West Bengal, Jammu and Kashmir, Punjab, Haryana, Uttarakhand, Chattishgarh, Bihar and Orissa.

Table 6.1. Major Peas producing countries of the world

Country	Area (in '000 ha)	Production (in '000 MT)	Productivity (in MT/ha)	% Share of world production
India	314	2299	7.3	29.5
China	241	2210	9.2	28.3
USA	86	859	10.0	11.0
France	31	360	11.7	4.6
Egypt	28	290	10.4	3.7
UK	32	133	4.2	1.7
Turkey	17	122	7.0	1.6
Algeria	24	110	4.5	1.4
Hungary	14	85	6.2	1.1
Netherlands	5	84	16.5	1.1
Others	303	1248	4.1	16.0
Total	**1094**	**7799**		

Source: FAO (except Indian data) (Indian data source: Indian Horticulture Database 2006)

6.1.1. Pests

6.1.1.1. Bean Fly, *Ophiomyia centrocematis*

(i) Damage: This species is specific to peas and lays eggs in the base of the stem. This causes drying of leaves, starting from the base. Hence, the first symptom is drying of the first primitive leaf at the base. This occurs after about 20 days of sowing. A database on bean flies has been developed by the Indian Institute of Horticultural Research, Bangalore (Anon, 2000).

Table 6.2. Area, production and productivity of leading peas growing states in India (2005–06)

State	Area (in '000 ha)	Production (in '000 MT)	Productivity (MT/ha)
Uttar Pradesh	139.0	1146.8	8.3
Madhya Pradesh	19.4	212.9	11.0
Himachal Pradesh	16.3	177.0	10.9
West Bengal	21.7	129.4	6.0
Jammu & Kashmir	7.3	112.2	15.4
Punjab	17.2	103.3	6.0
Haryana	11.1	84.9	7.6
Uttarakhand	'9.6	60.9	6.3
Chattishgarh	7.2	50.8	7.1
Bihar	8.1	. 50.5	6.2
Orissa	4.8	42.0	8.7
Others	52.4	128.0	2.4
Total	**314.0**	**2298.7**	**7.3**

(*Source*: Indian Horticulture Database 2006)

(ii) Management

- Apply neem cake at 250 kg per hectare to the ridges immediately after germination. This should be followed by sprays of pulverized neem seed powder extract 4% or neem soap 1% or neem formulation with more than 10000 ppm azadirachtin 2–3 ml/l at 20 and 30 DAS.
- Spray acephate at 0.75 ml per l or dimethoate 30 EC at 2 ml per litre or 0.05% phosphamidon at 20 and 30 DAS.
- Sow the crop in second fortnight of October to escape the damage from stemfly in Punjab.
- Application of 0.05% phosphamidon at 20 DAS gives effective control of the pest. Second spraying may be repeated after 10 days.

6.1.1.2. Lycaenid Pod Borer, *Lampides boeticus*

(i) Damage: This is a butterfly pest and commonly known as pea blue. The eggs are laid singly in flowers and flower buds. In peas, the larvae bore into young pods. The entry hole gets blocked after some time. When the larvae attain fourth instar, they migrate and cause extensive damage to flower buds and pods (Krishna Moorthy and Srinivasan, 1989). Hence, control measures have to be taken in the early stages of the pest incidence.

(ii) Management

- In peas, the first spray has to be given at flowering as soon as eggs are noticed with cypermethrin 25 EC at 0.5 ml/l or chloripyriphos 20 EC at 2.5 ml/l or indoxacarb 14.5 SC at 0.5 ml/l.
- Weekly sprays of 5% NSKE after flowering reduces the pest incidence to a great extent.

6.1.1.3. Gram Pod Borer, *Helicoverpa armigera*

(i) Damage: This polyphagous pest also attacks peas occasionally along with *Lampides boeticus*. The habit of egg laying and larval feeding are the same as pea blue butterfly.

(ii) Management

- In peas, the first spray has to be given at flowering as soon as eggs are noticed with indoxacarb 14.5 SC at 0.5 ml/l or novaluron at 0.75 g/l.

6.1.1.4. Aphids, *Aphis craccivora, Acrythosiphon pisum*

(i) Damage: Both nymphs and adults suck the sap from young plant parts like shoot tips and inflorescence. *A. craccivora* is greenish black and A. *pisum* is greenish in colour. *A. pisum* generally occurs on peas.

(ii) Management

- Clip infected shoots and spray neem soap or pongamia soaps at 1%.
- Spray acephate 75 WP at 0.75 ml/l or dimethoate 30 EC 2 ml/l or 0.05% phosphamidon or 0.05% monocrotophos.

Pesticide Residues

The persistence of terminal pesticide residues and waiting periods for peas and beans are presented in Table 6.3. (Awasthi, 2001).

Table 6.3. Terminal pesticide residues and their waiting periods for peas and beans

Pesticide schedule	Dosage/ Concentration (%)	Terminal residues (ppm/days)	Waiting period (days)
Endosulfan spray at pod formation stage	0.07	0.14/10	1
Monocrotophos spray at pod formation stage	0.05	0.13/7	6
Quinalphos spray at pod formation stage	0.05	0.06/15	8
Cypermethrin spray at pod formation stage	0.01	0.03/15	3
Fenvalerate spray at pod formation stage	0.01	0.07/15	3
Deltamethrin spray at pod formation stage	0.002	0.07/10	8
Carbofuran soil application at sowing	0.5 kg *a.i.*/ha	0.15/55	70
Phorate soil application at sowing	1.0 kg *a.i.*/ha	0.26/60	70

6.1.2. Diseases

6.1.2.1. White Rot, *Sclerotium rolfsii*

(i) Symptoms: It is a ubiquitous fungus with a broad host range. Initially wet rotting of bark is observed. Entire bark of the plant near collar region rots. The characteristic symptoms include white, cottony fungus growth is observed on affected portion as well as on parts in contact with soil. Gradually this hyphal mat is converted into small, mustard-like sclerotia that survive in the soil.

(ii) Management

Summer ploughing followed by irrigation in summer helps in reducing the initial inoculum of the fungus. Field sanitation by burning all plant debris after harvest is effective. Crop rotation with low land paddy, wheat/barley is effective in reducing the initial inoculum. Follow proper spacing between rows and plants. Avoid excessive irrigation from pre-flowering stage onwards. Stake medium to tall cultivars. Avoid injury to stem during picking. The late sown crop in November escape the disease to a great extent.

Soil drenching with 0.3% copper oxychloride followed by 0.1% carbendazim near the collar region in evening is recommended. Soil application of *Trichoderma* spp. supplemented with organic matter and green manuring are beneficial.

6.1.2.2. White Mold, *Sclerotinia sclerotiorum*

(i) Symptoms: The disease is commonly found on pea which is responsible for complete killing of plants. Symptoms observed include wet and soft rotting of the tissues. White fungus growth is observed on the rotted portion. Embedded sclerotia covered with white mycelium are formed on the infected portion as well as on inner portion of the pith and fruits.

(ii) Management

All infected parts must be carefully collected and burnt before drying of the plants. Deep ploughing and crop rotation with low land paddy with continuous stagnation of water reduces the sclerotial population. Close planting and dense canopy of plants should be avoided. Alternate sprays of 0.1% carbendazim and 0.25% mancozeb is essential at 7 days interval to check the disease. Reduction in white mold of pea was reported under field conditions through the application of *Trichoderma harzianum*.

6.1.2.3. Powdery Mildew, *Erysiphe polygonum*

(i) Symptoms: Mealy white patches appear on both sides of leaves, stems, branches, pods and tendrils (Fig 6.1). These patches originate as minute discoloured specks from which a powdery mass radiates on all sides and covers a large area of aerial parts. It causes considerable damage and may result in 20–30% losses in pod number and quality. If it attacks early, plants fail to bear fruit or the pods get chaffy.

Figure 6.1. Powdery mildew on pea leaves and pod

(ii) Management

Avoid delay in sowing which should be completed by first fortnight of October. Infected plant debris must be burnt after harvesting of pods. Spray 0.01% penconazole or dinocap or karathane or carbendazim or baycor and repeat the spray after 10–15 days, if necessary. Use tolerant *cvs.* like Ageta, Organ, Sugar and NDVP-4 which are also vegetable types (Pandey *et al.*, 1999).

In organically grown pea, biopriming of pea seeds with *Trichoderma harzianum* + *Pseudomonas fluorescens* was most effective in improving the seedling stand. Foliar application of bioagents significantly reduced the incidence of powdery mildew disease. Application of *P. fluorescens* + *T. harzianum* through seed and foliar spray was most effective in increasing the yield of organically grown pea (Table 6.4).

Table 6.4. Effect of different biocontrol agents applied through seed, FYM and/or foliar application on powdery mildew disease and yield of pea

Treatment	Seedling stand/m^2	Powdery mildew incidence (%)	Yield (kg/ha)
Seed biopriming with *Trichoderma harzianum* + sprays with *Pseudomonas fluorescens*	56	13	355
FYM colonized with *Trichoderma harzianum* + sprays with *Pseudomonas fluorescens*	63	16	352
Seed biopriming with *Trichoderma harzianum* + FYM colonized with *T. harzianum* + sprays with *T. harzianum*	69	12	388
Control	52	36	301
CD at 5%	**6**	**10**	**53**

6.1.2.4. Wilt, *Fusarium oxysporum* f. sp. *pisi*

(i) Symptoms: Infection leads to drooping and wilting of plants at early stage of plant growth (Fig 6.2). Vascular discolouration of stem and reddish appearance in the pith extends towards roots. The roots turn black and rot. Plant growth is checked, foliage turns yellow and downward curling of stipules and leaflets takes place. The entire plant wilts and the stem shrivels. Epinasty of affected plants is a characteristic symptom.

Figure 6.2. Fusarium wilt of pea

(ii) Management

Summer ploughing should be carried out to reduce the soil inoculum. Field sanitation and crop rotation are most important for controlling this soil-borne pathogen. All infected plants should be uprooted and burnt. Proper drainage should be maintained in the field.

Treat seeds with thiram at 0.3% or carbendazim at 0.2% before sowing and avoid early sowing. In badly infected areas, treat the soil with 0.1% carbendazim. Soil amendment with green manure and application of *Trichoderma harzianum* is most beneficial.

6.1.2.5. Rust, *Uromyces pisi, U. fabae*

(i) Symptoms: It occurs in severe form in cooler regions of India. The disease is characterized by yellow spots bearing asci in clusters with curling symptoms. Pustules present a light appearance on both surfaces of leaves and other parts. Teleuto pustules are dark brown or almost black in colour. In cool and wet climate of Northern India, the entire plant may be killed.

(ii) Management

Burning of infected crop debris will reduce the inoculum. Early bloom stage of crop 4 weeks before harvest is most critical period for rust and must be protected from the infection. Fungicide spray should be immediately started when average of 2 spots per leaf are observed in the field. The best fungicide is 0.025% flusilazole sprayed at weekly interval.

Spraying of hexaconazole or bitertanol or tridimefon or tebuconazole or karathane or carbendazim or baycor all at 0.05% at 10–15 days interval is also effective. The disease can also be controlled by sprays of Calixin, Bayleton and mancozeb. Pea *cvs*. NDVP-3 (vegetable type), Batri Brown and Pea-9 are reported to be resistant to rust disease (Katiyar and Ram, 1987).

6.1.2.6. Pea Mosaic

The disease is caused by marble pea mosaic strain of pea virus 2A.

(i) Symptoms: The diseased plants are pale, weak and dwarfed. The young leaves show a general mottling whereas the older ones display distinct irregular whitish areas. The tendrils are curled abnormally and the apical leaves remain folded. Invariably small specks of light brown discoloured areas are seen on petioles, stems and tendrils. The diseased plants flower later than the healthy ones and the flowers are smaller in size and fewer in number. Comparatively fewer pods are set on diseased plants and they are smaller in size and contain a few seeds (Srinivasan and Nariani, 1966).

(ii) Transmission: The virus is mechanically sap transmissible. Six species of aphids *viz.*, *Aphis craccivora, A. gossypii, A. evonymi, Myzus persicae, Rhophalosiphon pseudobrassicae* and *R. maidis* transmit the virus.

(iii) Management

Pea *cvs*. Canners Perfection, Horal, Hundred Fold and Little Marvel are immune to this virus. The pea varieties Dwarf and Kelvedon Wonder are found to be highly resistant.

6.1.3. Nematodes

Root-knot, reniform and lance nematodes are the important limiting factors in successful cultivation of pea crop.

6.1.3.1. Root-knot Nematodes, *Meloidogyne* spp.

M. incognita was responsible for 20.00 to 50.61% loss in pod yield of peas (Parvatha Reddy, 1985b; Upadhayay and Dwivedi, 1987; Sharma, 1985).

(i) Symptoms: The infested fields show patches in which peas are stunted, chlorotic and have few flowers which produce small and often empty pods. The root systems are reduced in size, exhibit large galls and poor nodulation. Senescence also tends to occur earlier.

(ii) Management

(a) Cultural Methods: Intercropping with marigold or mustard with pea reduced the damage of root-knot nematodes.

Peas escape nematode attack, in the subtropics when sowing is postponed to mid-autumn, when temperatures drop.

(b) Chemical Methods

Seed Treatment: Bhagawati and Phukan (1990) found that carbofuran at 3% w/w as seed treatment was effective in reducing galls and egg masses in roots of pea and increased yields. Seed treatment with phenamiphos, aldicarb and carbofuran all at 1 per cent concentration were effective in controlling the root-knot nematode (*M. incognita*) infecting peas and in increasing their pod yields (Parvatha Reddy, 1984).

Main Field Treatment: Aldicarb, carbofuran, ethoprophos and phenamiphos each at 1 to 2 kg *a.i.* per ha were found effective in reducing the root-knot nematode population and in increasing pod yields of peas (58 to 85 per cent) (Parvatha Reddy, 1985a; Rao and Singh, 1978).

Drill application of carbofuran at 1.5 kg *a.i.*/ha below the seed level reduces the incidence of *M. incognita* in pea.

(c) Host Resistance: Pea *cv.* Wando is reported to be highly resistant to *M. incognita* and moderately resistant to *M. javanica.*

6.1.3.2. Reniform Nematode, *Rotylenchulus reniformis*

Dalal and Vats (1998) reported that *R. reniformis* was responsible for 15.80% loss in pod yield of peas.

(i) Interaction with Other Pathogens: Vats and Dalal (1997) reported maximum and early wilting in pea due to prior inoculation of *R. reniformis* followed by *F. oxysporum* f. sp. *pisi*.

(ii) Management

(a) Cultural Methods: Intercropping with marigold or mustard with pea reduced the damage of reniform nematodes.

6.1.3.3. Lance Nematode, *Hoplolaimus uniformis*

(i) Interaction with Other Pathogens: In peas, early yellowing and root rot disease is dependant upon the presence of both *Hoplolaimus uniformis* and *F. oxysporum* f. sp. *pisi* race 3.

6.2. FRENCH BEAN

French bean (*Phaseolus vulgaris*), also known as common, haricot, kidney, string, salad, runner or snap bean, is an important leguminous vegetable. It is consumed as tender pods, shelled green beans and dry beans. It is a nutritious vegetable which contains proteins (1.7 g), calcium (50 mg), phosphorus (28 mg), iron (1.7 mg), carotene (132 mg), Thiamine (0.08 mg), Riboflavin (0.06 mg) and vitamin C (24 mg/100 g of edible pods). It is largely grown in hilly areas of Himachal Pradesh, Jammu and Kashmir and north-eastern states during summer, winter and autumn crop in parts of Uttar Pradesh, Maharashtra, Karnataka and Andhra Pradesh.

6.2.1. Pests

6.2.1.1. Bean Fly, *Ophiomyia phaseoli*

(i) Damage: This species is serious on French beans. Some times this species is also noticed on dolichos beans. This pest is particularly serious during dry periods of Kharif (June-August) and after January-February months when temperature increases above 30°C. At times more than 70% mortality is caused by this pest attack. The adults are small flies. These lay eggs in the unifoliate leaves that come immediately after germination (Fig 6.3). The oviposition is done by puncturing the leaf and eggs are laid under the leaf epidermis and these later turn into white spots, often confused with disease. These dried ovipositional cavities are often observed within 2–3 days of germination of seeds. The larvae mine the leaf lamina, veins, midrib, petiole and enter the stem. Larval feeding in stem results in mortality or reduction of the plant growth (Fig 6.3). Generally the incidence is high during prolonged dry spells between rains, particularly in June-July months. When the incidence is early and high, plants invariably die. This mortality is noticed by farmers only after about 25–30 days, when the plants are about to flower. Any control measure taken at that time does not control the mortality, as it is too late for the plant to recover.

Figure 6.3. Stem fly damage on stem of French bean

(ii) Management

- Apply neem cake 250 kg/ha immediately after germination. Do not delay, particularly during kharif period.
- Monitor the plants for adult activities, puncture marks and petiole mining soon after germination.
- As soon as a few adults are noticed hovering over the crop, spray endosulfan 35 EC at 2ml/l or acephate 75 WP at 0.75 g/l or PNSPE 4% or neem soap 1% or neem

formulation with 10000 ppm azadirachtin at 2 ml/l. However, please note that the botanicals get washed away by rain and they will be ineffective if it rains within 1–2 days of spray.

- Second spraying is given when on an average 5 leaves have petiole mining symptoms per 10 leaves (about 12–20 days of sowing).

As soon as few puncture marks are noticed on unifoliate leaves caused by oviposition by the adults inside leaf lamina, spraying of 0.07% endosulfan or 4% NSKE/4% pulverized NSPE/1% neem soap/1% pongamia soap is to be given. Second spray can be given after about 15 to 20 DAS. Soil application of neem cake at 200 kg/ha is also effective (Krishna Moorthy and Srinivasan, 1989, 1990, 1997).

6.2.1.2. Serpentine Leaf Miner, *Liriomyza trifolii*

(i) Damage: The accidentally introduced pest serpentine leaf miner has become a pest on French beans.

(ii) Management

NSKE 4% applied at 10 and 15 DAS French bean was able to reduce serpentine leaf miner incidence up to 21 DAS (15.20 mines as compared to 62.93 mines in control) and also increased pod yield (5.05 t/ha as compared to 1.05 t/ha in control) (Krishna Kumar, 1998).

6.2.1.3. Spider Mite

(i) Damage: Webbed colonies of spider mites feed on underside of leaves causing whitish grey patches. The leaves dry up when the infestation is high.

(ii) Management

The predatory mite, *Phytoseilus persimilis* gave excellent control of spider mites when released at 10 adults/plant under open field conditions (Krishnamoorthy and Mani, 1989b).

6.2.1.4. Aphid, *Aphis craccivora*

(i) Damage: Small soft bodied insects suck the sap from leaves and tender shoots and flower buds which curl, deform and dry up (Fig 6.4).

Figure 6.4. Aphid infestation on petiole and leaves of French bean

(ii) Management

Single spray of water suspension of *Fusarium pallidoroseum* (7×10^6 spores/ml) gave effective control of aphids under field conditions. Reinfestation of plant by aphids was not observed while the plants sprayed with quinalphos at 0.05% showed reinfestation necessitating repeated application of the insecticide (3 sprays).

6.2.1.5. Gujarat Hairy Caterpillar, *Amsacta moorei*

(i) Management

Moths can be effectively managed by installing UV light traps (Pest-O-Flash) at 1 light trap/ha. Spraying of 4% NSKE on bunds and field boundaries gives effective and economical control.

6.2.2. Diseases

6.2.2.1. Anthracnose, *Colletotrichum lindemuthianum*

(i) Symptoms: Most characteristic symptoms appear on immature pods, leaves (Fig 6.5) and cotyledons. Stems are also infected. Black, sunken cankers with light coloured or grey central area appear on pods. The central portion of the spot shows pinkish masses of the fungus in wet weather. It causes more damage in cool and temperate regions.

Figure 6.5. Anthracnose on leaf of French bean leaf and pod

(ii) Management

Use disease-free seed and treat seed with carbendazim at 2 g/kg seed before sowing. In disease prone areas, spray right from the beginning 0.2% chlorothalanil or zineb or 0.1% mancozeb.

6.2.2.2. Collar Rot or Web Blight, *Rhizoctonia solani*

(i) Symptoms: At emergence, the infection on hypocotyl and stem results in damping-off symptoms, but afterwards, elongated, sunken, reddish brown lesions are produced on the stem at ground level. On the foliage, circular to irregular brown spots having distinct borders appear. Leaf scald symptoms are common. In rainy season, the disease assumes serious proportions. Under continuous wet weather, along with high temperature, extensive damage results.

(ii) Management

Treat seed with 0.3% thiram 75 WP or 0.2% carbendazim. On disease appearance, spray 0.1% carbendazim. Soil application of *Trichoderma* spp. along with FYM before sowing the seed helps to control soil-borne problems.

Integration of *Glomus mosseae* with *T. viride* gave total protection.

6.2.2.3. Rust, *Uromyces phaseoli*

(i) Symptoms: Rust disease occurs throughout India where beans are grown. Disease symptoms are mostly confined to leaves but young stem and branches are also attacked. The reddish brown rust pustules appear abundantly on the lower surface of leaves (Fig 6.6). When the leaves become thoroughly infected, these shrivel and fall off from the plant. In severe cases, most of the vines are defoliated. Uredospores of the rust pathogen are the main source of infection.

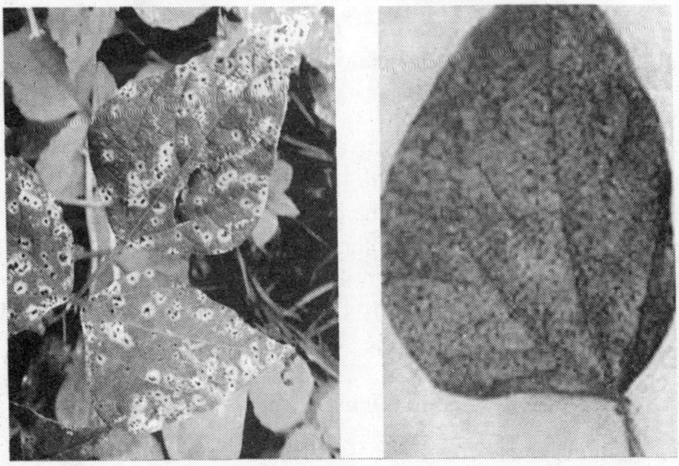

Figure 6.6. Rust on leaf of French bean

(ii) Management

Cultural operations like good field sanitation, destruction of infected plant material and crop rotation are important in minimizing the disease incidence. Fungicidal sprays with 0.2% mancozeb and 0.05% bayleton are effective for disease control.

T. viride and *T. koningii* as 2% foliar sprays were effective both in French bean and cowpea.

6.2.2.4. Angular Leaf Spot, *Isariopsis grisicola*

(i) Symptoms: Reddish-brown lesions appear mostly on lower surface of leaves. Later, symptoms also appear on pods which results in defoliation.

(ii) Management

Follow at least 2-year crop rotation with non-host crops in infected areas. On appearance of disease, spray chlorothalanil/dithionon/propineb/quantaf/carbendazim at 0.1%.

6.2.2.5. French Bean Common Mosaic Virus

This is one of the most important virus diseases among the legumes in India.

(i) Symptoms: The important symptoms of the disease are severe mosaic mottling of the leaves and various types of malformations. Trifoliate leaves affected with mosaic usually have irregular shaped, light yellow and green areas of various sizes, a type of mottling characteristic of mosaic in general. Infected leaves may be narrower than the normal ones, with downward cupping caused by unequal growth of the tissue. Infected plant produce very few, small and abnormal fruits.

(ii) Transmission: The virus is transmitted by mechanical sap inoculation, pollen and infected seeds. Several species of aphid vectors *viz., Aphis gossypii, A. medicaginis, A. rumicis, A. spiraecola, Brevicoryne brassicae, Macrosiphum pisi, M. gei* and *Myzus persicae* transmit the virus under natural conditions.

(iii) Management

- Select virus-free seeds for sowing.
- Rogue infected plants from the field.
- Remove weed hosts in and around the bean field.
- Apply carbofuran at 1.0 kg *a.i.*/ha at the time of sowing followed by 3 to 4 sprays of dimethoate or monocrotophos both at 0.1% at 10 days interval (Singh *et al.*, 1981).
- Use resistant French bean *cvs*. GM 108, 1140, Silver, PI 217179, Sinilac, Sea Farer and Idle Refugi (Sastry *et al.*, 1981).

6.2.2.6. French Bean Golden Mosaic Virus

This is a very widespread and serious disease of beans and other legumes. This disease was first reported by Capoor and Varma (1948) from India. As much as 70–80% losses in yield have been observed in case of early infection.

(i) Symptoms: The disease is characterized by scattered, slightly discoloured patches on leaves, which gradually turn bright yellow (Fig 6.7). At times, the whole leaf turns golden yellow. The chlorophyll of the leaves is partially or completely destroyed. Affected plants are usually dwarf and pod production is greatly reduced.

Figure 6.7. Yellow vein mosaic virus infection on French bean leaves

(ii) Transmission: The virus is readily transmitted by bud grafting. In nature, the disease is transmitted by the whitefly, *Bemisia tabaci.*

(iii) Management

- Remove the infected plants from the field.
- Soil application of carbofuan, fensulfothion or disulfoton or disyston or phorate all at 1.5 kg *a i.*/ha at the time of sowing followed by 3–4 foliar sprays of dimethoate or monocrotophos both at 0.05% at 10 days interval significantly controlled the whitefly population and the disease spread. Spraying of 2% mineral oil has proved very effective in checking the spread of the disease (Singh *et al.*, 1981).

Integrated Pest and Disease Management

- Seed treatment with *P. fluorescens* at 5 g/kg of seeds.
- Soil application of enriched FYM with *T. harzianum* at 1 tonne/ha.
- Spraying of 4% pulverized NSPE or 1% neem soap immediately after germination and repeated after 1 week to control stem fly.
- Spraying of *P. fluorescens* 2 times (20 and 45 DAP) to induce disease resistance.
- Spraying of 4% pulverized NSPE or 1% neem soap at flowering.
- Spraying of 1% neem soap or 1% pongamia oil on the under surface of the leaves, if mite/hopper incidence is noticed.

6.2.3. Nematodes

Root-knot and reniform nematodes are the major limiting factors in successful production of French bean crop.

6.2.3.1. Root-knot Nematodes, *Meloidogyne* spp.

Meloidogyne incognita and *M. javanica* appear to be the most common root-knot species of French beans and have been reported causing damage in tropics and subtropics. Lal and Ansari (1960) reported *M. arenaria* for the first time on beans from Bihar, India. *M. incognita* was responsible for 19.38 to 43.48% loss in pod yield of French bean (Das,

1994; Patel *et al.*, 2004; Parvatha Reddy and Singh, 1981), while *M. javanica* caused 30–40% loss in yield (Sharma *et al.*, 2002).

(i) Symptoms: Gall size on roots of French bean is variable and may be nearly undetectable. The only visible symptom on the roots is the presence of large egg-masses (Fig 6.8).

Figure 6.8. Root-knot nematode on French bean

(ii) Interaction with Other Pathogens: *M. incognita* in association with soil-borne fungi such as *R. solani* and *F. solani* was responsible for root-rot/wilt disease complex in French bean. In the presence of nematode, the root rot due to *R. solani* and wilt due to *F. solani* increased considerably (Parvatha Reddy *et al.*, 1979; Singh *et al.*, 1981).

(iii) Management

(a) Physical Methods: Root-knot nematodes can be effectively controlled by a 4–8 weeks solarization, assuming that the land will remain uncropped during summer.

(b) Cultural Methods: Intercropping of French bean with finger millet, chilli and groundnut helps in the management of *M. incognita* Race 2 (Ramappa, 1988). Growing of chilli or groundnut or finger millet in rotation with French bean significantly reduced the root-knot nematode population (Ramappa, 1988).

The root-knot nematodes would be unable to initially invade bean roots if the crop is sown at the end of winter or early in the spring, when soil temperatures are below 15°C. Escape from early root penetration would give the plant a head start. Yield would increase because the large root system can withstand the damage caused by delayed nematode invasion. Moreover, the beans would be harvested by the end of spring or early in summer, thus limiting the number of generations produced (often to only one) and overall population densities.

Botanicals: Spot application of neem cake at 16.6 g/spot 15 days before sowing was found effective in reducing root galls (64.4 compared to 122.9 in control), number of egg masses (16.0 compared to 44.4 in control) and final nematode population (253 compared to 437 in control) and in increasing plant growth and yield (3.375 t/ha compared to 0.715 t/ha in control) (Ahmed and Choudhury, 2004).

(c) Chemical Methods: Seed treatment with phenamiphos, aldicarb and carbofuran all at 1 per cent concentration were effective in controlling the root-knot nematode (*M. incognita*) infecting French bean and in increasing their pod yields (Parvatha Reddy, 1984). Seed treatment with oxamyl at 3-10% prevented development of *M. incognita*.

Aldicarb, carbofuran, ethoprophos and phenamiphos each at 1 to 2 kg *a.i.* per ha were found effective in reducing the root-knot nematode population and in increasing pod yields of French bean (100 to 112 per cent) (Parvatha Reddy, 1985a; Rao and Singh, 1978).

(d) Host Resistance: French bean *cvs*. Banat, Blue Lake Stringless, Bountiful Flat, Brown Beauty, Cambridge Countess, Gallaroy, Kenya-3, Pinto W5-114, Seafarer and Sutton's Masterpiece were reported to be resistant to *M. incognita* (Singh *et al*. (1981). Wyatt *et al*. (1983) released the first snap bean *cv*. Nemasnap resistant to *M. incognita*. In Kenya, the *cvs*. Kahuti, Red Haricot, Rono, Saginaw and Kiburn were resistant to *M. incognita* and *M. javanica* (Ngundo, 1977). Contender has been reported to be moderately resistant to *M. javanica*. French bean *cvs*. Kibbu, Manoa Wonder, Red Haricot, Rono and Saginaw were reported to be resistant to both *M. incognita* and *M. javanica*.

6.2.3.2. Reniform Nematode, *Rotylenchulus reniformis*

The reniform nematode also damages French bean.

(i) Interaction with Other Pathogens: Vadhera *et al*. (1995) demonstrated that the incidence of root rot was maximum in simultaneous inoculation of *R. reniformis* and *F. solani* on beans. The combination treatment resulted in maximum reduction in plant growth parameters.

(ii) Management

(a) Cultural Methods: Neem and karanj oil cakes at 2 tonnes per ha were most effective in reducing *R. reniformis* infecting French bean.

(b) Chemical Methods: Satisfactory nematode control has been obtained with six foliar sprays of oxamyl at 0.56 kg *a.i.*/ha combined with a soil drench of 2.24 kg *a.i.*/ha of the same chemical. Furrow application of 2.5 kg *a.i.*/ha of carbofuran was also found effective.

6.2.3.3. Root-knot Nematode, *Meloidogyne incognita* and Root Rot, *Macrophomina phaseolina* Disease Complex

(i) Management

(a) Cultural Methods: The beneficial effects of neem cake were observed not only against *M. incognita* and the fungus *M. phaseolina* individually, but also when both the pathogens formed a disease complex on French bean.

6.3. COWPEA

Cowpea (*Vigna unguiculata*) also known as southern pea and black-eye pea, is one of the most important vegetables. It is cultivated for its long, green or purplish pods to be cooked as vegetable or for dry seeds used as pulse. Its foliage is used as fodder or green manure. In India, cowpea is grown almost throughout the country.

6.3.1. Pests

6.3.1.1. Spotted Pod Borer, *Maruca testulalis*

(i) Damage: This is a serious pest on cowpeas (Fig 6.9). The pyralid moth lays eggs on small flower buds as soon as flowering starts. Flower buds can be examined for holes caused by young larvae. Therefore, control measure has to be taken at flower bud stage itself to prevent the larvae from developing and feeding on pods.

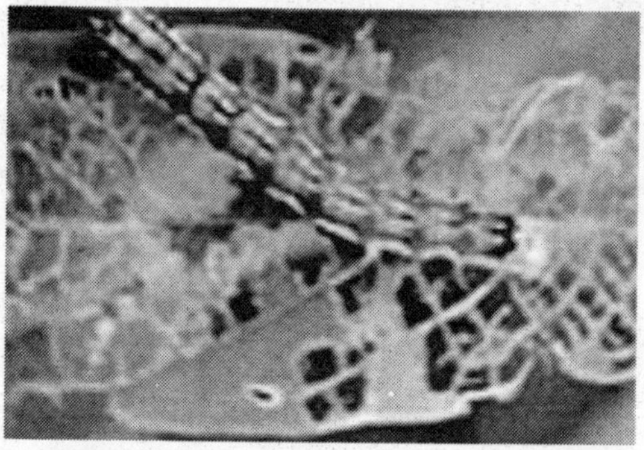

Figure 6.9. Pod borer damage on cowpea

(ii) Management

- Apply neem cake at 250 kg/ha at flowering and the first spray has to be given at flower bud formation with pulverized neem seed powder extract (PNSPE) 4% or neem soap 1%. Repeat after one week.
- Spray cypermethrin 25 EC at 0.5 ml/l or chloripyriphos 20 EC at 2.5 ml/l or indoxacarb 14.5 SC at 0.5 ml/l.
- *Crotalaria* sp. as a trap crop has been proposed in cowpea fields against pod borer in Nigeria (Jackai and Singh, 1981).

6.3.1.2. Bugs

(i) Damage: Many bugs are known to attack leguminous crops. Of these, *Nezara viridula* and *Coptosoma cribraria* are often serious, mostly on cowpeas and lablab. Eggs are laid on tender plant parts. Nymphs and adults suck the sap from foliage, flower parts and pods and emit characteristic bad smell.

(ii) Management

- Apply neem cake at 250 kg per ha immediately after germination and repeat at flowering. Spray neem soap 1% or pulverized neem seed powder extract 4% at 10 days interval.
- Dust endosulfan 4% or spray with indoxacarb at 0.5 ml/l if the incidence is high.

6.3.1.3. Vertebrate Pests

(i) Damage: Small green barbet, Red vented bulbul and Rose ringed parakeet, *Bandicota bengalensis* and *Funambulus palmarum* fed on cowpea flower buds and seeds.

(ii) Management

Intercropping of cowpea with ridge gourd and erecting pandal to 0.5 m height reduced damage, resulting in higher pod yields (Anon., 1992).

6.3.2. Diseases

6.3.2.1. Leaf Blight, *Pseudocercospora cruenta, Ccercospora cruenta*

(i) Symptoms: *Pseudocercospora cruenta* is more destructive pathogen than *C. cruenta.* Symptoms appear as black, sooty and mouldy growth of fungus observed on all foliar parts (Fig 6.10). Initially the spore masses are olivaceous dark brown and spread very fast in the field. Quick defoliation is characteristic feature of the disease which leads to naked stem and twigs on the plant. However, *C. cruenta* is less destructive and form small circular brown spots with reddish margin.

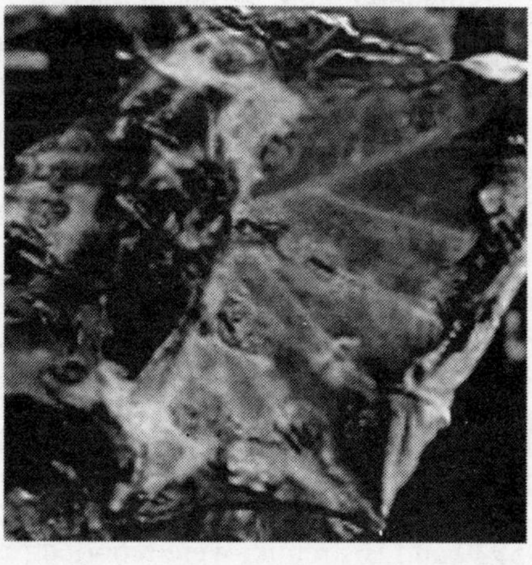

Figure 6.10. Leaf blight on cowpea leaves

(ii) Management

Management practices include crop rotation with non-host crops and field sanitation by burning the infected crop debris. Avoid delay in planting particularly during kharif season.

Hexaconazole or bitertanol or triadimefon or tebuconazole all at 0.05% are highly effective (Pandey and Pandey, 2002).

6.3.2.2. Bacterial Blight, *Xanthomonas campestris* pv. *Phaseoli/vignicola*

Bacterial blight is the major disease particularly in cowpea. Presently, the disease is prevalent in most of the cowpea growing areas in India. The disease is responsible for losses in yield up to 90% (Kishun, 1989). The disease appears during the months of July to September in rainy season.

(i) Symptoms: Infected cotyledons become red and shriveled. In case of severe infection, the cotyledons, primary leaves and growing points are killed in the initial stage. Small, translucent, water-soaked spots appear on leaves. As the spots develop, the centre turns yellow to brown and dries with a narrow, yellow hallow around it. Reddish streaks and cankers appear on stem, reducing plant vigour and yield.

(ii) Epidemiology: The pathogen is seed-borne. The bacterium survives in seed up to 250 days (Shah *et al.*, 1989). The disease spreads up to 62% plants with an initial inoculum load of 1% infected seed at 20–80% RH and temperature range of 25–34°C (Shekhawat and Patel, 1977). Intermittent rain has been found favourable for the disease development (Shah *et al.*, 1989). Cowpea plants were found most susceptible up to the age of 30 days (Kishun, 1989).

(iii) Management

Use of healthy disease-free seed for sowing and clean cultivation helps to reduce the disease incidence. Proper drainage should be maintained in the field to avoid water stagnation. Three years crop rotation with non-host crops is essential and sufficient to dispose off over wintering pathogen.

Soak seed in a solution of 0.01% Streptomycin and 0.25% copper oxychloride for 4 hr before sowing followed by spraying the crop during the rainy season. Pre-soaking of seeds at 15°C for 20 hr and then immersing them in water maintained at 30–60°C for 25 min and clean cultivation gave effective control of the disease.

Spraying of 1% Bordeaux mixture; 0.1% carbendazim + 250 ppm agrimycin-100 followed by 250 ppm agrimycin-100 are effective (Chakravarti *et al.*, 1978). Kishun *et al.* (1980) obtained good control by seed treatment with 100 ppm plantomycin followed by 2 sprays of 100 ppm plantomycin or 0.3% copper oxychloride.

Cowpea *cvs*. S-288 (Sreekantaradhya *et al.*, 1977), Iron (Singh and Patel, 1977), TVU 200, 347, 401, 456, 580, 726, 1460, 2460 and 4540 (Kishun *et al.*, 1980), Iron Grey (Gour *et al.*, 1981), Pundulea 68-5939 and Bistremiki were reported resistant.

6.3.2.3. Cowpea Mosaic Virus

(i) Symptoms: The virus causes mosaic, mottling, vein banding of leaves and stunting of plants (Govindaswamy *et al.*, 1970). At times, blisters are also seen on the infected leaves.

(ii) Transmission: The virus is easily transmitted through sap. Seed transmission is up to 17.5%. Aphids mainly *Myzus persicae, Aphis fabae, A. craccivora* and *A. gossypii* transmit the virus in stylet-borne manner.

(iii) Management

- Since the disease is seed-borne, use seeds collected only from healthy plants for sowing.
- Apply carbofuran at 1.0 kg *a.i.*/ha at the time of sowing followed by 3 to 4 foliar sprays of dimethoate or monocrotophos both at 0.1% at 10 days interval to control aphids.
- Spraying mineral oils help in reducing the spread of the virus.
- Use resistant cowpea varieties such as P-340, P-1093, T-2, C-2509, C-42 and EC 78.

6.3.2.4. Cowpea Yellow Fleck

(i) Symptoms: This is a whitefly transmitted disease caused by Gemini virus. Yellow and green mosaic symptoms are seen on the apical leaves. Later, most of the leaves turn completely yellow in color. Infected pods turned yellow from normal green colour. Pod setting is also reduced.

(ii) Transmission: The virus is readily transmitted by grafting. In nature, the whitefly, *Bemisia tabaci* transmit the disease up to 20%.

(iii) Management

Rouging should be practiced at the initial stage. Soil application of carbofuran at 1.5 kg *a.i.*/ha is recommended. Foliar spray of 0.05% monocrotophos at 10 days interval up to flowering stage is effective. Application of mineral oils has been reported effective in reducing the spread of the disease.

Cowpea *cvs*. Arka Garima, Pusa Parvati, Contender, Kentucky Wonder are reported to be resistant.

6.3.3. Nematodes

Root-knot, reniform and pigeonpea cyst nematodes are the major limiting factors in successful production of cowpea crop.

6.3.3.1. Root-knot Nematode, *Meloidogyne incognita*

Parvatha Reddy and Singh (1981) reported that *M. incognita* was responsible for 28.60% loss in pod yield of cowpea.

(i) Symptoms: Symptoms of damage induced by root-knot nematode include patches of stunted and yellowed plants. Severe damage can lead to reduced numbers of leaves and buds. At high densities severe root galling occurs (Fig 6.11). Visual symptoms of damage first occurred at 1,000 and 10,000 juveniles/500 g of soil.

Figure 6.11. Heavy galling of cowpea roots infected with *Meloidogyne incognita* (Courtesy: F.E. Caveness)

(ii) Interaction with Other Pathogens: The presence of heavy infestations of *M. javanica* on a cowpea *cv.* tolerant to wilt incited by *Fusarium oxysporum* f. sp. *tracheiphilum* caused increased wilting when compared to the susceptible *cv.*

(iii) Management

(a) Cultural Methods: Summer ploughing was found effective for the management of root-knot nematodes and increased the cowpea yield over non-ploughed field.

The root-knot nematodes associated with cowpea can be effectively managed by the application of neem or eupatorium leaves at 15 t/ha two weeks before sowing. Neem cake incorporation in the previous crop, caused reduction in density of all nematodes in the soil on the following cowpea crop (Jain and Hasan, 1986). Cocoa pod husks incorporated at 6 t/ha caused 28% reductions in galling and 6.7% increases in yield (Egunjobi, 1985).

Rotation of cowpea with graminaceous crops or *Crotalaria spectabilis* is recommended. Populations of root-knot decreased greatly when compared to fallowed plots when *C. spectabilis* was grown as a weed free cover crop. Mulching with cowpea foliage was also highly effective in suppressing root-knot nematode populations.

M. javanica populations were lower when cowpea and maize were grown under mixed rather than under sole crop cropping systems. One crop of rice was sufficient to effectively reduce root-knot nematode infestations in succeeding susceptible cowpea crop. The reduction was greater than with rotations with non-host crops.

(b) Chemical Methods: Seed treatment with phenamiphos, aldicarb and carbofuran all at 1 per cent concentration were effective in controlling the root-knot nematode (*M. incognita*) infecting cowpea and in increasing their pod yields (Parvatha Reddy, 1984). Carbosulfon at 1,000 ppm used as seed treatment reduced juvenile penetration in cowpea plants. Posse and monocrotophos at 1,000 ppm reduced root-knot index (2.2 and 2.8, respectively) as compared to untreated check (3.5) when cowpea seeds were dipped in these chemicals for 12 hr.

Aldicarb, carbofuran, ethoprophos and phenamiphos each at 1 to 2 kg *a.i.* per ha were found effective in reducing the root-knot nematode population and in increasing pod yields of cowpea (40 to 50 per cent) (Parvatha Reddy, 1985a; Rao and Singh, 1978). An increase in seed yield to the tune of 61.8% was reported with application of phenamiphos at 2.5 kg *a.i.*/ha (Jain and Hasan, 1984).

(c) Biological Methods: Cowpea seed treatment with *Paecilomyces lilacinus* spores was reported to be effective in reducing the root-knot infestation level (Midha, 1985). Application of certain organic materials, *viz.* subabool leaves, neem cake or saw dust prior to the application of *P. lilacinus* not only enhanced the growth and multiplication of the fungus resulting into high percentage of egg mass infection but also effectively reduced the soil incidence in cowpea (Hasan, 1988). Application of *P. lilacinus* at 15 kg/ha along with certain organic materials effectively reduced the root-knot nematode incidence in cowpea (Hasan and Jain, 1992).

(d) Integrated Methods: Hasan and Jain (1992) reported that soil application of *P. lilacinus* cultured on sorghum seeds together with certain organic matter effectively reduced the incidence of *M. incognita* and increased the crop yield of cowpea.

Summer ploughing along with seed treatment with carbosulfon 3% w/w or seed soaking in monocrotophos at 0.1% for 6 hr gave effective control of root-knot nematodes and increased the cowpea yield.

6.3.3.2. Reniform Nematode, *Rotylenchulus reniformis*

The emergence of cowpea seedlings was delayed by 7 days and seedling stand was reduced to the tune of 6 to 11% due to *R. reniformis* at one nematode/g of soil (Nanjappa *et al.*, 1978). *R. reniformis* was responsible for 13.20 to 32.00% loss in pod yield of cowpea (Palanisamy and Sivakumar, 1981; Hasan and Jain, 1998).

(i) Histopathology: *R. reniformis* feeds on cortical tissue of cowpea.

(ii) Races: The nematode has two races based on their ability to parasitize cowpea, castor or cotton with race a reproducing on all three hosts and race B only on cowpea (Dasgupta and Seshadri, 1971).

(iii) Management

(a) Physical Methods: Under field conditions, soil solarization significantly reduced soil population of the reniform nematode up to 15 cm deep and increased the yield of cowpea.

(b) Cultural Methods: Soil application of neem seed kernel at 2.5 t/ha was most effective in reducing the number of females/plant (46.2 compared to 80.7 in check), number of egg masses/plant (26.0 compared to 45.5 in check), number of eggs and larvae/egg mass (46.2 compared to 80.7 in check) and final soil population/200 cc soil (143.5 compared to 831.0 in check) of the reniform nematode (Ram and Baheti, 2004).

Summer ploughing was found effective for the management of reniform nematodes and increased the cowpea yield over non-ploughed field. Solarization was considered as an effective method for reducing the nematode densities to a depth of 15 cm.

The reniform nematodes associated with cowpea can be effectively managed by the application of neem or eupatorium leaves at 15 t/ha two weeks before sowing.

The reniform nematode densities were suppressed when cowpea was grown intercropped with maize.

(c) Host Resistance: Cowpea *cvs.* V-16 and Pusa Phalguni are reported to be resistant to *R. reniformis* (Thaker and Patel, 1984).

(d) Integrated Methods: Summer ploughing along with seed treatment with carbosulfan 3% w/w or seed soaking in monocrotophos at 0.1% for 6 hr gave effective control of reniform nematodes and increased the cowpea yield.

6.3.3.3. Pigeonpea Cyst Nematode, *Heterodera cajani*

The cyst nematode has been found associated with cowpea in a number of regions of India.

(i) Symptoms: The cyst nematode retarded emergence of leaves and reduced the number of flowering buds, flowers, growing pods and yield.

(ii) Management

(a) Cultural Methods: Cowpea rotated with paddy rice may be less affected by the cyst nematode, because of the negative effect of flooding on nematode densities.

(b) Host Resistance: Cowpea *cv.* CO 6 was found resistant to the pigeonpea cyst nematode. The cowpea *cv.* Barsati Mutant has been reported to be tolerant to the cyst nematode (Sharma and Sethi, 1976).

6.4. PIGEONPEA

6.4.1. Pests

6.4.1.1. Pod Borer, *Helicoverpa armigera*

(i) Damage: Pod borer is one of the major pests in tropical and subtropical regions. The adult moths have yellowish to orange fore wings in females and greenish grey in males, with a light dark transverse band in the distal third. The kidney-shaped marking is slightly distinct and smoky. Hind wings are pale grey with a broad, darker marginal band and small brown marking near the base. Females lay several hundred eggs on all parts of plant. On hatching, the larvae feed for short time on the tender leaflets, flower buds and tender roots. Slowly it enters and feed on the seeds inside the pods (Fig 6.12). The half portion of larva remains inside while feeding on the developing seeds.

(ii) Management

Bhatnagar *et al.* (1982) reported that intercropping pigeonpea with sorghum resulted in reduction of *Helicoverpa armigera* damage to pigeonpea and an increase in the combined yield.

Soil incorporation of neem cake at 250 kg/ha at the time of flowering gave significant reduction in borer incidence in pigeonpea to 6% as compared to more than 20% in other non-IPM plots (Krishna Moorthy and Krishna Kumar, 2002).

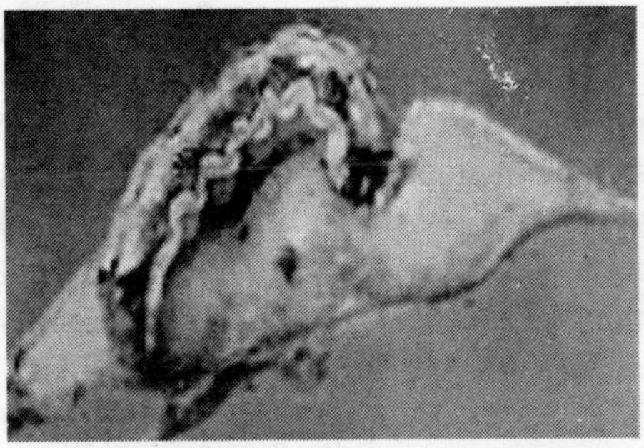

Figure 6.12. Pod borer on pigeonpea

Integrated management of pigeonpea pod borer using *Ha* NPV at 2×10^6 PIB's/ml + nematode DD-136 (*Steinernema feltiae*) at 3×10^3 infective juveniles/ml was attempted. The least population of *H. armigera* was observed in virus + nematode treated plots (0.2) when compared to control (12.6). The percentage of pod damage was significantly lower in virus + nematode (1.07) when compared to control (40.31). Significant increase in pod yield was obtained in virus + nematode treated plots (10.7 kg) when compared to control (3.2 kg) (Narayanan and Gopalakrishnan, 1987).

IPM

Validated IPM Technology (Karnataka)

- Fall ploughing to expose pupae to hot sun and natural enemies.
- Mixing of sorghum and mesta seeds at 250 g/ha with pigeonpea seeds, which act as live bird perches.
- Installation of pheromone traps at 5/ha to monitor the pests.
- Erecting the branched twigs at 20/ha which act as bird perches.
- First spraying with ovicide indoxacarb or methomyl at 300 g/ha or profenophos at 2 litres/ha.
- Second spraying with 5% NSKE or commercial neem formulation (1500 ppm Azadirachtin) at 2 litres/ha.
- Third spraying with *Ha* NPV at 250 LE/ha.
- Fourth spraying with indoxacarb at 300 ml/ha or chlorpyriphos at 2.5 litres/ha or quinalphos or endosulfan at 2 litres/ha.
- If necessary, spraying of synthetic pyrethroids at 500 ml/ha.
- In case of scarcity of water, dust 4% endosulfan followed by 1.5% quinalphos or 0.4% fenvalerate at 25 kg/ha.

Validated IPM Technology (Maharashtra)

Pre-sowing

- Deep ploughing and exposure of soil to hot summer to kill pupating larvae and fungal propagules.

- Use of FYM at 5 MT/ha (if possible mixed along with neem cake) to reduce the disease as well as the cyst nematode population.
- In areas where termite and cut worm is a problem, seed treatment with chlorpyriphos at 8 ml/kg or pre-sowing mixing of soil with chlorpyriphos dust is recommended as they are relatively low cost and initially protect the seed and seedlings from cut worms and seed-borne pathogens.
- Synchronized sowing of multiple pest resistant varieties such as Sharad, Asha, Maruthi, Bahar, Abhaya, BSMR-736 along with fungicidal seed treatment.
- Ridge planting to prevent incidence of blight caused by *Phytophthora drechsleri*.

Sowing Time

- Intercrop sorghum to reduce wilt, conserve beneficial insects and serve as bird perches.
- Early planting around mid–June in north west plain zone to avoid *Helicoverpa armigera*.
- Use marigold as a trap crop on border or interspersed with crop for pod borer control.

Post-Sowing

- Monitoring through regular field scouting along with pheromone traps (5–10/ha) to assess population build up of borers and their management. A threshold level of 5–6 moths/trap/day is indicative of its peak activity period and warrants initiation of management practices.
- Rogue out and destroy sterility mosaic affected plants.
- Mechanical shaking of plants and collection of larvae for preparation of *Ha* NPV locally to cut down the cost.
- Spray 5% NSKE or neem based formulation against *H. armigera* and other Lepidopterous pests.
- Spray with *Ha* NPV at 450 LE/ha along with UV retardant.
- Spray eco-friendly pesticide like endosulfan if pest complex exceeds ETL.

Validated IPM Technology (Uttar Pradesh)

- Deep summer ploughing to destroy immature stages and pathogen propagules.
- Prior to sowing, soil application of *Trichoderma harzianum* at 10 g/kg of FYM for controlling the pigeonpea wilt.
- Sowing on ridges to control *Phytophthora*.
- High yielding varieties with pest/disease tolerance—Asha and Maruthi against wilt and PSM for Andhra Pradesh and Karnataka, and Bahar for Uttar Pradesh (Varanasi).
- Installation of pheromone traps at 10/ha in the month of September.
- Erection of bird perches at 25/ha for facilitating predation of *Helicoverpa* larvae.
- One spray with 2% neem oil.
- Two applications of 5% NSKE in September and October.
- Spray *Ha* NPV at 500 LE/ha (1.5×10^{12} POB)/ha in September and October—when larvae are small.

- Shaking of plants 5 times a day starting from October for short duration pigeonpea.
- Need based spray with endosulfan at 2 litres/ha.

6.4.2. Diseases

6.4.2.1. Wilt, *Fusarium udum*

(i) Symptoms: The disease is characterized by slow wilting of the plant. The symptoms can be observed after a month of sowing or at the flowering or pod formation stage. The affected plants become yellow in color followed by drooping and finally the whole plant dries up. The symptoms resemble as if the plant is suffering from the drought. The disease can be diagnosed whenever the affected stem is cut and opened where the browning of the xylem vessels could be clearly seen. It has been recorded that wilting can be partial or total. However, it has been observed that partial wilting is mainly associated with lateral root infection, whereas tap root infection may cause complete wilting.

(ii) Management

Bacillus subtilis brought about a marked reduction in the incidence of pigeonpea wilt. Amendment of soil with roots of certain leguminous crops (sweet clover), molasses and oil cakes (groundnut cake) markedly increased the antibiotic (bulbiformin) production by *B. subtilis* and a reduction of 88% in the incidence of wilt. Seedlings gained resistance to *Fusarium* infection when the seeds were bacterized with *B. subtilis* before sowing. It was suggested that bulbiformin became systemic in the plant and provided a protective zone around the roots of pigeonpea seedlings.

Trichoderma harzianum effectively suppressed *Fusarial* wilt of pigeonpea.

6.4.3. Nematodes

6.4.3.1. Cyst Nematode, *Hetrodera cajani*

(i) Symptoms: *Hetrodera cajani* exhibit sexual dimorphism. The males are vermiform and live in the soil. The females are sedentary and remain in the root system. Usually one generation is completed in each season. During the off-season, the nematode survives as cysts which contains eggs and larvae. Economic threshold level for economic damage to pigeonpea was reported at 2 larvae/g of soil. The plant exhibit fewer leaves, reduced tillering, delayed emergence of ears, reduced number of spikelets and grains. Rains between vegetative and flowering stages activate the nematodes, while the sandy loam soil favour their activity.

6.5. CLUSTER BEAN

Cluster bean (*Cyamopsis tetragonoloba*) is also popularly called as 'guar' in India. Its tender pods being edible and are used in preparation of curry. The green pods are rich in vitamins A and C. Since it is draught tolerant, it can be grown in summer. It is also grown as a forage and green manure crop. Its seeds are reported to be good cattle feed.

6.5.1. Pests

6.5.1.1. Pod Gall Midge, *Asphondylia cyamopsii*

(i) Damage: Cluster bean is found severely infested with pod gall midge causing severe reduction in yield. The pest remained active during July to November. However, maximum damage was found in September-October. Summer crop grown in March-April remained completely free from gall midge infestation. The total life cycle of *Asphondylia cyamopsii* is completed in 26 days.

(ii) Management

The parasitism by parasitoids (*Dioxybracon* sp.) varied from 47 to 61%. Carbaryl at 0.2% and endosulfan at 0.1% spray gave effective control of the pest (Chari, 1970).

6.5.2. Diseases

6.5.2.1. Root Rot, *Sclerotium rolfsii*

(i) Symptoms: It is a ubiquitous fungus with a broad host range. Initially wet rotting of bark is observed. Entire bark of the plant near collar region rots. The characteristic symptoms include white, cottony fungus growth is observed on affected portion as well as on parts in contact with soil. Gradually this hyphal mat is converted into small, mustard-like sclerotia that survive in the soil.

(ii) Management

Control of root rot of cluster bean was achieved with *Streptomyces nigrifaciens*.

6.5.2.2. Bacterial Blight, *Xanthomonas campestris* pv. *cyamopsidis*

At present, the disease is prevalent at Delhi, Haryana, Punjab, Rajasthan and Uttar Pradesh. Loss in yield due to this disease is reported between 19 to 85% (Gupta, 1978; Gandhi and Chand, 1985b; Lodha and Gupta, 1981, 1986).

(i) Symptoms: Two types of symptoms *viz.*, leaf spot and blight are produced by the single pathogen (Srivastav and Rao, 1963). Several small, circular, water-soaked spots are formed on leaflets. They are oily and olive coloured on the lower surface. Spots, when fully developed become jet black in color and several of them coalesce to form necrotic patches, surrounded by chlorotic halo. The infection from blighted leaves spread to stem through petiole and in advance cases, stem gets cracked. The pods also show heavy spotting.

(ii) Epidemiology: Pathogen is seed-borne (Srivastav and Rao, 1963). The disease spreads to new areas by infected seeds and is destructive to rainy season crop (Srivastav and Rao, 1970). The intensity of the disease is increased with increase in amount of rain fall and number of rainy days (Karwasra *et al.*, 1981, 1984) and with increase in the nitrogen level from 20–40 kg/ha (Gandhi and Chand, 1985a). Younger plants are reported to be more susceptible (Singh and Swarup, 1987; Gandhi and Chand, 1988).

(iii) Management

- Use of healthy seed from summer grown irrigated crop (Srivastav and Rao, 1970).
- Seed treatment with hot water at 56°C for 10 min (Singh and Swarup, 1986).
- Cluster bean *cvs*. G 85, 102, 126, and 155 are reported to be highly resistant; whereas HG 75, 182, HFG 119, 340 (Gandhi and Chand, 1987), G 28, 89, 98, 100, 119, 127, 139, 158, 166, 172, 223, 246, 253 and 254 (Chahal *et al*., 1979) are moderately resistant.
- Gupta (1978) reported good control of the disease by spraying with 100–250 ppm streptocycline and 100–500 ppm agrimycin-100. Spraying of 100 ppm streptocycline followed by 100 ppm agrimycin-100 or 2000 ppm ceresan wet was also found effective (Singh and Swarup, 1986).

6.5.2.3. Cluster Bean Mosaic Virus

(i) Symptoms: Mosaic pattern, leaf distortion, blisters and shortening of leaves, stunted growth and dark green appearance of the plants are the typical symptoms of the disease.

(ii) Transmission: The virus is easily sap transmissible. Several aphid species transmit the virus and *Aphis craccivora* is the most efficient vector. The virus is seed-borne up to 40%.

(iii) Management

- Use disease-free seeds for sowing.
- Seed treatment in hot water at 55°C for 15 min followed by incubation at 25°C for 4 days.

6.5.3. Nematodes

Root-knot nematode and *Fusarium* wilt disease complex is the major limiting factor in successful production of cluster bean crop.

6.5.3.1. Root-knot Nematode, *Meloidogyne javanica* and Wilt, *Fusarium solani* Disease Complex

(i) Management

(a) Integrated Methods: *Pseudomonas aeruginosa* and *Paecilomyces lilacinus* when used together significantly reduced infection of the disease complex on cluster bean (Perveen *et al*., 1998).

6.6. FIELD BEAN

6.6.1. Pests

6.6.1.1. Pod Borer, *Adisura atkinsoni*

(i) Damage: This is the major pod borer in field bean. The larvae of this pest resembles *H. armigera*. The eggs are laid on tender pods. The young larvae bore into tender pods. They develop inside the bored pod and come out after attaining fourth instar, which is a migratory stage. It causes heavy damage by way of feeding the developing pods and reducing the marketable yield. Hence, control measures have to be taken at tender pod stage and again at flat pod stage (when pods have small developing seeds) of the crop.

(ii) Management

- Apply neem cake 250 kg/ha at flowering and the first spray with monocrotophos has to be given at flowering and repeat after 20 days. This should be followed by sprays of pulverised neem seed powder extract (PNSPE) 4% or neem soap 1% at tender pod formation. Repeat sprays at 10 days interval. The borer incidence was 14% due to neem cake application as compared to more than 36% in control (Krishna Murthy and Krishna Kumar, 2002).
- Spraying monocrotophos at tender pod stage based on peak egg laying and again when on an average 1.5 flat pods are damaged per inflorescence was found effective (Krishna Moorthy *et al.*, 1988).
- Spray cypermethrin 25 EC at 0.5 ml/l or chloripyriphos 20 EC at 2.5 ml/l or indoxacarb 14.5 SC at 0.5 ml/l at peak egg laying stage or tender pod stage and repeat after 15 days.
- 'Biotrol' (*Bacillus thuruingiensis*) at 1 g/l reduced the pod borer infestation on field beans (Surulivelu *et al.*, 1977).
- Crop rotation with cereals (jowar, maize) reduced the incidence of pod borer.
- Spraying of NPV at 250 LE/ha in combination with endosulfan at 0.035%, twice at fort-nightly intervals significantly reduced both larval population and pod damage by *Adisura atkinsoni* and *Sphaenanches anisodactylus* (Narayanan, 1987).

6.6.1.2. Lycaenid Pod Borer, *Lampides boeticus*

(i) Damage: This is a butterfly pest and commonly known as pea blue. The eggs are laid singly in flowers and flower buds. In lablab, the larvae bore into flower buds. The pest can be monitored by the presence of the bored holes. When the larvae attain fourth instar, they migrate and cause extensive damage to flower buds and pods. Hence, control measures have to be taken in the early stages of the pest incidence.

(ii) Management

- In lablab, spraying has to be done at flower bud stage, when on an average about 1 flower bud is damaged per inflorescence. Spraying with any synthetic pyrethroid or 0.05% chlorpyriphos or 5% NSKE after flowering is effective.

6.6.1.3. Bugs

(i) Damage: Many bugs are known to attack leguminous crops. Of these, *Nezara viridula* and *Coptosoma cribraria* are often serious, mostly on cowpeas and lablab. Eggs are laid on tender plant parts. Nymphs and adults suck the sap from foliage, flower parts and pods and emit characteristic bad smell.

(ii) Management

- Apply neem cake at 250 kg per ha immediately after germination and repeat at flowering. Spray neem soap 1% or pulverized neem seed powder extract 4% at 10 days interval.
- Dust endosulfan 4% or spray with indoxacarb at 0.5 ml/l if the incidence is high.

Table 6.5. Management of insect pests in leguminous vegetable crops

Crop	Pest	Symptoms	Pesticide/Bioagent/Dose
French beans	Stem fly	Puncture marks in leaves, mining in petiole and stem	Endosulfan-2 ml/l -10 & 15 DAS; Monocrotophos-1.25 ml/l -10 & 15 DAS; NSKE 5%-50 gm/l -7, 10 & 15 DAS
	Jassids	Swarming of adults while walking	Phosphamidon-0.5 ml/l or Monocrotophos-1.25 ml/l at seedling stage
	Serpentine leaf miner	Mines generally in basal leaves, rarely more in trifoliate leaves	NSKE 4% or Triazophos-1.5 ml/l at 10 & 20 DAS
	Tetranychus spp.		*Phytoseiulus persimilis,* 10 adults/ plant or release 1–6 leaves with predatory mites. Once, 30 days after germination or as and when infestation is noticed.
Peas	Stem fly	Yellowing of basal leaf	Phosphamidon-0.5 ml/l or Monocrotophos-1.25 ml/l at 10 & 20 DAS
	Pod borer	Boring of pods with holes or with out holes	Fenvalerate/Cypermethrin-0.5 ml/L at the time of flowering
Pigeonpea	*H. armigera*		*Ha* NPV at 250 LE/ha
Chickpea	*H. armigera*		*Ha* NPV at 250 LE/ha

Table 6.6. Leguminous vegetable crop varieties resistant to biotic stresses

Vegetable crop	Pest/Disease	Resistant varieties
French bean	Rust	Arka Bold, Swarna Priya, Swarna Latha.
	Angular leaf spot, Mosaic	Pant Anupama.
	Rust, *Alternaria* leaf spot	Arka Bold.
	Rust, Bacterial blight	Arka Anoop.
Field bean	Viral diseases, jassid, aphid, fruit borer	Pusa Sem-2, Pusa Sem-3.
Cowpea	Bacterial blight	Pusa Komal.
Pea	Powdery mildew	Pusa Pragati, JP-4, JP-93, FC-1, PRS-4, NDVP-4.
	Powdery mildew, Rust	Arka Ajit, Arka Karthik, Arka Sampoorna.
	Fusarium Wilt	JP Batri Brown 3, JP Batri Brown 4.

7 LEAFY VEGETABLES CROPS

7.1. AMARANTHUS

7.1.1. Pests

Amaranth is susceptible to damage by foliar insects such as leaf miners, leaf roller caterpillars, cutworms, aphids, flea beetles, and mites. An effective method of controlling insect pests is to cover the bed with a fine screen or nylon mesh netting (32-mesh or finer).

7.1.1.1. Leaf Webber/Caterpillar, *Hymenia recurvalis*

(i) Damage: Eggs are laid in and around the leaf veins and the larvae joins leaves together and after enclosing itself inside, scrape the green part thus making leaves unfit for consumption (Fig 7.1). Green slender caterpillar feed on leaves.

Figure 7.1. Leaf caterpillar on amaranthus leaf

(ii) Management

The population of leaf webber caterpillars and the extent of leaf damage in yellow oleander, *Thevetia nerrifolia* plant extract (2% seed and fresh leaf extract) treated plots were significantly low compared to control plots. Seed extract sprayed plots gave an yield of 4.10 kg/plot as compared to 2.75 kg/plot in control.

'Thuricide' dust (3×10^6 spores/g) at 20 to 25 kg/ha, 'Thuricide' WP (3×10^6 spores/g) at 0.1 and 0.2% and 'Thuricide' EC (15×10^9 spores/g) at 0.1 and 0.2% were found to be highly effective in controlling the leaf caterpillar (Sundarababu, 1968).

The leaf webber was found to be parasitized to the tune of 62% by the parasitoid, *Apanteles* sp.

7.1.1.2. Stem Weevils, *Hypolixus truncatulus, Hypolixus haerens*

(i) Damage: Several species of weevils feed on amaranth. Adult weevils feed on leaves, but the larvae (grubs) are more damaging because they bore into roots and stems, causing rotting and potentially lodging and predisposition to diseases. Stem-boring weevils such as the pigweed weevil (*Hypolixus haerens*) are the most damaging causing plants to wither and lodge. The adult weevil lays its eggs in leaf veins, stem or branches, and the larvae bore through stems to the root collar hollowing the stems (Fig 7.2). Feeding by larvae results in stems that are more susceptible to wind breakage resulting in increased crop losses. The larvae pupate in the stem. Attack by this weevil has been associated with extensive tissue discolouration, decay and cankers in branches, stems, and root collars of *Amaranthus hybridus*. This weevil has been found to be associated with fungi (mainly *Fusarium* spp.) that cause tissue decay and a canker disease.

Figure 7.2. Weevil larvae and adult feeding in amaranth stem (Courtesy: A.M. Varela)

(ii) Management

Uproot and destroy attacked plants to reduce number of weevils and prevent damage to healthy plants.

7.1.1.3. Aphid, *Lipaphis erysimi*

(i) Damage: Aphids suck plant sap (Fig 7.3) resulting in stunted growth of the plant.

(ii) Management

Spraying 4% NSKE gives effective control of aphids.

Figure 7.3. Aphid infestation on amaranthus leaf

7.1.1.4. Bugs

(i) Damage: Bugs can cause severe damage to flowering head and seeds, and may be particularly damaging to grain amaranth when present in large numbers during the critical seed fill stage (Fig 7.4). They are usually of minor importance in vegetable amaranth, and no control measures are needed.

Figure 7.4. Bugs feeding on amaranth flowering head (Courtesy: A.M. Varela)

7.1.1.5. Stem Borer, *Lixus truncatulus*

(i) Damage: The larvae of the stem borer *Lixus truncatulus* may cause much damage, sometimes already in the seedling stage. The basal part of the plant containing the pupae swells and the plant growth is much retarded.

7.1.2. Diseases

7.1.2.1. Damping-off, *Pythium aphanidermatum, Rhizoctonia solani, Aphanomyces* sp.

(i) Symptoms: Damping-off is a serious disease in seedlings. The disease is caused by *Pythium aphanidermatum, Rhizoctonia solani* and *Aphanomyces* sp. Seeds may rot in the soil before emergence (pre-emergence damping-off) or seedlings may exhibit stem canker above the soil line and/or root necrosis. Affected seedlings eventually wilt (post-emergence damping-off) (Fig 7.5). The disease is favoured by high soil water content and low soil temperatures. Also dense planting without sufficient aeration enhances disease development.

Figure 7.5. Damping-off of amaranthus seedlings

(ii) Management

- Use disease-free seeds.
- Avoid over watering.
- Avoid dense sowing and planting.
- Provide good drainage.
- Spray dithiocarbamates.

7.1.2.2. Stem Rot/Blight, *Choanephora cucurbitarium*

(i) Symptoms: Stem rot is caused by the fungus *Choanephora cucurbitarium*. It causes wet rot of stems and leaves. Affected plant parts have hairy appearance (silk-like threads) consisting of fungal spores. Infection is predisposed by injuries. During rainy season it can cause heavy defoliation. The disease is spread by air currents and infected seeds. Warm, and moist conditions favour disease development.

(ii) Management

- Plant certified disease-free seeds.
- Avoid dense planting to allow sufficient aeration.
- Practice good field sanitation.
- Wider spacing reduces the rate of infection.

7.1.2.3. White Rust, *Albugo bliti*

(i) Symptoms: White rust pathogen forms white blisters on the leaves (Fig 7.6).

7.1.2.4. Downy Mildew, *Peronospora* spp.

(i) Symptoms: Downy mildew causes discolouration and drying of the leaves (Fig 7.7).

Figure 7.6. White rust on amaranthus leaf

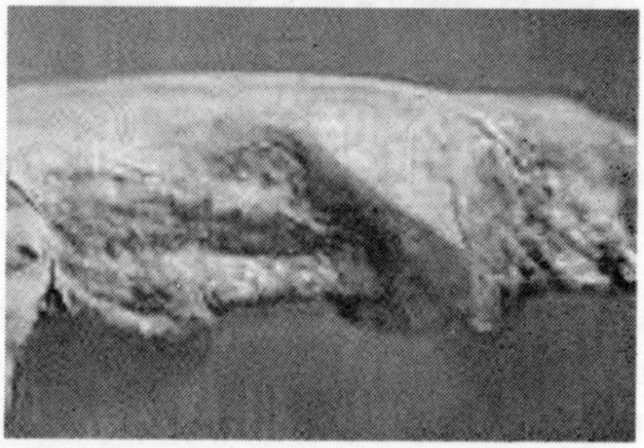

Figure 7.7. Downy mildew on amaranthus leaf

7.1.2.5. Amaranthus Mosaic Virus

(i) Symptoms: The diseased plants exhibit slightly smaller leaves, showing crinkling and mosaic mottling of dark green areas interspersed with light green patches. The leaves are also reduced in size (Ramakrishnan *et al.*, 1971). The virus attacks *Amaranthus gangeticus, Achyranthes aspera, Celosia cristata, Gompherina globosa, Nicotiana glutinosa* and *Petunia hybrida.*

(ii) Transmission: The virus is sap transmissible. *Aphis gosypii* acts as the vector of this disease.

(iii) Management

The disease incidence can be minimized by 3 to 4 sprays of dimethoate or monocrotophos both at 0.5% at weekly intervals. The spraying should be stopped at least 20 days before first harvest.

7.2. LETTUCE

Lettuce (*Lactuca sativa*) is a very common cool season salad crop. Its leaves are rich in vitamin A (900 IU), C (10 mg), choline (178 mg) and minerals–calcium (50 mg) and phosphorus (28 mg).

7.2.1. Nematodes

Root-knot and reniform nematodes are recognized as the major limiting factors in successful cultivation of lettuce crop.

7.2.1.1. Root-knot Nematodes, *Meloidogyne* spp.

The root-knot nematode cause large decreases in yield of lettuce.

(i) Management

(a) Physical Methods: The most important nematode pests controlled by greenhouse steaming are root-knot nematodes on lettuces.

(b) Cultural Methods: Planting of 4 and 5 seedlings of *Tagetes erecta* in a pot around a lettuce plant reduced root galling by 56 and 72% and nematode population by 73 and 94%, respectively (Pervez *et al.*, 1988).

(c) Chemical Methods: Soil fumigation with 98% methyl bromide + 2% chloropicrin (Dowfume MC-2) at 1 lb/100 cu. ft. of soil was best, both in terms of nematode control (0 galls as compared to 65 galls in control) and yield increase (82.5 g/plant as compared to 46.4 g/plant in control).

7.2.1.2. Reniform Nematode, *Rotylenchulus reniformis*

(i) Management

(a) Physical Methods: Under field conditions, soil solarization significantly reduced soil population of the reniform nematode up to 15 cm deep and increased the yield of lettuce.

7.3. SPINACH

7.3.1. Diseases

7.3.1.1. Spinach Mosaic

This disease is caused by a strain of cucumber mosaic virus.

(i) Symptoms: The disease is characterized by light and dark green patches and mosaic mottling of infected leaves. The leaves are reduced in size, distorted and invariably show chlorotic sectors which are almost white. The infected plants show stunted growth (Naqvi and Mahmood, 1975).

(ii) Transmission: The virus is easily transmitted by sap inoculation. In nature, the disease is transmitted by aphid vector, *Myzus persicae* in a non-persistent manner.

(iii) Management

The disease incidence can be checked by 1 or 2 sprays of dimethoate or monocrotophos both at 0.05%. The spraying should be stopped at least 20 days before the first harvest.

7.4. CURRY LEAF

Curry leaf (*Murraya koenigii*) is a fast-growing deciduous shrub or small tree with deep roots and scented leaves. It is cultivated all over South and Southeast Asia. The leaves are an important ingredient extensively used in curry in south India and Sri Lanka. They are particularly used in south India cooking to provide a flavouring for curries, vegetable, fish and meat dishes, soups (rasams), pickles, butter milk preparations, chutneys, scrambled eggs and curry powder blends.

Curry leaf is used in South Asian traditional medicine to treat the digestive system, skin conditions and diabetes. Its anti-diabetic properties are supported by scientific research.

7.4.1. Pests

7.4.1.1. Lemon Butterfly, *Papilio demoleus*

(i) Damage: Caterpillars of lemon butterfly feed on leaves (Fig 7.8).

Figure 7.8. Larva and adult of lemon butterfly

(ii) Management

Hand pick and destroy caterpillars.

7.4.1.2. Scale Insect

(i) Damage: Scale insects affect branches of curry leaf plant (Fig 7.9).

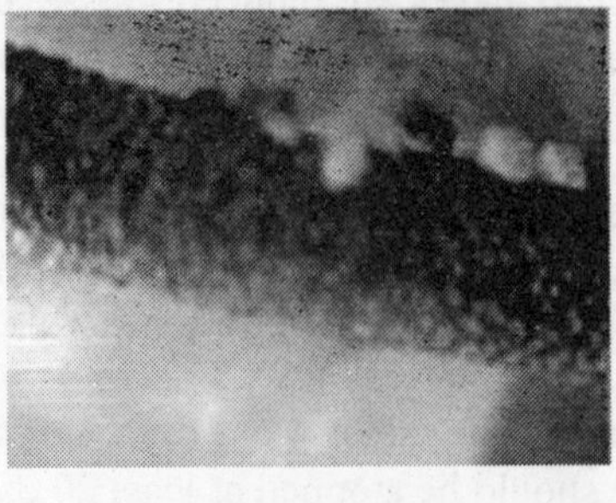

Figure 7.9. Scale infestation on stem of curry leaf plant

(ii) Management

Spraying 4% NSKE gives effective control f the pest.

7.4.2. Diseases

7.4.2.1. Leaf Spot, *Cercospora* sp.

(i) Symptoms: The pathogen causes leaf spots on curry leaves (Fig 7.10).

Figure 7.10. *Cercospora* leaf spots on curry leaf

(ii) Management

Spraying carbendazim at 0.25 ml/litre of water gives effective management of leaf spot disease.

7.5. DRUMSTICK

7.5.1. Pests

7.5.1.1. Bud Worm, *Noorda moringae*

(i) Damage: Larvae bore into flower buds (Fig 7.11) and cause shedding of flowers.

Figure 7.11. Bud worm infestation on drumstick

(ii) Management

Dusting 10% carbaryl at 25 kg/ha gives effective control of the pest.

7.5.1.2. Leaf Caterpillar, *Noorda blitiealis*

(i) Damage: The leaf caterpillar causes defoliation of trees (Fig 7.12).

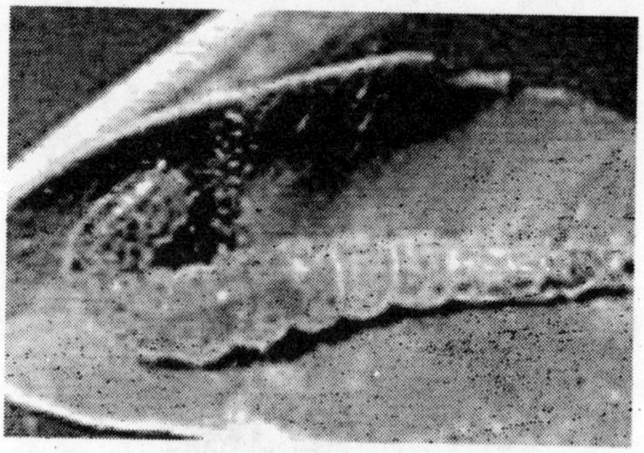

Figure 7.12. Leaf caterpillar infestation on drumstick

(ii) Management

Dusting 10% carbaryl at 25 kg/ha gives effective control of the pest.

7.5.1.3. Hairy Caterpillar, *Eupterote mollifera*

(i) Damage: The hairy caterpillar causes defoliation of trees (Fig 7.13). Larvae collect on branches in a gregarious mass.

Figure 7.13. Hairy caterpillar infestation on drumstick

(ii) Management

Destroy larvae by torching.

7.5.1.4. Fruitfly, *Gitonia* sp.

(i) Damage: Fruitfly maggots cause fruits to dry from the tip.

(ii) Management

Spraying dichlorvos at 1 ml/litre of water gives effective control of fruitfly.

Pesticide Residues

The persistence of terminal pesticide residues and waiting periods for leafy vegetables are presented in Table 7.1. (Awasthi, 2001).

Table 7.1. Terminal pesticide residues and their waiting periods for leafy vegetables

Pesticide schedule	Dosage/ Concentration (%)	Terminal residues (ppm/days)	Waiting period (days)
DDVP spray on foliage	0.05	0.25/3	3
Malathion spray on foliage	0.06	0.03/3	Nil
Phosphamidon spray on foliage	0.05	0.24/15	11
Dimethoate spray on foliage	0.05	0.19/21	10
Endosulfan spray on foliage	0.07	0.14/22	8
Fenvalerate spray on foliage	0.01	0.28/15	20

Table 7.2. Leafy vegetable crop varieties resistant to biotic stresses

Vegetable crop	Pest/Disease	Resistant varieties
Amaranth	White rust	Arka Arunima, Arka Suguna (MR).
Palak	*Cercospora* leaf spot	Arka Anupama.

8 CUCURBITACEOUS VEGETABLES CROPS

8.1. PUMPKIN

Pumpkin (*Cucurbita moschata*) occupies a prominent place among vegetables owing to its high productivity, nutritive value, good storability, long period of availability, better transport qualities and extensive cultivation in sub-tropical and tropical parts of the world. Fruits are rich in vitamin A. It is also consumed as processed and stock feed. The fruits can be used in preparing sweets, candy, fermented into beverages. Yellow or orange-fleshed pumpkins are rich in carotene. In India, it is grown mainly in Assam, West Bengal, Tamil Nadu, Karnataka, Madhya Pradesh, Uttar Pradesh, Orissa and Bihar.

8.1.1. Pests

8.1.1.1. Red Pumpkin Beetle, *Aulacophora foveicollis* and *A. lewisii*

(i) Damage: The adults are small, elongated yellow beetles. The yellow brown adults flying near plants defoliate the leaves, immediately after germination (Fig 8.1). The larvae feed on roots and plant parts touching the ground.

Figure 8.1. Red pumpkin beetle on pumpkin leaves

(ii) Management

Mechanically collect and destroy the pest if incidence is low. If the incidence is high, spray indoxacarb 14.5 SC at 0.5 ml/l or 0.2% carbaryl or 0.05% quinalphos or 0.05% chlorpyrifos.

The red pumpkin beetle was found to be highly susceptible to *Bacillus thuringiensis* (Sundarababu, 1969). Spraying with *Beauvaria bassiana* (2×10^6 cfu/g) is effective against red pumpkin beetles.

8.1.1.2. Serpentine Leaf Miner, *Liriomyza trifolii*

(i) Damage: This is an introduced pest occurring on many cucurbit vegetables. Heavy incidence is noticed in watermelon, pumpkin, cucumber etc. However, bitter gourd seems to be resistant. Mines having serpent shape with small larvae are noticed all over the leaves (Fig 8.2). The incidence starts after about 15 days of germination of seeds. Left alone, it gets suppressed by natural enemies.

Figure 8.2. Serpentine leaf miner damage on pumpkin leaf

(ii) Management

- Left alone, the leaf miner gets suppressed by natural enemies. A native larval parasitoid, *Hemiptarsenus vericornis* has been observed to be the major parasitoid on this pest.
- Soil application of neem cake at 250 kg/ha immediately after germination.
- Destroy cotyledon leaves with leaf mining at 7 days after germination.
- Spray 4% pulverized NSPE/NSKE or neem soap 1% or neem formulation with 10,000 ppm or more (2 ml/l) 15 days after sowing and repeat after 15 days if necessary.
- If the incidence is high, first remove all severely infected leaves and destroy. Then mix neem soap 5 g and hostothion 1 ml per litre and spray. After one week spray neem soap 1% or pulverized NSPE or neem formulation with 10,000 ppm or more (2 ml/l).
- Never repeatedly spray any one insecticide again and again.

8.1.1.3. Fruit Fly, *Bactrocera cucurbitae*

(i) Damage: This is the major pest of cucurbits and is widely distributed. The damage by maggots results in rotting of young and ripened fruits or drying and shriveling of fruits before maturity. Some times even the base of the plant gets attacked and plants start wilting. The incidence is more in wet climate.

Eggs are white and are laid in groups and are 1.0–1.5 mm long and taper at both ends. Full grown larvae are 5–10 mm in length. Pupae are brown and 5–8 mm long. The life cycle may take about 10–18 days, but gets prolonged during winter in north India (Butani and Jotwani, 1984).

(ii) Management

- Use of resistant varieties *e.g.* Arka Suryamukhi (pumpkin).
- Soil application of neem cake at 250 kg/ha immediately after germination and repeat at flowering followed by sprays of neem soap 1% or pulverized 4% NSPE/NSKE at 10 days interval after flowering were found effective.
- Crush pumpkin 1 kg and add 100 gm jaggery and 10 ml malathion and keep in the plot (10–15 places per ha). This will attract the adults to the fermenting pumpkin and they lay eggs and get killed. Repeat the process 2–3 times in a cropping season.
- Erect cue lure (para pheromone trap) 8 per ha to attract and trap male fruit flies.
- Spray carbaryl 50 WP at 3 g/l or indoxacarb 0.5 ml/l. Foliar spray of 0.05% fenthion with 5% jaggery at fruit formation/ripening stage is very effective.

The fruits of cucurbits can be protected from fruit flies by planting maize as a trap crop. The maize is then treated with insecticides to kill the adult flies. This way, susceptible cucurbit crops escape damage and do not have any insecticide residues. Since the trap area encourages the population of natural enemies, fruit flies will be further reduced by these natural enemies (Matthews, 1984).

8.1.1.4. Thrips, *Thrips palmi*

(i) Damage: Both nymphs and adults feed on the young shoot tips and floral parts and are suspected vector of tospo virus. They are serious during summer months and are suspected to transmit bud necrosis disease in watermelon.

(ii) Management

- Soil application of neem cake (once immediately after germination and again at flowering) followed by NSPE at 4% and neem soap 1% alternately at 10–15 days interval.
- Spray any systemic insecticides like 0.05% acephate or dimethoate 30 EC at 2 ml/l.

8.1.1.5. Leaf Eating Caterpillar, *Diaphania* (= *Margaronia*) *indica*

(i) Damage: Long shining caterpillars feed on leaves and fruits.

(ii) Management

- Soil application of neem cake (once immediately after germination and again at flowering) followed by NSPE at 4% and neem or pongamia soap 1% alternately at 10–15 days interval.
- Spray carbaryl 50 WP at 3 gm/l or indoxacarb at 0.5 ml/l.
- Dusting the plants with ash temporarily repels the beetles. Sow the crop in November to avoid the damage from the pest. Pumpkin leaf beetles and the leaf feeding caterpillars on gourds were found to be highly susceptible to *B. thuringiensis*.

8.1.1.6. Red Spider Mite, *Tetranychus neocaledonicus*

(i) Damage: This pest is serious on cucurbits during warmer climate. Colonies of mites within silk web can be observed on ventral leaf surface when the incidence is high.

(ii) Management

- Spraying of neem or pongamia soap at 1%. Spray lower surface thoroughly.
- Alternately, spray dimethoate 30 EC at 2 ml/l or ethion 50 EC at 1 ml/l or wettable sulphur 80 WP at 3 g/l.

8.1.1.7. Vertebrate Pests

(i) Damage: Damage to cucurbits was noticed at sowing, flowering and fruiting stages. At sowing, Bluc rock pigeon (*Columba livia*), squirrel (*Funambulus palmarum*) and rabbit (*Orictolagus cuniculus*) damaged the seedlings. An estimation revealed that on an average 17.2 to 30.5% seedlings are lost due to these vertebrates. Birds even consumed petals, detach androecium and damage the reproductive parts. The birds preferred flowers of cucumbers to that of snake gourd, ridgc gourd and bitter gourd that are less tasty with high cucurbitacins (Srihari and Chakravarthy, 1998).

IPM

- Soil application of neem cake at 250 kg/ha after germination and repeat once at flowering.
- Removal of cotyledon leaves infected with leaf miner one week after germination, followed by spraying of neem seed powder extract at 4% or neem soap at 1% reduces the incidence of leaf miner substantially.
- Spray neem seed powder extract 4% or neem soap or pongamia soap at 1% after flowering at 10 days interval.
- Erect pheromone traps cue lure at 8 traps/ha and change after 60 days.

Pesticide Residues

The persistence of terminal pesticide residues and waiting periods for gourds and melons are presented in Table 8.1. (Awasthi, 2001).

8.1.2. Diseases

8.1.2.1. Downy Mildew, *Pseudoperonospora cubensis*

(i) Symptoms: This is the most important disease in successful cultivation of cucurbitaceous crops. Characteristic symptoms are angular yellow spots often restricted by veins appearing on upper surface, giving purplish growth on the lower surface. In moist weather, such leaves die quickly. Young fruits and tips of vines are affected and fail to grow (Fig. 8.3).

Table 8.1. Terminal pesticide residues and their waiting periods for gourds and melons

Pesticide schedule	Dosage/ Concentration (%)	Terminal residues (ppm/days)	Waiting period (days)
Quinalphos spray at fruiting stage	0.05	0.55/5	7
Monocrotophos spray at fruiting stage	0.05	0.62/7	10
Dimethoate spray at fruiting stage	0.05	0.33/10	6
Cypermethrin spray at fruiting stage	0.01	0.17/7	1
Fenvalerate spray at fruiting stage	0.01	0.13/10	2
Metalaxyl spray at fruiting stage	0.15	0.21/10	11
Carbofuran soil application at planting	1.5 kg *a.i.*/ha	ND*/60	60

* ND—Not detectable

Figure 8.3 Downy mildew on cucurbits

(ii) Management

Removal of badly infected leaves, followed by spraying fungicides prevents spread of the disease. Spray 0.3% copper oxychloride or 0.2% mancozeb or propineb or chlorothalanil or dithionon.

If it is a seed crop, the spray of 0.2% metalaxyl mancozeb or cymoxanil mancozeb or AL-Fosetyl at fortnightly interval gives very good control.

8.1.2.2. Anthracnose, *Colletotrichum lagenarium*

(i) Symptoms: This disease is serious on fruits, young stems and vines. Small, yellowish or water-soaked areas that enlarge rapidly and turn brownish appear on the foliage of most cucurbits. Growing leaves may be destroyed by the disease and coalescing spots may cause death of the entire leaf. On fruits, the spots are roughly circular, sunken, and water-soaked with dark borders. In the centre of these spots, point-wise blank stromata bearing pink spore masses are produced in humid weather.

(ii) Management

Spray 0.25% mancozeb or chlorothalanil or dethionon or 0.1% carbendazim at 15 days interval, especially during rainy season.

8.1.2.3. Powdery Mildew, *Erysiphe cichoracearum, Sphaerotheca fuligenea*

(i) Symptoms: This disease is widely occurring on musk melon, bottle gourd, cucumber, pointed gourd, water melon and pumpkin. The disease becomes rampant during dry and humid weather. The disease appears as mealy-white to dirty grey spots on the foliage (Fig 8.4) and green stems. Later, these become powdery and ultimately cover the entire green surface of the plant, causing premature withering, drying and defoliation. Under severe infection, the fruits may also be covered with white powdery mass. Young fruits turn pale brown and drop off. The quality of fruit is adversely affected.

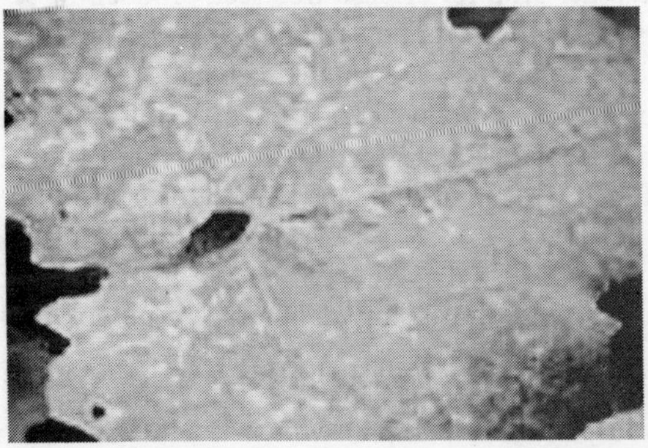

Figure 8.4. Powdery mildew infection on pumpkin leaf

(ii) Management

Destruction of infected plant material and weeds harbouring the pathogen reduces the disease incidence. For effective control of the disease, 0.025% penconazole, 0.1% tridemorph, 0.15% benomyl, 0.05% karathane, 0.5% elosal, 0.2% wettable sulphur, 0.1% carbendazim have been recommended.

Muskmelon *cvs.* Arka Rajhans, Florigold, Gulfstream, Jacumba, Edisto and Cantaloupe No. 45 are reported to be resistant to powdery mildew disease.

Integrated Pest and Disease Management

- Summer ploughing to expose resting stages of insects and nematodes followed by irrigation in first week of May and then again subsequent ploughing after green manuring in the month of June and July.
- Soil application of FYM enriched with *Trichoderma harzianum* at 3–5 MT/ha within a week of ploughing of sunhemp.
- Always prefer to grow all the cucurbits on bower system particularly in kharif/rainy season.
- Seed treatment with 0.25% carbendazim or *T. harzianum/T. viride* at 4 g/kg of seed.
- Restrict the use of exotic hybrids in view of their susceptibility to gummy stem blight and viral diseases.
- Soil application of neem cake enriched with *T. harzianum* to minimize gummosis, nematodes, root rot and collar rot.
- Soil application of neem cake at 250 kg/ha at 15 DAS followed by 3 sprays of either 4% NSPE or 1% pongamia soap for effective management of leaf miner and fruitfly. No incidence of leaf eating caterpillar was observed in such fields.
- Rouging of virus infected plants from time to time.
- Give 2 foliar sprays of 0.25% mancozeb at initial infection of downy mildew.
- Spray 0.1% carbendazim or 0.25% chlorothalanil for controlling foliar diseases such as anthracnose and *Cercospora* leaf spot.
- Always use summer crop for disease-free seed production.
- Give enough spacing to prevent overlapping of the crop.
- Collection and destruction of red pumpkin beetles.
- Stirring the soil causes mortality of pupal stages of fruitfly, *Dacus cucurbitae*.
- Removal and destruction of damaged fruits from time to time.
- Installation of methyl euginol sex traps to monitor fruitfly.

8.1.3. Nematodes

The root-knot nematode is the most important pest on pumpkin.

8.1.3.1. Root-knot Nematodes, *Meloidogyne* spp.

The very broad leaved plant like pumpkin which is infected with root-knot nematodes may show day time wilting and develop much larger galls. In cucurbits, the roots react to the presence of *Meloidogyne* spp. by the formation of large, fleshy galls.

(i) Management

(a) Physical Methods: Pulverizing the planting pit soil and exposing to sunlight by repeated raking and heaping 10 to 15 cm thick dry trash and burning it before application of compost helps in reducing the nematode population. Frequent cultivation may be given in May-June to expose the soil to sun.

(b) Cultural Methods: Application of 250 g of fresh neem or karanj cake at the time of sowing reduces the nematode population in soil. Crop rotation with rice, oats, wheat and taramira is also effective.

(c) Chemical Methods: Seed treatment with carbosulfan (Marshall 25 ST) at 6% w/w is effective against root-knot nematodes. Application of carbofuran at 0.2g *a.i.*/pit at sowing is also effective against root-knot and reniform nematodes.

(d) Host Resistance: Pumpkin *cvs.* Jaipuri and Dasna were found resistant to root-knot nematodes.

(e) Integrated Methods: Addition of 250 g fresh neem cake or karanj cake enriched with *Trichoderma harzianum/Paecilomyces lilacinus* per planting pit effectively controlled the nematodes.

8.1.3.2. Root-knot Nematode, *Meloidogyne javanica* and Root Rot/Wilt, *Macrophomina phaseolina, Fusarium oxysporum, F. solani* Disease Complex

(i) Management

(a) Integrated Methods: *Pseudomonas aeruginosa* and *Paecilomyces lilacinus* when used together significantly reduced infection of the disease complex on pumpkin (Perveen *et al.*, 1998).

8.2. CUCUMBER

India is considered to be the home of cucumber (*Cucumis sativus*). Cucumber is cultivated for fresh consumption or as pickling cucumber for preservation, marinated with vinegar, salt, dill or other spices. It is an important salad crop cultivated both in north and south and lower as well as higher hills in India.

8.2.1. Pests

8.2.1.1. Fruit Fly, *Dacus cucurbitae*

(i) Damage: Fruitfly infestation causes oozing of resinous fluid from fruits which become distorted and malformed. Maggots feed on fruit pulp causing premature dropping of fruits.

(ii) Management

Application of neem cake or sprays of 4% NSKE were very effective in controlling fruit fly in cucumber (Table 8.2). Soil application of neem cake reduced the incidence of fruit fly to 6%, whereas insecticide sprayed plots recorded 15% incidence.

Table 8.2. Efficacy of neem cake and NSKE on fruit fly in cucumber

Treatment	Dose (%)	Incidence of fruit fly (%)
NSKE	4.00	11.08
Carbaryl	0.15	15.17
Metasystox	0.05	22.77
Monocrotophos	0.05	20.84
Phosphomidon	0.05	22.36
Neem cake	100 g/pit	6.26
Pongamia cake	100 g/pit	21.45
Control	—	48.96

8.2.1.2. Serpentine Leaf Miner, *Liriomyza trifolii*

(i) Damage: Mines having serpent shape with small larvae are noticed all over the leaves. The incidence starts after about 15 days of germination of seeds. Leaf miners make small tunnels in the leaves of plants. Other symptoms include discoloration of the leaves. Left alone, it gets suppressed by natural enemies.

(ii) Management

NSKE 4% was effective in reducing the serpentine leaf miner, *Liriomyza trifollii* damage and increasing the fruit yield of cucumber (9.1 kg/12 sq. m.) compared to control (2.4 kg/ 12 sq. m.) (Krishna Kumar, 1998).

8.2.1.3. Vertebrate Pests

(i) Damage: Small green barbet caused a loss of about 38% fruit in cucumber, a local delicacy when grown as a sole crop on wooden platforms called pandals in South Canara district, Karnataka.

(ii) Management

The damage was reduced to 8.5% when cucumber was grown on raised basins (2 m diameter, 0.3 m height) mixed with pumpkin, bottle gourd and ridge gourd (Srihari and Chakravarthy, 1998).

8.2.2. Diseases

8.2.2.1. Powdery Mildew, *Sphaerotheca fuliginea, Erysiphe cichoracearum*

(i) Symptoms: This is a fungal disease that is spread by spores carried by the wind. White to gray fungal deposits are observed on the leaves and stems. As the mildew spreads, the leaves may begin to drop off. White powdery coating is produced on the leaves. Affected plants turn yellow and wither.

(ii) Management

Make sure to keep the field free of debris and destroy any infected plant parts. Crop rotation with paddy in low lying areas is very effective against powdery mildew. Application of *Ampelomyces quisqualis* to the foliage, starting shortly after colonies of *S. fuliginea* began to appear, efficiently suppressed powdery mildew and increased yields of marketable cucumbers. Water spray of *A. quisqualis* gave acceptable control of powdery mildew on cucumber in the field if applied with 2% paraffin oil.

Both *A. quisqualis* and *Sporothrix flocculosa* are effective in controlling powdery mildew on cucumber under high humidity conditions, especially in greenhouses. The pycnidial hyperparasite, *A. quisqualis* has sticky conidia that are splash dispersed, as are the spores of yeast like antagonist, *S. flocculosa*. Biocontrol of powdery mildew therefore, depends on maintaining high humidity and occasional water sprays to ensure splash dispersal of antagonist spores. *Tilletiopsis washingtonensis* and *T. pallescens* when applied at 1×10^8 spores/ml, could significantly reduce powdery mildew on cucumber in greenhouse.

8.2.2.2. Downy Mildew, *Pseudoperonospora cubensis*

(i) Symptoms: This devastating disease is caused by the fungus *Pseudoperonospora cubensis*. The disease is most important in the northeast on muskmelon, pumpkin, and winter squash, though all cucurbits are susceptible. Because the fungus does not overwinter in the north, the disease is reintroduced each season by airborne conidia blown into the area from southern states. If appropriate conditions are present, plantings can be infected so rapidly that they appear to have been frosted. Foliar symptoms vary depending on how quickly infection occurs. Upper surfaces of leaves may show angular pale-green areas bounded by leaf veins that give the impression of mosaic. These areas change to yellow angular spots and become necrotic as the disease progresses. When leaves are infected by a heavy spore shower, small individual necrotic and chlorotic flecks appear on the upper surface. In a moist environment, sporulation occurs on the lower leaf surface. The chlorotic flecks appear as light brown areas when held up to the light. Areas with active sporulation give the undersurface a purplish or dark brown sooty appearance. The fungal growth consists of large lemon-shaped spores called sporangia that are borne on branched structures called sporangiophores. The sporangia are wind-borne and can survive over long distances when the air is moist. They are also readily spread to adjoining plants by splashing rain.

Although most downy mildew fungi are considered cool, wet weather diseases, cucurbit mildew can infect plants over a wide temperature range (50–80°F) with the optimum at 61–72°F. Periods of heavy dew lasting into midmorning are sufficient for infection to occur. Mid-August is a critical time to scout for mildew because appropriate environmental conditions usually exist then in the Hudson Valley and on Long Island. Practice cleanliness in the garden by removing garden debris. Try and keep the foliage as dry and possible.

(ii) Management

Because disease development is dependent upon outside inoculum, rotation cannot prevent the disease. Select resistant varieties if they are available. Downy mildew can be controlled with the application of protectant and systemic fungicides.

8.2.2.3. Angular Leaf Spot, *Pseudomonas syringae* pv. *lachrymans*

(i) Symptoms: This bacterial disease can occur on most cucurbits, including cucumber, muskmelon, pumpkin, and winter squash. The disease is less common on cucumber because resistant varieties are widely grown. Leaf spots are variable in size and may be angular in shape because leaf veins limit enlargement of spots. Initial symptoms appear as brown water-soaked spots, as illustrated on butternut squash (Fig 8.5). Under moist conditions, especially in early morning, small droplets of bacterial ooze can come from the water-soaked areas; which, when dry, appear as a white residue. The initial brown spots may be surrounded by a yellow halo, become white, and, as they dry, tear away from the healthy tissue, producing irregular holes in the affected leaves. Fruit spots are smaller than leaf spots and are usually circular. These spots, as shown on the processing squash Delicious, are usually superficial but can develop into deeper rots if secondary soft-rot bacteria invade.

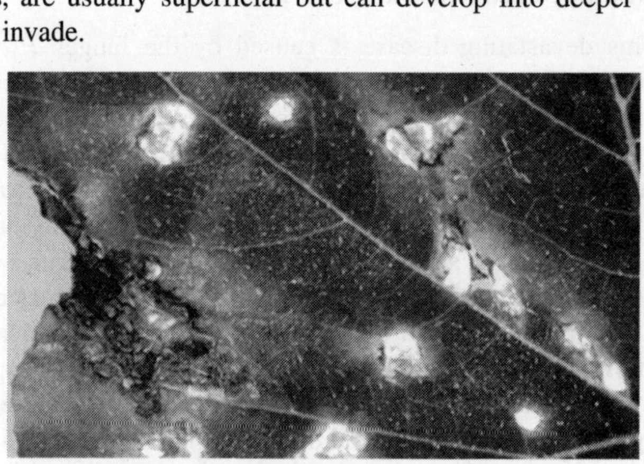

Figure 8.5. Angular leaf spot on cucumber

The bacteria can survive for possibly 2 years in soil or debris from diseased plants. It is also known to be associated with seed. Bacteria can enter through wounds or stomata. Moisture in the form of rainfall or relative humidity greater than 95 percent for several days is required for infection and later disease development. Dry weather for 2 weeks can arrest the disease.

(ii) Management

Controls include the selection of resistant cucumber varieties, use of disease-free seed, and a 2-year rotation out of all cucurbits. Check with your seed supplier for the current list of resistant cucumbers. Copper fungicides can be applied to slow disease spread during particularly wet periods but can be dropped if dry weather continues for 2 weeks. The best way to avoid this disease is to keep the foliage dry. Drip irrigation and soaker hoses work great to accomplish this.

8.2.2.4. Cucumber Mosaic Virus (CMV)

(i) Symptoms: This disease infects almost all the cucurbits. The young leaves develop greenish-yellow areas and they are more translucent than those in the remaining parts of the leaf. Yellow mottle is seen on the leaves (Fig 8.6). This is followed by leaf distortion and stunting of plants. Infected fruits are mottled with yellowish-green colour and blistered. Mosaic virus causes deformation of the cucumber plants, with small leaves and underdeveloped plants. Discoloration and spotting on the leaves are observed. This disease is spread by aphids.

(ii) Transmission: CMV is transmitted by sap, seed and several species of aphids, especially *Aphis gossypii* and *Myzus persicae* so make sure to control these pests (Kennedy *et al.*, 1982). Remove the infected plant and destroy it. Do not compost plants infected with the virus. The fruits from infected plants should also be destroyed.

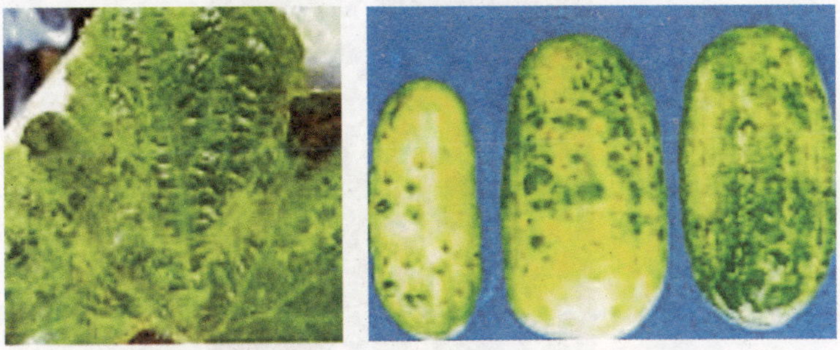

Figure 8.6. Cucumber mosaic virus on leaves and fruits

(iii) Management

- Select seeds from absolutely virus-free plants.
- Remove weed hosts from the field.
- Rouging of infected plants from time to time.
- Use barrier crops like sunflower, sorghum and pearl millet all round the cucumber crop to prevent the vectors entering the field.
- Mulch the soil with saw dust or wood chips or with aluminium foil which repels the aphid vectors and reduce the CMV infection up to 87.9%.
- Spray with either dimethoate or monocrotophos each at 0.1% at weekly interval to control aphid vectors. Spraying of mineral oils at 0.75 to 1.0% at weekly intervals has been found promising.

IDM

- Seed treatment with *P. fluorescens* at 5 g/kg of seeds.
- Soil application of enriched FYM with *T. harzianum* at 1 MT/ha.
- Soil application of fresh neem cake at 250 kg/ha immediately after germination and repeated at the time of flowering.
- Spraying of *P. fluorescens* 2 times (20 and 45 DAP) to induce disease resistance.

8.2.3. Nematodes

The root-knot nematode is the most important pest on cucumber that is of some economic importance.

8.2.3.1. Root-knot Nematodes, *Meloidogyne* spp.

(i) Symptoms: The root-knot nematodes cause severe galling on cucumber roots (Fig 8.7).

(ii) Management

(a) Physical Methods: The most important nematode pests controlled by greenhouse steaming are root-knot nematodes attacking cucumbers.

(b) Chemical Methods: Drill application of carbofuran at 1.5 kg *a.i.*/ha below the seed level reduces the incidence of *M. incognita* in cucurbitaceous crops.

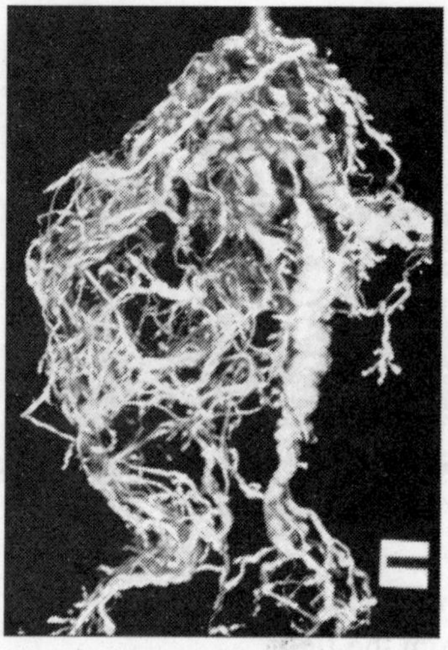

Figure 8.7. Cucumber roots severely galled due to root-knot nematode infection

(c) Integrated Methods: Combined inoculation of AMF and *Pseudomonas fluorescens* had positive effect on root-knot nematode control on cucumber (Jakobsen, 1999).

8.2.3.2. Dagger Nematode, *Xiphinema americanum* and Tobacco Ring Spot Virus Disease Complex

(i) Management

(a) Chemical Methods: Benomyl application inhibited feeding of *X. americanum* on cucumber, which was reflected in a marked reduction of its transmission ability of tobacco ring spot virus (McGuire and Good, 1970).

8.3. POINTED GOURD

Pointed gourd (*Trichosanthes diocea*) or parwal is widely cultivated in Bihar, Uttar Pradesh, West Bengal, Orissa, Assam, Madhya Pradesh and Gujarat. Recently, it has been introduced in and around Hyderabad and Bangalore.

8.3.1. Nematodes

8.3.1.1. Root-knot Nematode, *Meloidogyne incognita*

(i) Symptoms: The root-knot nematode is the major limiting factor in successful production of pointed gourd. Verma (2001) reported that *M. incognita* was responsible for 30–40% loss in yield of pointed gourd.

(ii) Management

(a) Cultural Methods: Selection of healthy looking roots for planting is recommended. Planting of marigold, *Tagetes patula* at 3 to 5 plants/pit will manage attack of root-knot nematodes on pointed gourd.

(b) Chemical Methods: Rooted cuttings should be given a 6 hr dip in 0.1% carbosulfan 25 EC or triazophos 40 EC before planting. In the ratoon crop, carbofuran/phorate should be applied at 0.2g *a.i.*/plant.

(c) Integrated Methods: Integration of *Paecilomyces lilacinus* (50 g) + *Trichoderma harzianum* (100 g) + neem cake (250 g) + marigold (3 plants/pit) increased plant growth parameters and yield (6.6 kg/plant compared to 1.4 kg/plant in control), and reduced root galling (14.3/5 g roots compared to 75.0/g roots in control), number of egg masses (40/5 g roots compared to 150/g roots in control) and final nematode population (66.6/100 g soil compared to 86660/100 g soil in control) (Verma *et al.*, 2005).

8.4. BOTTLE GOURD

Bottle gourd (*Lagenaria siceraria*) is grown for immature fruits used for culinary purposes. Tender fruits are rich in vitamin B. It is also used for preparation of different types of sweets. Fruit pulp is very good source of fibre-free carbohydrates and fruit pericarp for crude fibre. The oil extracted from kernels of seed, is a fine cooking medium. In India, it is cultivated in Uttar Pradesh, Punjab, Gujarat, Assam, Meghalaya and Rajasthan.

8.4.1. Nematodes

8.4.1.1. Root-knot Nematodes, *Meloidogyne* spp.

The root-knot nematode is an important limiting factor in successful cultivation of bottle gourd.

(i) Management

(a) Cultural Methods: Application of press mud at 10 t/ha (ICBR 1: 18.95) or mustard cake at 3.333 t/ha (ICBR 1: 16.10) or neem cake at 4.286 t/ha one week in advance of sowing gave effective management of root-knot with increased yield of bottle gourd (Dahiya *et al.*, 1998).

(b) Chemical Methods: Seed treatment with aldicarb and carbofuran at 3 per cent was effective against root-knot nematodes infecting bottle gourd (Kandasamy and Sivakumar, 1981). Carbosulfan and benfuracarb at 3% w/w as seed treatment increased yield and reduced root galling in bottle gourd.

8.5. RIDGE GOURD

Ridge gourd (*Luffa acutangula*) or ribbed gourd is a monoecious viny vegetable. It has some medicinal value and has all the nutrients. The fibre from well grown fruits has commercial value, as it can be used for making brushes, for cleaning. It is mainly cultivated in Andhra Pradesh, Tamil Nadu, Karnataka, Gujarat, Assam, West Bengal and Konkan region of Maharashtra.

8.5.1. Nematodes

8.5.1.1. Root-knot Nematodes, *Meloidogyne* spp.

The root-knot nematode is the major limiting factor in successful production of ridge gourd.

(i) Management

(a) Host Resistance: Ridge gourd *cvs*. Panipati and Meerut Special were found resistant to root-knot nematodes.

8.6. BITTER GOURD

8.6.1. Pests

8.6.1.1. Fruit Fly, *Batocera cucurbitae*

(i) Damage: Fruitfly infestation causes oozing of resinous fluid from fruits which become distorted and malformed. Maggots feed on fruit pulp (Fig 8.8) causing premature dropping of fruits.

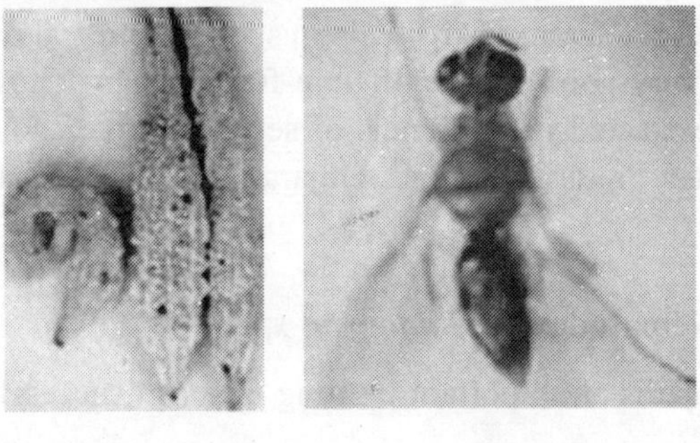

Figure 8.8. Fruitfly damage on bitter gourd fruits

(ii) Management

The incidence of fruit fly was reduced to less than 9% by sprays of neem/pongamia soap as compared to 32% incidence in unsprayed control. Sprays of 4% pulverized neem seed powder recorded the lowest (0 to 3.60%) fruit fly incidence (Table 8.3). Soil application of neem cake followed by foliar sprays of neem seed powder extract is effective for managing fruit fly in bitter gourd (Krishna Murthy and Krishna Kumar, 2002).

Table 8.3. Management of fruit fly in bitter gourd

Treatment	Dose (%)	Incidence of fruit fly (%)
Neem seed powder extract	7.5%	0.00
Carbaryl	0.15%	21.31
Methyl demeton	0.05%	24.09
Monocrotophos	0.05%	26.33
Phosphamidon	0.05%	39.97
Neem cake	100 g/pit	8.33
Pongamia cake	100 g/pit	18.66
Control	—	49.51
CD (P = 0.05)		**14.98**

Bitter gourd *cvs.* Hissar II and Ghoti possessed field resistance to fruitfly attack.

8.6.2. Diseases

8.6.2.1. Bitter Gourd Mosaic Virus

(i) Symptoms: The symptoms on the naturally infected bitter gourd plants are confined mostly to leaves. The leaves in the secondary branches produced at the apical end of the plant show the symptoms of the disease. Scattered, small, irregular yellowish patches are seen on the leaves. Some leaves show vein clearing in one or two lobes of the leaf. Severely infected plants show reduction in their size and elongation and/or suppression of one or two lobes. Young developing leaves are completely distorted, filiform and malformed with considerable reduction in their size. Some of the leaves show marked reduction in the development of lamina resulting in a shoe string effect (Nagarajan and Ramakrishnan, 1971).

(ii) Transmission: The virus is transmitted by 5 species of aphids, *viz., Aphis gossypii, A. malvae, A. nerii, Brevicoryne brassicae* and *Myzus persicae.* The virus can infect *Cucurbita moschata, C. maxima, C. pepo, C. melo, Citrullus vulgaris, Lagenaria vulgaris* and *Trichosanthes anguina.*

(iii) Management

- Spray the crop just after germination with nuvacron or dimecron both at 0.05% at 10 day intervals to prevent aphid vectors.
- Spraying of 1.5% mineral oil minimizes the virus spread.

8.6.2.2. Bittergourd Witches' Broom

The disease was first reported by Singh (1985) from Bangalore, Karnataka. When the plants get infection in the early stages, they do not bear any fruits and the loss is 100%. However, in late infections, the yield loss is about 30–50%. The disease also attacks other cucurbits like *Lagenaria vulgaris, Trichosanthes anguina* and *Luffa acutangula.*

(i) Symptoms: The characteristic symptoms are malformation and proliferation of auxiliary buds. Diseased plants show many abnormally small leaves which fail to attain full size. The diseased plant bears many flowers and blossom earlier than healthy plants. Internode bud sprouts give rise to many bud-like chlorotic leaves. Flowers on infected plants show characteristic green phyllody symptoms. A variety of abnormalities ranging from near abnormal flowers showing green patches on the yellow petals or greenish yellow flower to a completely vegetative green leafy structure can be observed. Fruits from these flowers are very small, thin, cylindrical, deformed and completely seedless. Plants showing severe witches' broom symptoms do not bear fruits.

(ii) Transmission: The disease is easily transmitted by grafting. Leaf hoppers have been suspected as vectors of this disease.

(iii) Management

- Application of carbofuran at 1.5 kg *a.i.*/ha at the time of sowing seeds followed by 5–6 sprays of dimecron or nuvacron or metasystox all at 0.05% at 10 days interval keeps the vector population under check.

- Spraying of oxytetracycline hydrochloride at 500 ppm at weekly intervals is effective in suppressing the symptoms of the disease.

8.7. WATERMELON

Watermelon (*Citrullus lanatus*) is an important cucurbitaceous summer vegetable grown for ripe fruits. Though it can be grown in garden land, it is a major river-bed crop of Uttar Pradesh, Rajasthan, Gujarat, Maharashtra and Andhra Pradesh. An excellent desert fruit, it is relished by rich as well as poor. The fruit juice makes an excellent refreshing and cooling beverage after adding a pinch of salt and black pepper. The fruits contain 92% water, 0.2% protein, 0.3% minerals and 7.0% carbohydrates in a 100 g edible flesh.

8.7.1. Pests

8.7.1.1. Thrips, *Thrips palmi*

(i) Damage: The nymphs and adults are found moving on young shoot tips. If the tip of the plant is tapped, many young ones and adults can be collected on a white sheet of paper. Thrips are very serious during summer months and are suspected to transmit bud necrosis disease in watermelon.

(ii) Management

Spraying of systemic insecticides like 0.05% acephate reduces the pest incidence. Sprays of 1% neem soap/1% pongamia soap were found to reduce thrips damage in watermelon. Sprays of 4% NSKE/4% pulverized NSPE were also found very effective. Soil application of neem cake once immediately after germination and again at flowering also reduces the pest incidence.

8.7.2. Diseases

8.7.2.1. Anthracnose, *Colletotrichum* sp.

(i) Symptoms: Sunken spots with raised margins are produced on leaves and fruits (Fig 8.9).

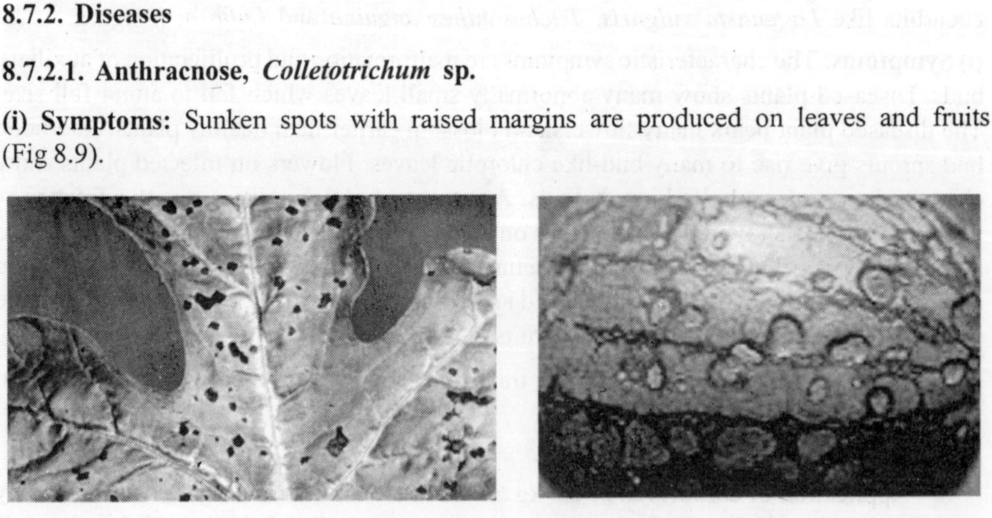

Figure 8.9. Anthracnose on water melon leaf and fruit

(ii) Management

Spraying carbendazim at 2 g/litre of water gives effective management of the disease.

8.7.2.2. Powdery Mildew, *Erysiphe cichoracearum*

(i) Symptoms: White powdery coating is produced on the leaves (Fig 8.10). Affected plants turn yellow and wither.

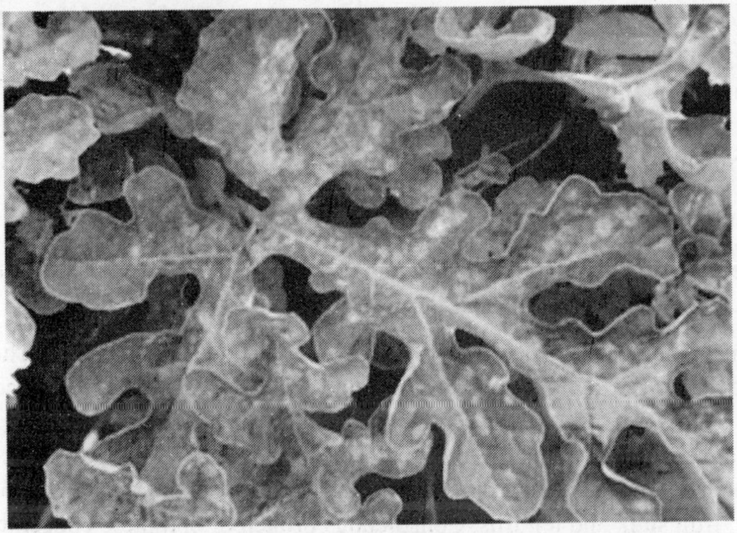

Figure 8.10. Powdery mildew on water melon leaf

(ii) Management

Spraying carbendazim at 2 g/litre of water is effective for the control of the disease.

8.7.2.3. Watermelon Mosaic virus

The disease was first reported from the USA in 1932. In India, the virus was first reported by Bhargava and Joshi (1960).

(i) Symptoms: Diseased vines are conspicuous because the tips of the runners and a proliferation of shoots from around the crown protrude stiffly above the general level of the vines, giving a 'petunia-like' appearance. The tips and shoots have shorter than normal internodes which produces crowding of the young leaves stunted and rolled. The characteristic symptoms are mosaic, mottling and blistering of leaves and stunted growth of the infected plants. Flowers on severely infected plants show abnormalities of the petals which are of unequal size and greenish in color. Stamens are distorted and unevenly developed and the flowers may fall of prematurely. Melons produced on diseased vines are mottled and sometimes deformed. Diseased vines bear few fruits of small size.

(ii) Transmission: The virus is easily transmitted by a mechanical means. Insect vectors include several species of aphids *viz*., *Aphis gossypi, A. craccivora, A. cirthosiphose, A. spiraecola, Myzus persicae, Brachycaudatus relichtysi* and *Cavariella aegopodii.*

(iii) Management

- Rogue out infected plants from the field as soon as they are noticed.
- Control weed hosts mainly *Melothria pendula*, which act as a source of inoculum.
- Spray the crop 4 to 5 times with either dimethoate or monocrotophos both at 0.1% at 10 days interval to kill aphid vectors.
- Spread of the virus is significantly reduced by spray of miner oils.
- Grow 5 to 6 rows of barrier crops such as sorghum, maize or pearl millet around the watermelon crop 50 to 60 days before sowing to prevent the entry of vectors.

8.7.2.4. Watermelon Bud Necrosis

This disease was first observed at Hessaraghatta, Bangalore, Karnataka, during 1990 on watermelon.

(i) Symptoms: The initial symptom of the disease is necrosis of the buds. The buds turn brown to black and die. This is followed by longitudinal streaks on the vine commencing from the tip downwards. These streaks may vary in size from 10 to 50 mm in length. The streaks first appear brown and later turn to black colour. The leaf symptoms include brown or black spots covering the entire leaf lamina and leaf petiole. Streaks are also noticed on leaf petioles, fruit stalk and tendrils. Conspicuous broken or complete concentric rings are produced on fruits. The rings vary in size from 5 to 30 mm in diameter. The margins of these rings are corky in texture and raised. The disease causes severe losses in yield of fruits.

(ii) Transmission: The virus is transmitted by mechanical sap inoculation. Thrips transmit this virus under natural conditions.

(iii) Management

The control measures used are same as described under tomato spotted wilt.

8.7.3. Nematodes

The root-knot nematode is the major limiting factor in successful production of watermelon.

8.7.3.1. Root-knot Nematodes, *Meloidogyne* spp.

Hasan and Jain (1998) reported that *M. incognita* was responsible for 18–33% loss in yield of watermelon.

(i) Management

(a) Host Resistance: Watermelon *cvs.* Shehjanpuri was found resistant to *M. incognita*, while Dixie Queen was found resistant to *M. javanica.*

8.8. MUSK MELON

8.8.1. Pests

8.8.1.1. Red Pumpkin Beetle, *Raphidopalpa fovecollis;* Fruit Fly, *Bactrocera cucurbitae*

(i) Damage: Adult beetles damages the cotyledonary leaves. Sometimes severe damage results in death of plants. Grubs feed on roots. Damage by beetle to grown up leaves does not cause economic damage.

(ii) Management

Soil application of carbofuran at 200 to 300 g *a.i.*/ha at sowing, followed by 4 sprays with 250 to 500 g *a.i.*/ha of carbaryl at 10 days interval (beginning 10 days after sowing) gave effective control of both red pumpkin beetle and fruit fly even under high insect pressure. Spraying of endosulfan at 15 ml or carbaryl at 2 g/litre of water at cotyledonary leaf stage gives effective management of the pest.

Musk melon *cv.* Kharda is reported to be resistant, while Lakhazda is resistant to fruit fly.

8.8.2. Diseases

8.8.2.1. Alternaria Leaf Blight, *Alternaria cucumerina*

(i) Symptoms: Alternaria leaf blight caused by the fungus *Alternaria cucumerina* is a common disease of muskmelon (cantaloupe) but can also infect watermelon, squash, and cucumber. Leaf blight usually occurs from midseason or when the canopy closes, can reduce late season fruit production, and can result in poor quality if the disease is severe. Symptoms first appear on the upper surface of crown leaves as small spots 1 to 2 mm in diameter although other lesions can be more than 10 mm in diameter (Fig 8.11). These larger spots can show a target-like pattern of rings which is typical of most *Alternaria* spp.

Figure 8.11. Alternaria leaf blight on musk melon

Leaf blight overwinters as mycelium or chlamydospores in diseased plant debris and can probably survive for more than a year. Conidia are produced in the spring and act as the primary inoculum. Conidia produced on infected plants provide inoculum for repeating secondary cycles during the season.

(ii) Management

A 2-year rotation out of cucurbits is the easiest way to break this cycle. There are few Alternaria–resistant varieties. Protectant fungicides should be applied in mid–July when vines run and fruit have been set. Infection and disease development are favoured by lengthy periods of high relative humidity (18 hr) over a broad range of temperatures (68 to 90°F).

8.8.2.2. Wilt, *Fusarium solani*

(i) Management

T. harzianum and *T. viride* when applied as seed treatment (5 g/kg) followed by soil application along with FYM/neem cake at 1 kg/basin were effective. Soil application of neem cake/mustard cake at 2% reduced the wilt by 80% and 65%, respectively.

8.9. GHERKEN

8.9.1. Pests

8.9.1.1. Fruit Fly

(i) Damage: Fruitfly infestation causes oozing of resinous fluid from fruits which become distorted and malformed. Maggots feed on fruit pulp causing premature dropping of fruits.

(ii) Management

The incidence of fruit fly was reduced to less than 9% by sprays of both neem and pongamia soaps as compared to 32% incidence in unsprayed control (Table 8.4). Spraying of monocrotophos and phosphamidon recorded more than 10% fruit fly incidence (Krishna Murthy and Krishna Kumar, 2002). The soap sprays were also effective on leaf miner.

Table 8.4. Management of fruit fly in Gherkins (2001)

Treatment	Dose	Fruitfly incidence
NSPE	7.5%	2.25
Carbaryl	0.15%	10.77
Metasystox	0.05%	12.28
Monocrotophos	0.05%	11.95
Phosphamidon	0.05%	28.68
Neem cake	100 g/pit	5.31
Pongamia cake	100 g/pit	9.50
Neem soap	1%	9.44
Pongamia soap	1%	9.40
Control	—	32.30
CD (P = 0.05)		**4.07**

Table 8.5. Management of insect pests in cucurbitaceous vegetable crops

Crop	Pest	Symptoms	Pesticide/Bioagent/Dose
Cucurbits	Red pumpkin beetle	Leaves damaged and with holes	Sevin 50 EC-4 gm/l immediately after germination
	Fruit fly	Rotting of fruits	Folithion 1000 EC/fenthion-1 ml/l spray with 5% jaggery at fruit formation
	Serpentine leaf miner	Mines on the leaves	NSKE 4%-40 g/l; Triazophos-1.5 ml/l once in 20 days

Table 8.6. Cucurbitaceous vegetable crop varieties resistant to biotic stresses

Vegetable crop	Pest/Disease	Resistant varieties
Muskmelon	Powdery mildew	Arka Rajhans, Pusa Madhuras (MR), RM-43.
	Powdery mildew; Downey mildew	Punjab Rasila.
	Fusarium wilt	Pusa Madhuras, Durgapura Madhu, Arka Jeet, Punjab Sunehari (MR), Harela.
Watermelon	Powdery mildew; Downy mildew; Anthracnose	Arka Manik
Pumpkin	Fruitfly	Arka Suryamukhi.
Ridge gourd	Powdery mildew; Downy mildew	Swarna Uphaar.
Bottle gourd	Blossom end rot	Arka Bahar (T).
Cucumber	Powdery mildew	Swarna Poorna.
	Powdery mildew; Downy mildew	Poinsette
	Cucumber Green Mottle Virus	DURM-1

9 ROOT VEGETABLES CROPS

9.1. RADISH

Radish (*Raphanus sativus*) is a root vegetable which can be grown as an intercrop with other crops. Its leaves are rich in vitamins A and C. It can be eaten raw and also cooked. It has some medicinal value. Radish is cultivated throughout India. West Bengal, Bihar, Uttar Pradesh, Punjab, Assam, Haryana, Gujarat and Himachal Pradesh are major radish-growing states.

9.1.1. Pests

9.1.1.1. Mustard Sawfly, *Athalia lugens proxima*

(i) Damage: The activity of mustard sawfly is found from last week of July to end of March. The egg, larval, prepupal and pupal period last for 3, 10, 3 and 10 days, respectively. The pest hibernates in the pre-pupal stage in an earthen cocoon (Patel, 1967). Besides attacking radish (Fig 9.1), the pest also attacks mustard, cabbage and cauliflower.

Figure 9.1. Mustard sawfly damage on radish leaf

(ii) Management

Spraying 5% NSKE showed higher antifeedant and larvicidal effect. Repellin at 1%, margocide at 0.1% and neemark at 1% showed their higher oviposition deterrent, antifeedent and larvicidal action, respectively on sawfly. Synthetic insecticides like triazophos + decamethrin at 0.036% and quinalphos at 0.05% gave higher larval mortality (Patel, 1995).

9.1.1.2. Aphids, *Brevicoryne brassicae, Lipaphis erysimi, Myzus persicae*

(i) Damage: Colonies of aphids are seen clustered on tender stems (Fig 9.2) and ventral side of leaves. They suck the plant sap. The affected plants curl and show yellowing symptoms. The seed crop of radish suffers comparatively more as compared to crop grown for edible root.

Figure 9.2. Aphids on radish stem

(ii) Management

Three sprays of monocrotophos or endosulphan at 2 ml/litre of water directed on attainment of 50 aphids (*Lipaphis erysimi*)/plant provided the highest cost: benefit ratio (Bath and Singh, 1989).

Soil application of aldicarb, phorate or carbofuran all at 1.5 kg *a.i.*/ha as and when aphid infestation is noticed gave effective control of the pest.

9.1.1.3. Diamondback Moth (DBM), *Plutella xylostella*

(i) Management

Effective control of *P. xylostella* on radish was reported with 'Thuricide' 90 TS flowable at 2.25 l/ha (Narayanan *et al.*, 1970).

9.1.2. Diseases

9.1.2.1. White Rust, *Albugo candida*

(i) Symptoms: Prominent white pustules appear on the surface of leaves, stem and inflorescence. Marked swelling and distortion of the affected plant parts are common. Due to hypertrophy and hyperplasia of the tissues, the floral parts show swellings and distortions. Infected plants are stunted in growth.

(ii) Epidemiology: The pathogen survives in infected plant debris and on other hosts. Low temperature, high humidity and high moisture favour the disease development.

(iii) Management

Spraying with Dithane M-45 or Phytoalexin/Macimin all at 0.3% is effective.

9.1.2.2. Bacterial Root Rot, *Erwinia rhapontici*

The disease was first reported from Bangalore, Karnataka, during 1983 and its incidence ranged from 15 to 50% (Kishun, 1983b).

(i) Symptoms: The characteristic symptoms of the disease are rotting of pith tissues resulting in cavity formation and wilting of plants. The disease is more severe in seed crop grown from cut roots.

(ii) Epidemiology: The pathogen is soil-borne and the disease spreads through implements, irrigation water, insects, etc.

(iii) Management

Root dip treatment with 100 ppm of agrimycin-100 is found most effective in reducing the disease incidence (Kishun, 1983b).

9.1.2.3. Radish Mosaic Virus

The disease is caused by a strain of turnip mosaic virus. The disease was first reported by Thakur and Dube (1981) and later by Ahlawat and Chenulu (1984). The disease incidence ranges from 7 to 63% (Ahlawat and Chenulu, 1984). The virus causes losses ranging from 3.3 to 100.0% in root weight and up to 95.8% in seed weight depending on the time of infection.

(i) Symptoms: The affected plants are conspicuous with stunting, reduced leaf lamina and root size. Diseased leaves show interveinal chlorosis, vein banding and dark green patches on light green leaves. Curling and puckering of leaves are not uncommon. Mosaic pattern is also seen on the stem, petioles and pods. Seeds from diseased plants show poor germination. The virus has a wide host range.

(ii) Transmission: The virus is readily sap transmitted. The virus is not transmitted through seeds in radish. Several species of aphids, *Aphis gossypii, A. spiraecola, Brevicoryne brassicae, Lipaphis erysimi, Myzus persicae, Rhapalosiphum maidis, R. padi, Schizaphus graminum, Toxoptera citricidus* and *T. odinae* have been found to transmit the virus in a non-persistent manner (Ahlawat and Chenulu, 1982). *Lipaphis erysimi* and *Myzus persicae* are reported to be efficient vectors (Thakur and Dube, 1981).

(iii) Management

The disease spread can be minimized by soil application of carbofuran at 1.5 kg *a.i.*/ha at the time of sowing seeds, followed by 2–3 foliar sprays of dimethoate or monocrotophos both at 0.5% at 10 days interval.

9.1.3. Nematodes

The stunt nematode is recognized as the major limiting factor in successful production of radish crop.

9.1.3.1. Stunt Nematodes, *Tylenchorhynchus* spp.

(i) Management

(a) Cultural Methods: The mustard (rabi), radish (summer), sesame (kharif) rotation sequence considerably reduced the population of *Tylenchorhynchus* spp. (Haque and Gaur, 1985).

9.2. BEET ROOT

9.2.1. Diseases

9.2.1.1. Seedling Blight, *Rhizoctonia solani*

(i) Management

Spraying 0.2% mancozeb gives effective control of seedling blight disease. *Streptomyces distaticus* was effective in suppressing seedling blight of beet root.

9.2.1.2. Leaf Spot, *Cercospora* sp.

(i) Symptoms: The pathogen causes small circular spots in large number on older leaves. The centre of the spot turn brown which later turn grey with red margin.

(ii) Management

Spraying of 1% Bordeaux mixture as soon as the disease incidence is noticed gives effective control of the disease.

9.2.1.3. Storage Rot, *Sclerotium rolfsii*

(i) Management

T. harzianum was highly effective in suppressing *S. rolfsii* in beet root under field conditions. The conidial mixture of 2 bioagents *T. pseudokoningii* (effective in reducing the mycelial growth) and *T. virens* (good colonizer of sclerotia) was effective in reducing the incidence of storage rot.

9.2.1.4. Purple Leaf virus

The virus was first recorded at Pune during 1958 (Capoor and Sharma, 1965).

(i) Symptoms: The infected plants are stunted and leaves have a tendency to stand erect and come closer, unlike in healthy plants where the leaves are broad, long and profuse. The leaves of infected plants show an unusual intense purple colour, while the young emerging leaves show it prominently. Few leaves develop minute necrotic lesions all over the lamina.

The virus also infects *Cassia tora, Nicotiana glutinosa, Chenopodium amaranticolor, C. album* and *Gompharena globosa.*

(ii) Transmission: The virus, a strain of TMV, is readily transmitted by mechanical sap inoculation.

(iii) Management

Rouging of virus-infected plants and removal of weed hosts may help in minimizing the disease incidence.

9.2.2. Nematodes

9.2.2.1. Root-knot Nematode, *Meloidogyne* spp.

The root-knot nematode causes knotting on lateral roots. Knots have pigment similar to the tuber (Fig 9.3). The nematode is responsible for stunting of plants and chlorosis of leaves.

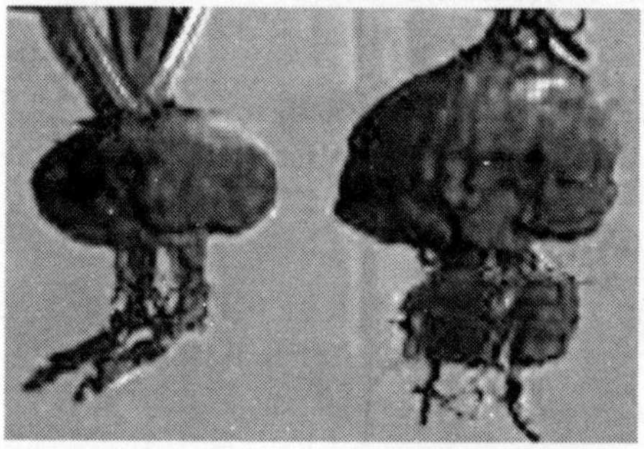

Figure 9.3. Roots of beet root infected with root-knot nematodes

9.3. CARROT

Carrot (*Daucus carota*) is another root vegetable rich in vitamin A and some minerals. It is grown all over India. Main carrot growing states are Uttar Pradesh, Assam, Karnataka, Andhra Pradesh, Punjab and Haryana. It is taken raw as well as in cooked form. It is made into pickles and sweetmeat. Carrot juice is a rich source of carotene (a precursor of vitamin A) and contains appreciable quantities of thiamine and riboflavin. It is sometimes used for colouring butter and other foods. Green carrot leaves are highly nutritive, rich in protein, minerals and vitamins and used as fodder and also for preparation of poultry feed.

9.3.1. Pests

9.3.1.1. Thrips, *Thrips tabaci*

(i) Damage: Adults and larvae suck the cell sap from the leaves of the plants (Fig 9.4). White blotches appear on leaves. Tips of leaves wither and turn brown.

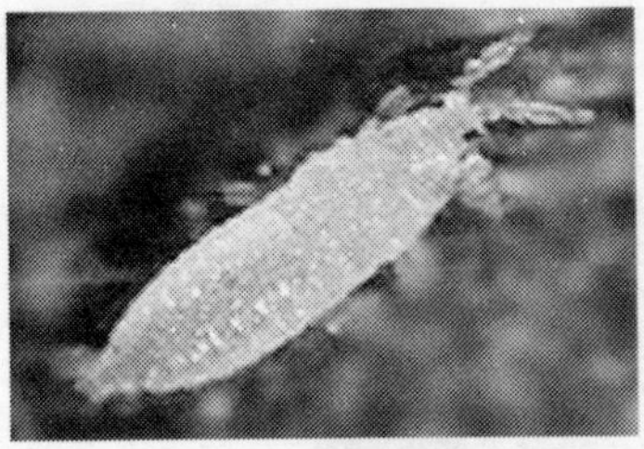

Figure 9.4. Nymph and adult of thrips on carrot

(ii) Management

Spraying monocrotophos or phosphamidon or malathion at 5 ml/10 litres of water gives effective control of the pest.

9.3.1.2. Rust fly, *Psila rosae*

(i) Damage: Larva mines into the root and causes holes, which rots and leads to secondary infection (Fig 9.5). Drooping and discoloration of foliage from green to rusty brown are the other symptoms.

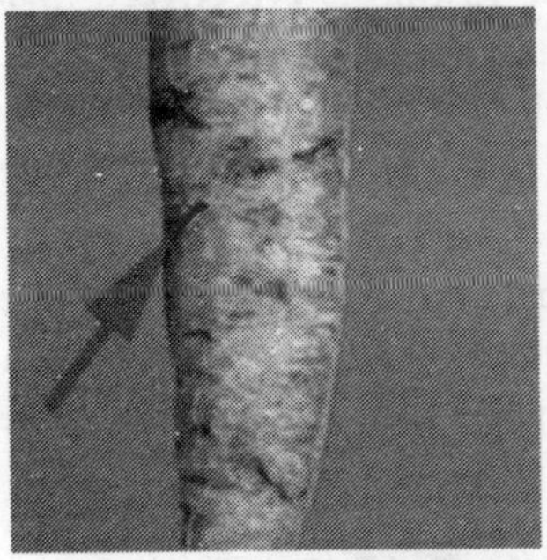

Figure 9.5. Rustfly damage on carrot and adult of rustfly

(ii) Management

Deep summer ploughing, rotation of crops, destruction of wild hosts are the cultural methods that can be used to reduce the pest population. Application of carbofuran at 1 kg *a.i.*/ha at a depth of 10–15 cm during land preparation is effective.

9.3.2. Diseases

9.3.2.1. *Cercospora* Leaf Blight, *Cercospora carotae*

(i) Symptoms: Lesions are formed on leaves, petioles, stems and floral parts. On leaves, brown to black spots appear and later may shrivel and break (Fig 9.6). Floral parts when attacked also shrivel. The pathogen survives on and in the seed and also in diseased plant refuse in the soil. The fungus is active under moderately warm conditions.

Figure 9.6. Cercospora leaf blight on carrot leaf

(ii) Management

Seed treatment with hot water at 50°C for 15 min is effective to manage the disease. Crop rotation and destruction of infected plant material in the field will minimize infection. Spraying of 0.2% foltaf or 0.3% copper oxychloride is effective to control leaf blight.

9.3.2.2. *Alternaria* Leaf Blight, *Alternaria dauci*

(i) Symptoms: Dark brown to black spots appear mostly along the leaf margin and petioles (Fig 9.7). In severe cases, leaflet is shriveled and killed. The pathogen survives in and on the seed and also in diseased material in the soil. The disease is severe under high moisture conditions.

Figure 9.7. *Alternaria* leaf blight on carrot leaf

(ii) Management

Seed treatment with hot water at 50°C for 15 min is effective to manage the disease. Crop rotation and destruction of infected plant material in the field will minimize infection. Spraying of 0.2% foltaf or 0.3% copper oxychloride is effective to control leaf blight.

9.3.2.3. Powdery Mildew, *Erysiphe cichoracearum*

(i) Symptoms: White powdery growth appears on leaves and petioles causing the leaves to turn brown and wilt (Fig 9.8).

Figure 9.8. Powdery mildew on carrot leaves

(ii) Management

Spray 0.2% wettable sulphur fungicides 4 weeks after sowing and repeat the spray at 7 and 10 weeks after sowing. Carrot *cv.* Arka Suraj is resistant to powdery mildew disease.

9.3.2.4. Soft Rot, *Erwinia carotovora*

(i) Symptoms: Water soaked irregular lesions appear on tubers (Fig 9.9). Mostly the disease occurs after the harvest.

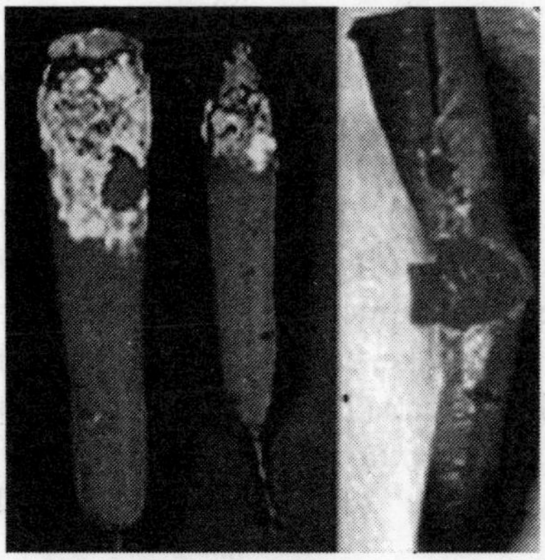

Figure 9.9. Soft rot on carrot

(ii) Management

Spraying carbendazim at 2 g/litre of water at an interval of 8–10 days is effective for the management of the disease.

9.3.3. Nematodes

The root-knot nematode is the major limiting factor in successful cultivation of carrot crop.

9.3.3.1. Root-knot Nematodes, *Meloidogyne* spp.

Devi (1993) reported that *M. incognita* was responsible for 56.64% loss in yield of carrots. Root-knot nematodes develop characteristic forking of the roots in carrots (Fig 9.10).

Figure 9.10. Forking of carrots incited by root-knot nematodes (Courtesy: AICRP on Nematodes, Hisar)

(i) Management

(a) Cultural Methods: The population of *M. incognita* larvae decreases in combination of spinach-marigold-carrot and spinach-radish-carrot (Hasan and Jain, 1998). Late sown carrots suffer less damage due to *M. incognita* in California, USA.

Botanicals: The oil cakes (neem, groundnut, mustard and castor) were effective in reducing parasitic nematode population (*Hoplolaimus*, *Tylenchorhynchus, Meloidogyne* and *Helicotylenchus* species) and in increasing yields of carrot.

In a field trial, saw dust, neem cake, poultry manure at 1.0 and 1.5 t/ha and water hyacinth at 1 and 2 t/ha increased yield of carrot (28–66%) by reducing the infestation of *M. incognita*. The maximum reduction of *M. incognita* in soil was recorded in neem cake (38.5%) followed by saw dust (23%), poultry manure (22%) and water hyacinth (14%) at higher dose over control (Devi, 1993).

(b) Biological Methods: Application of *P. lilacinus* (containing 1×10^6 spores at 10 g/2.5 sq. m.) recorded least gall index (1.6 as against 5.0 in control), reduced nematode population by 45.23%, reduced forking by 92.4% and increased the yield by 147.6% (Sivakumar, 1998).

(c) Host Resistance: Carrot *cv.* Arka Suraj is reported to be tolerant to the root-knot nematode (Swamy and Sadashiva, 2004). Among 103 germplasm screened, 10 genotypes namely New Kuroda, Selendro, Perfecta Elita, Zeno, Nantes Strong Top, Nantes, F-524, Danvers Half Long, Avenger and HYDC-4 were found resistant to *M. incognita* (Veere Gowda *et al.*, 2001).

(d) Integrated Methods: A soil fumigant at the beginning of two consecutive carrot crops, followed by two year onion production and at last a cover crop sudax in the course of 5 year rotation can control *M. hapla* problem (Bird, 1981).

Seed treatment with *T. harzianum* at 10 g (with 1×10^6 cfu/g)/kg and *P. fluorescens* at 10 g (with 1×10^9 cfu/g)/kg and subsequent field application of 5 tonnes of enriched FYM with *T. harzianum* (with 1×10^6 cfu/g) and *P. fluorescens* (with 1×10^9 cfu/g) per ha significantly reduced root-knot and reniform nematodes in carrot roots by 79 and 75%, respectively. The yield increase was to the tune of 29.8% with a cost: benefit ratio of 1 : 13.6.

9.3.3.2. Lesion nematode, *Pratylenchus* sp.

The lesion nematode (Fig 9.11) feeds on roots, leading to yellowing and drying of the leaves. Empty patches are seen after the leaves dry in the field (Fig 9.11). Roots become forked with rosette formation.

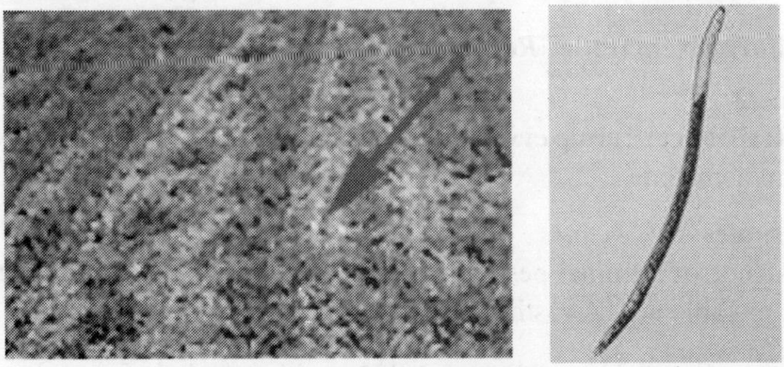

Figure 9.11. Carrot crop damaged by the lesion nematode

Management

Soil application of neem cake at 500 kg/ha or carbofuran at 1.5 kg *a.i.*/ha at a depth of 10–15 cm during land preparation helps in reducing the nematode population in soil.

9.4. TURNIP

Turnip (*Brassica rapa*) is grown in temperate, sub-tropical and tropical regions of India. It is extensively cultivated in Bihar, Haryana, Himachal Pradesh, Punjab and Tamil Nadu. Turnip is rich in vitamins A, B, C and calcium and iron. It can be eaten raw.

9.4.1. Diseases

9.4.1.1. Bacterial Soft Rot, *Erwinia carotovora* pv. *carotovora*

In India, the disease was first noticed in Pusa, Bihar during 1931 (Prasad, 1931b).

(i) Symptoms: In early infection, foliage appears to be normal, but in advanced stage the leaves turn yellow and drop off. On roots, soft, water-soaked patches, paler than healthy roots are visible. Inside the soft patches, tissues are reduced to white pulp from which a putrid smell emanates.

(ii) Management

Crop rotation with non-host crops is effective in reducing the disease incidence (Prasad, 1931b).

9.4.2. Nematodes

Root-knot and reniform nematodes are recognized as the major limiting factors in successful production of turnip crop.

9.4.2.1. Root-knot Nematodes, *Meloidogyne* spp.

(i) Management

(a) Cultural Methods: The oil cakes (neem, groundnut, mustard and castor) were effective in reducing parasitic nematode population (*Hoplolaimus*, *Tylenchorhynchus*, *Meloidogyne* and *Helicotylenchus* species) and in increasing yields of turnip.

9.4.2.2. Reniform Nematode, *Rotylenchulus reniformis*

(i) Management

(a) Host Resistance: Turnip *cvs.* Purple Top and White Globe are reported to be resistant to the reniform nematode.

Pesticide Residues

The persistence of terminal pesticide residues and waiting periods for root vegetables are presented in Table 9.1 (Awasthi, 2001).

Table 9.1. Terminal pesticide residues and their waiting periods for root vegetables

Pesticide schedule	Dosage	Terminal residues (ppm/days)	Waiting period (days)
BHC dust soil application at sowing	5.0 kg *a.i.*/ha	ND*/120	105
Aldrin dust soil application at sowing	3.0 kg *a.i.*/ha	0.005/90	90
Carbaryl dust soil application at sowing	1.5 kg *a.i.*/ha	ND*/60	60
Methyl parathion dust soil application at sowing	1.0 kg *a.i.*/ha	ND*/60	60
Carbofuran soil application at planting	1.0 kg *a.i.*/ha	ND*/55	60

* ND–Not detectable

Table 9.2. Management of insect pests in root vegetable crops

Crop	Pest	Symptoms	Pesticide/ Dose
Radish	Aphid	Clustering of aphids in young leaves	Phosphamidon-0.5 ml/l or Monocrotophos-1.25 ml/l as and when observed
	Mustard saw fly	Leaves with holes & black larvae are found crawling around plants	Endosulfan/ Methyl Parathion-2 ml/l; Quinalphos-1.25 ml/l as and when observed

10 MUSHROOMS

Mushrooms have achieved significant importance due to their nutritive and medicinal values. Mushrooms, the edible fleshy fungi, represent one of the few valuable sources of nutritive and palatable food. They are rich in proteins (35–48%), vitamins and minerals. They are weight reducers and useful for diabetic patients. Though, these fungi have been naturally occurring and being utilized in India since time immemorial, the commercial cultivation is only of recent times. It was only in 1964 that the first successful crop of mushroom was grown at Solan in Himachal Pradesh. Since then, research on various aspects of mushroom cultivation has been strengthened and today it is one of the important cottage industries, not only in the state of Himachal Pradesh but its commercial cultivation has also been initiated in most of the other states and union territories of India.

At present, it is estimated that 40,000 tonnes of mushrooms are produced annually. The major share (85%) is contributed by button mushroom (*Agaricus bisporus* and *A. bitorquis*). Oyster mushroom (*Pleurotus* spp.) is popular in southern states and Maharashtra and milky mushroom (*Calocybe indica*) in Tamil Nadu, Andhra Pradesh and Karnataka (Table 10.1).

Table 10.1. State-wise production of cultivated mushrooms

State	Production in MT	Type of mushroom
Tamil Nadu	10000	Agaricus/Volvariella
Maharashtra	7000	Agaricus/Pleurotus
Haryana	4000	Agaricus
Punjab	4000	Agaricus
Uttar Pradesh	4000	Agaricus/Pleurotus
Andhra Pradesh	2500	Agaricus/Pleurotus
Madhya Pradesh	2000	Agaricus/Pleurotus
Goa	1000	Agaricus
North eastern states	1000	Agaricus/Lentinus/
Orissa	700	Pleurotus/Volvariella
Karnataka	600	Agaricus/Volvariella
Himachal Pradesh	500	Agaricus
Jammu & Kashmir	400	Agaricus

Table. 10.1. (Contd.)

State	Production in MT	Type of mushroom
West Bengal	250	Agaricus/Pleurotus
Kerala	100	Pleurotus/Volvariella
Rajasthan	50	Pleurotus
Others	1900	Agaricus/Pleurotus
Total	**40000**	

Source: NCMR&T, Solan (1996–97)

10.1. PESTS

Pests are a constant threat to the successful commercial production of mushrooms. Uncontrolled populations of any of the main mushroom pests can result in substantial losses in yield due to both direct larval action and associated disease spread by adult flies.

Flies found in cropping houses cause reduction in yield by:

- Damaging the compost.
- Feeding on the mycelium.
- Act as vectors of diseases.
- Nuisance to pickers.
- They can cause crop rejection at market in pre-packs or chips.

10.1.1. Sciarids—*Lycoriella castanescens, L. ingenua*

(i) Damage: Sciarids are compost feeders and really prefer unspawned compost to that colonised by *Agaricus* mycelium. The adult flies are attracted by volatiles given off by the compost so are a danger at any time after pasteurisation when the compost is cool. The adults are usually 3–4 mm long (although some species recently found associated with mushrooms have been slightly bigger) and with the naked eye can be distinguished from the other fly pests of mushrooms by their long antennae (Fig 10.1). The females deposit between 50 to 200 eggs in the compost and the developing larvae pass through four moults varying in size from 1 to 8 mm long before pupating. They are easily recognised by their black shiny heads. They feed on rotting vegetation, so mushroom compost is an ideal substrate for them but the accumulation of their waste (called frass) renders it unsuitable for mushroom growth. Although exudates produced by the growing *Agaricus* mycelium inhibit the larvae, they have been observed browsing on the growing tips and when there is a large infestation the larvae will burrow up into the mushroom stipes.

As well as direct damage to the crop the flies can also spread mites and disease, especially *Verticillium* as the spores of this disease are sticky they easily adhere to the bodies of the flies.

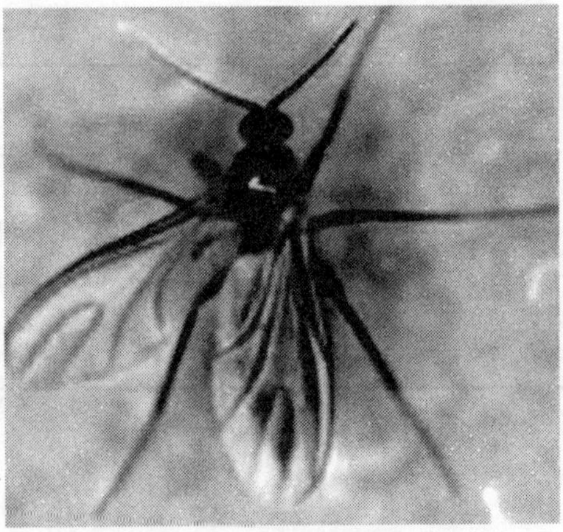

Figure 10.1. Adult sciarid fly

(ii) Management

- Efficient composting and peak heating.
- Exclusion of adults by filters and tight fitting doors. It is important that doors are not left open as an aid for ventilation.
- Disposal of crops debris at a distant location from the farm.

Sciarid larvae can be treated with the growth regulator Dimilin and with Nemasys M– an insect parasitic nematode. The adult flies of both sciarid and phorid can be sprayed with pyrethrums although the flies have often laid at least some of their eggs before leaving the beds.

10.1.2. Phorids, *Megaselia halterata*

(i) Damage: Phorid larvae are obligate mycelial feeders therefore the adult flies are not attracted to oviposit in the crop until after spawning. The larvae are white, 1 to 6 mm long, are stubby at one end and have a pointed head at the other. They feed on the growing mushroom mycelium but rarely feed on the fruiting body itself, although different species of phorid are known to do so. They can be distinguished from sciarid larvae by the absence of the black head and they develop more rapidly into a pupa. The adult fly is 2 to 3 mm long, has very short antennae and a characteristic hump-back. It can be distinguished from the sciarid fly by the short antennae and by its rapid, jerky, running movement. They are unable to fly when the temperature falls below 12°C and are therefore unlikely to re-infest mushroom houses between late autumn and early summer.

The female can lay 50 eggs close to growing mycelium, and at a temperature of 24°C these eggs will reach the adult stage inside two weeks. The larvae are white legless maggots with no black head and tend to have a blunt tail region. These larvae feed entirely on growing mycelium although some species can tunnel into mushrooms. Adults are usually annoying to pickers. In addition these flies can spread disease such as *Verticillium*.

(ii) Management

- Efficient composting and peak heating.
- Exclusion of adults by filters and tight fitting doors especially during the first three weeks after spawning.
- Disposal of stems immediately from the growing house, these should be dumped at a distant location. When the bags are emptied they should also be discarded well away from the farm.

Since the withdrawal of diazinon, there is no pesticide currently available to treat an infestation of phorid larvae.

10.1.3. Cecids, *Heteropeza pygmaea*

(i) Damage: The three most common species of cecids which infect the mushroom crop are *Heteropeze pygmaea, Mycophila speveri* and *Mycophilia barnesi*. Cecid flies are rarely seen as the larvae reproduce paedogenetically—*i.e.*, new generations are produced within the body of a 'mother' larva without sexual reproduction- and the flies are very small. It is usually the larvae which are about 1 mm long that are first noticed through the plastic bag. The commonest species is *Heteropeza pygmaea* which has white larvae whereas the *Mycophila* species have orange larvae. Each larva becomes a 'mother larva' which can give birth to 12–20 daughter larvae within a week of its own birth, without any adult midge being present. This ability to multiply so rapidly results in enormous numbers of larvae in the compost and when the food supply becomes limiting within the compost the larvae swarm and go onto the mushrooms, especially after the crop has been watered. Large numbers of larvae especially of the orange type make the mushrooms unsaleable. The larvae also carry a bacterium which causes increasing losses due to the degradation of mushrooms.

When there is a bad infestation the larvae clump together and flow over the sides of the mushroom beds onto the floor where they can be inadvertently transported to healthy crops on feet and equipment.

(ii) Management

(a) Efficient composting and peak heating: Cecids and their larvae are only killed when temperatures throughout the compost reach 55°C for 5 hr or 60°C for 3 hr.

(b) Strict hygiene: If all larvae are killed during pasteurization, infestations at cropping usually indicate that pests have entered the bags after filling. If there is a cecid problem on the farm it is vital to isolate infested houses by disinfecting tools, knives, buckets etc. Clean overalls should be worn in each house. Filters should be replaced at the end of each crop and doors should be tight fitting and not left open as an increased aid for ventilation. As per usual the area where casing is to be unloaded should be disinfected. Any bags that are visibly infected with cecids during cropping should be sprayed with a disinfectant, sealed and dumped at a distant location; the same applies for crop debris.

(c) Efficient cook out and disinfection of premises: Before the bags are emptied they should be sprayed with disinfectant, after emptying the interior of the house should be washed out.

10.1.4. Mites

Many mites are commonly found in straw and manure, most species are beneficial to mushroom growing as they feed on eelworms and other mites, although some can cause damage, these are as follows:

Tarsonemid mite (*Tarsonemus mycetophagus*)
Tyroglyphid mites (*Tyrophagus* spp.)
Red Pepper mites (*Pygmephorus* spp.)

10.1.4.1. Tarsonemid Mite, *Tarsonemus mycetophagus*

(i) Damage: These mites are pale brown and are so minute that they are only visible with the aid of a microscope. They cause damage by feeding entirely on hyphae of mushrooms and the grower will know if he has these mites present, as the base of the stem of the mushroom will show a reddish brown discolouration. Where severe infestations occur the whole base of the mushroom may be detached from the growing surface.

(ii) Management

- As with eelworms little can be done when mites are present in the growing house, therefore efficient composting and peak heating must take place to ensure that they are killed during the pasteurisation process.
- Good hygiene should be practiced around the farm, especially in the clearance of crop debris.

10.1.4.2. Tyroglyphid Mites, *Tyrophagus* spp.

(i) Damage: These mites can be identified as they are slow moving, translucent, with long hairs on their bodies. If these mites are present in abundance they eat small pits in the caps and stalks. These pits then suffer from bacterial decomposition, which breaks down tissues just below the surface. This results in the skin collapsing which leaves an open pit. Tyroglyphids may also feed on mushroom mycelium, where they are present in large numbers, crop reductions can be caused. Mites usually gain entry into the compost by clinging onto Sciarid flies when the mites are the migratory stage. These migratory stages are normally produced when mites become overcrowded.

(ii) Management

- The mites should not be a problem where efficient composting and peak heating takes place.
- Organic debris should not be allowed to accumulate around the farm as it provides a breeding ground for mites.

10.1.4.3. Red Pepper Mites, *Pygmephorous* spp.

(i) Damage: These mites are not regarded as primary pests; their presence is usually an indicator that *Trichoderma* (green mould) is present in the compost. These mites feed on various weed moulds but not mushrooms, thus their presence indicates that the compost is unsatisfactory. The mites are yellowish-brown in color, 0.25 mm in length and have a flattened appearance; they also are capable of rapid rates of reproduction. As already stated these mites are secondary pests and they often swarm on the casing and mushroom surfaces. Where this happens their presence makes the mushrooms unsaleable. These mites can also spread spores of *Trichoderma* from bag to bag.

(ii) Management

- Efficient composting and peak heating.
- Good husbandry around the farm.
- Where mites appear on individual bags during cropping, these should be sealed up and disposed of.

10.2. DISEASES

10.2.1. Wet Bubble, *Mycogone perniciosa*

It is one of the serious diseases causing severe crop losses. It is the commonest and geographically most widespread disease of mushrooms. This disease is occurring in almost all the countries wherever mushrooms are grown.

(i) Symptoms: The most characteristic symptom is the development of distorted masses of mushroom false tissue, which are initially white and fluffy but turn brown with age and decay. These distorted masses are known as 'Sclerodermoid masses' and they may be as big as 10 cm in diameter. Small amber to dark brown drops of liquid are seen on the surface of the undifferentiated tissue under high moisture/RH conditions. The name of the disease is derived from this typical condition. In addition to distortion symptom, small fluffy white patches of mycelium may occur on the surface of the casing, following the infection of a developing mushroom below the casing surface.

The base of the stalk shows brown discolouration when mature fruit bodies are attacked, and a white fluffy mycelial overgrowth appears on the affected portions of the mushroom. The stipe sometimes develops brown streaks, and the cap develops symptoms which appear as a sector of affected and stunted gills, covered by the contamination of the casing medium.

(ii) Epidemiology: The pathogen produces two types of spores: single celled, thin walled and relatively short lived conidia; and two celled aleurospore (terminal chlamydospore) with very thick-walled terminal cell for longer survival under adverse conditions. The conidia are light and air borne. Both aleurospores and conidia are spread by water splash.

This fungus is commonly found in soils, and the disease outbreak often follows unrestricted soil movement on the farm. The primary source of the pathogen on most farms is through contaminated casing. Under temperate climate, spores may survive in crop debris and hence can contaminate subsequent crops. Once the pathogen is established in the crop beds, the main means of secondary spread is through water splash and by excess water running off the beds. Pickers and workers may also carry and spread the pathogen on their hands, tools, boxes, clothing, etc.

(iii) Management

The most important and viable means of control of wet bubble disease is the elimination of the primary source of the pathogen by paying strict attention to hygiene at the farm. The casing medium should be freed from spores of *Mycogone perniciosa* which can be achieved by either steam pasteurization or formalin treatment of the casing medium and its careful storage in a clean room free from dust and debris from cropping rooms.

The following practices may be adopted to manage the disease:

- Disinfect tools/clothes of workers regularly to prevent the spread of inoculum.
- Avoid over-watering of cropping beds.
- Carefully remove all affected mushroom under a powerful light source.
- Pickers should use gloves and disinfectants at regular intervals.
- All diseased mushrooms should be removed before the crop is picked.
- Cover the affected mushroom with common salt or pots (push the pots well into the casing) to prevent the spread of the disease.
- Watering should be done only after all the diseased mushrooms have been removed.

Benzimidazole fungicides (benomyl, carbendazim and thiobendazole) are quite effective in controlling the disease. Prochloraz manganese also gives good control of wet bubble disease when mixed in casing soil. Dithiocarbamates (zineb and maneb) also control the disease when inoculum levels are low.

10.2.2. Dry Bubble, *Verticillium fungicola* var. *fungicola*

This is the most serious fungal disease of mushrooms causing serious economic losses. This fungus is a mycoparasite with necrotrophic behaviour and kills the host fungus on colonization. It is geographically most widespread disease of white button mushroom in the world. This disease is reported from UK, USA, Ireland, Germany, The Netherlands, Italy, India, Canada, Australia, etc. If this disease left uncontrolled, the pathogen can totally destroy the mushroom crop within 2 to 3 weeks.

(i) Symptoms: When the mushrooms are infected at an early stage of their development (primordial/pin head stage), the typical symptoms are small undifferentiated mass of tissue (up to 2.5 cm in diameter). When mushrooms are affected at later stage of development, they are often imperfectly formed with partially differentiated caps, or with distorted shaggy stipes and with tilted caps. Such affected mushrooms are covered with a fine grey white mycelial growth, discoloured, dry and do not rot as in wet bubble. When fully grown mushrooms are affected, these show small pimple-like outgrowths on top of pileus, or blue grey spots (1–2 cm in diameter) on the cap surface. Such spots have yellow or bluish-grey halo around them.

The growth of the pathogen on mushrooms is in the form of fine, grayish brown felt. The infected mushrooms are dry at first, but may undergo bacterial rot if not removed soon from beds. The diseased sporophores when cut longitudinally, often showed a grey-brown zone of variable thickness immediately below the surface, others were completely discoloured internally. The cavities found in sporophores were lined with spore-bearing conidiophores.

(ii) Epidemiology: Contaminated casing and not the compost is the most common initial source of inoculum on a mushroom farm. Air-borne spores can also serve as primary source of inoculum, or spores carried by flies, mites and pickers. The conidia of *Verticillium*, produced in clusters surrounded by sticky mucilage, help in effective dispersal of spores to new sites from infected beds. The mucilage enables the spores to get attached to dust particles, flies, mites, debris and pickers. Watering is another important means of dispersal, as excess water running off the infected beds carry spores to the lower beds and floor.

The optimum temperature for disease development is about 20°C. The time from infection to symptom appearance is 10 days for distortion symptoms and 3–4 days for cap spotting at 20°C. The pathogen grows best at 24°C.

(iii) Management

The most effective way to control this disease is the prevention of entry of this pathogen into crop beds. Strict attention to hygiene will go a long way in keeping this disease at bay in mushroom houses. Ensure that casing soil is free from infection of *V. fungicola*. Pasteurize or treat the casing medium with steam/formalin and store in a clean area, free from dust and debris. The affected mushroom should be removed as quickly as possible from the beds to avoid the danger of dispersal of inoculum through insects/pests, tools, pickers, etc. The affected mushrooms can also be covered with common salt or pots. Fly and mite infestation should be controlled to prevent dissemination of the inoculum. The diseased beds should be effectively cooked out with steam or chemically.

Prochloraz manganese gives good control of dry bubble disease.

10.2.3. Trichoderma Disease, *Trichoderma harzianum*

Trichoderma has been reported as one of the serious disease causing fungal pathogen on mushrooms. As high as 30% crop losses have been reported by the attack of this fungus on mushrooms. It is most commonly associated with mushroom farms and thrives best in high humidity and on carbohydrate materials. Both these parameters are excellent for the growth of *Trichoderma*.

(i) Symptoms: Green patches of sporulation, the main symptom of attack by *Trichoderma*, are clearly visible in bags, trays or base of shelves. When mature, these spores turn dark green and become less visible. These are then washed off into the casing below and the mushrooms emerging through this casing are badly spotted, often distorted and with a very short shelf life (Seaby, 1989).

There is drastic fall in yield with the appearance of blank patches on the bed, and casing becomes densely covered with red pepper mites or contaminated by nematodes.

(ii) Epidemiology: The spores of *Trichoderma* are transmitted to freshly peak heated compost by wind, attached to spent compost dust particles, by mites who have fed on *Trichoderma* mycelium in infected compost, on the hands and clothing of workers and by machinery. The infection persists in wooden shelf by colonization of wood which is most of the times wet. When new compost is filled, it gets contaminated from the fungus from the wood, unless effective cook out has been carried out at termination of cropping.

(iii) Management

The most effective and economic control measures are by giving maximum attention to farm hygiene. Benomyl spray is also reported to be effective but spot treatment of affected areas with common salt is often used in the mushroom farms. Cook out of the compost beds at termination of cropping with steam or formalin treatment reduces the infection. Clean all tools and equipment with formalin solution periodically. Use foot dips effectively while entering the mushroom house. Control mites effectively to prevent dissemination of infection.

10.2.4. Bacterial Blotch or Brown Blotch, *Pseudomonas tolaasii*

This is most common and serious bacterial disease of white button mushrooms which is widespread in nature. This disease causes heavy crop losses in marketable mushrooms, especially during the months when temperature and relative humidity are high.

(i) Symptoms: The characteristic symptom of this disease is the occurrence of dark brown, irregularly shaped spots on surface of the pileus. These spots may initially be light brown in colour but eventually become dark brown. The spots are typically slightly sunken and with a silken sheen. Severely affected mushrooms may become distorted and the entire crop may turn dark brown in colour with splitting of affected areas. In very severe cases, stalk may also be affected. The intensity of symptom development is higher due to more wetting of cap during watering.

(ii) Epidemiology: In favourable conditions for disease development like high temperature, high humidity and frequent wetting of caps, epidemic levels of blotch disease can occur. In between the crops, the pathogen survives on surfaces, in debris, on tools and on various structures. The chief means of spread of bacteria is through flies, mites and moving equipments in cropping rooms. The bacteria can also be dispersed on debris, on mushroom spores, in water droplets, by flies and on hands of pickers. Watering will disperse bacteria very readily when infection is present in the beds.

(iii) Management

The treatment of casing material with formalin or steam before use is essential. Frequent and prolonged wetting of caps should be avoided during watering. Use of bleaching powder (sodium hypochlorite) as a routine spray to the beds will help to keep the bacterial population below critical threshold levels of infection. The recommended effective concentration for the use of chlorine at each watering is 150 ppm.

10.2.5. Mushroom Viruses (Virus-1, Virus-2, Virus-3, Virus-4, Virus-5)

The viruses are widely distributed all over the world wherever mushrooms are grown. Severe outbreaks of virus diseases on many mushroom farms have been reported with severe crop losses.

(i) Symptoms: Varied symptoms such as distortion of sporophore, elongation of the stalk, tilting of caps and reduction in size of caps on normal sized stalks are caused by virus infection. Affected beds may show patchy appearance, associated with death of mushroom mycelium. Premature opening of sporophores and reduction in cropping are other symptoms of viral diseases.

(ii) Epidemiology: Most of the mushroom viruses are spread by mushroom spores and mycelium. The viruses can remain viable for 6 to 9 years in mushroom spores, depending upon the prevailing temperature of the growing areas. Mushroom spores become air-borne, but spread can also be by workers and flies. Infection of the mycelium at spawning or during spawn run and before casing results in maximum yield reduction. The compost may also become contaminated by infected mushroom spores during cool down after pasteurization. Mushroom viruses can also survive in mycelium which is left attached to wooden trays at the end of the crop. Mushroom spawn is one of the potential sources of virus infection.

(iii) Management

Maintenance of strict hygiene on a mushroom farm is very essential. The entry points to points of pasteurization chamber and growing rooms should be protected by fungal filters effectively. Cook out the beds after completion of cropping with steam, especially when wooden shelves/trays are used. Use mushroom spawn free from viruses.

Agaricus bitorquis is resistant to mushroom virus infection and can hence be used to break the disease chain in monoculture system on a mushroom farm.

10.3. NEMATODES

An intensive survey of mushroom farms in India revealed that 84.4% nematode infestation, which led to extremely poor flushes in initial stages followed by total crop failures (Seth, 1984). Nematodes are most dangerous pests of mushrooms which once entered into the crop cannot be eradicated without completely destroying the crop. Their presence in the beds simply means very poor yields or total crop failures.

In India, so far eight species of *Aphelenchoides* besides *Ditylenchus mycetophagus* have been found to damage mushroom mycelium (Table 10.2).

Table 10.2. Fungal feeding nematodes associated with mushrooms from India

Nematode Order	Fungal feeding nematodes	References
Tylenchida	*Ditylenchus mycetophagus*	Thapa *et al.*, 1981
Aphelenchida	*Aphelenchoides agarici*	Seth & Sharma, 1986
	A. asterocaudatus	Bahl & Prasad, 1985
	A. composticola	Khanna & Sharma, 1988
	A. neocomposticola	Khanna & Sharma, 1988
	A. minor	Seth & Sharma, 1986
	A. mycetophagus	Seth & Sharma, 1986
	A. sacchari	Sharma *et al.*, 1981
	A. swarupi	Seth & Sharma, 1986

10.3.1. *Aphelenchoides agarici*

This species was recorded from mushroom farms in Himachal Pradesh (Seth & Sharma, 1986). It is one of the most destructive nematode species and multiplies freely on mushroom mycelium. An initial inoculum level of 10 nematodes has been shown to cause 100% mycelial depletion within 25 days at 28°C resulting over 800 times multiplication rate under laboratory conditions (Khanna & Sharma, 1988). It completes its life cycle within 8 days at 28 ± 1°C and the ratio between female to male was 1. 5:1 (Seth, 1984).

10.3.2. *Aphelenchoides composticola*

It is one of the most widely distributed species present throughout the mushroom growing countries of the world. In India, it is known to occur in Himachal Pradesh and Punjab. This nematode also has a short life cycle of 8 days at 23°C and multiplies very fast on mushrooms (Cayrol, 1967). An initial inoculum of 10 nematodes completely depleted the mycelium within 40 days resulting in about 1000 times multiplication rate (Khanna & Sharma, 1988). *A. composticola* was responsible for 35–60% loss in yield of mushroom (Laqman Khaṇ, 2001).

10.3.3. *Aphelenchoides mycetophagus*:

It is recorded from Solan district in Himachal Pradesh. It is also one of the highly destructive nematode species infesting mushroom showing the pathogenicity status of the level of *A. composticola* (Khanna & Sharma, 1988).

10.3.4. *Aphelenchoides neocomposticola*

It was reported from Shimla district of Himachal Pradesh. The nematode, though highly pathogenic in nature, was found to be less damaging than *A. agarici*, *A. composticola* and *A. mycetophagus* (Khanna & Sharma, 1988).

10.3.5. *Aphelenchoides sacchari*

It was reported by Sharma *et al.* (1981) and found to be highly pathogenic to white button mushroom, *Agaricus bisporus*. An average of 50–100% yield loss in sporophores in white button mushroom has been reported due to this nematode species (Sharma *et al.*, 1984). It is quite prevalent on mushrooms in India (Thapa *et al.*, 1981). Singh *et al.* (2003) reported that *A. sacchari* was responsible for 40.6 to 100% loss in yield of mushroom.

10.3.6. *Ditylenchus mycetophagus*

This nematode is widely distributed in almost all mushroom growing belts of the world. In India, it is found in Himachal Pradesh, Jammu and Kashmir and Punjab (Thapa *et al.*, 1981). It is one of the highly destructive species. At an inoculum level of 10–1000 individuals could cause 15–70% mycelial depletion within 60 days (Sharma *et al.*, 1985). However, it is considered to be less pathogenic than *A. composticola* and *A. sacchari* (Sharma *et al.*, 1981).

(i) Nature of Damage

The mycophagus nematodes thrust their hollow stylet into the hyphal cell. The mycelium bleeds and its sap is sucked upon by to and pro movement of the stylet. After feeding upon one cell the nematode shifts to another cell with the help of the moisture film present in the compost.

(a) Symptoms: Initially whiteness of spawn slowly starts turning brown followed by sparse and patchy appearance of the mycelium. This mycelium is stingy in nature. The compost surface sinks. Alternately, medium and low quantity of sporophores (fruiting bodies) is obtained in succession of flushes. There are extremely poor yields in the later flushes finally leading to total failure of sporophore production thus reducing the total yields as well as duration of the crop.

(b) Losses: Economic losses due to decline in yields are generally very high. Extent of losses mainly depends upon the time of nematode infestation, nematode species involved and population of the nematodes. Losses caused by *A. sacchari* are only so far worked out and have been given in Table 10.3 (Sharma *et al.*, 1984).

Table 10.3. Yield losses in mushroom due to *Aphelenchoides sacchari*

Time of inoculation	Initial inoculum	Yield loss (%)
Spawning	1×10^1	98.2
	1×10^2	100.0
	1×10^3	100.0
	1×10^4	100.0
Casing	1×10^1	90.6
	1×10^2	92.1
	1×10^3	87.3
	1×10^4	100.0

(ii) Management

(a) Physical Methods: Use of heat has been the most successful method of nematode control in mushroom cultivation during composting and casing process as well as at the end of cropping. The air as well as bed temperature should be maintained at 60ºC for at least 2 hr during peak heating of compost. Thorough pasteurization of casing soil with steam for at least 4 hr gave good control of nematodes. Treatment of used trays and handling tools in boiling water for 1–2 min or in disinfectants like formalin is recommended.

An effective method of casing pasteurization, based on the use of solar heat, was developed wherein the thinly spread casing is exposed to the sun under a polythene cover (150 gauge) for 5–6 hr (Grewal and Grewal, 1988).

Wood and Goodey (1957) found that *r*-rays in between 48,000 and 96,000 *r* inactivated *Ditylenchus mycetophagus* in mushroom compost.

(b) Cultural Methods: Incorporation of neem/pongamia/coconut cake at 2 kg/100 kg of compost during first or second turning is recommended. Dry leaf powder of neem can be mixed with compost at 2–5% concentration. This has been found useful not only to check nematode multiplication in mushroom growing trays but also helps in proper mycelial growth and improved sporophore yields (Khanna and Sharma, 1994). Addition of oil cakes like neem, karanj, coconut, castor and groundnut have been reported to control nematodes and increase button production (Rao *et al.*, 1992). Further, incorporation of fresh leaves of Pedilanthus, eucalyptus and pongamia in the compost during first or second turning was observed to reduce the population of mushroom nematodes and to increase the yield of button mushrooms. The incorporation of dried leaves of neem, castor and *Cannabis sativa* resulted in the suppression of *A. composticola* below economic injury levels and increased mushroom yields by 6.5 to 19% (Grewal, 1989).

(c) Chemical Methods: Thionazin, applied at 80 ppm concentration, is quite effective in controlling the myceliophagous forms without producing any adverse effect on mycelial growth and also does not leave any residue in sporophores.

(d) Biological Methods: Nematode trapping fungus, *Arthrobotrys robusta* commercially available as 'Royal 300' (strain Antipolis) in France has been recommended for the management of mushroom nematodes. *Arthrobotrys oligospora* and *A. superba* have also been found to reduce nematode populations and increase the yields. In India, *Arthrobotrys irregularis* and *Candelariella musciformis* isolated from spent compost have shown promising results in checking nematode population (Khanna and Sharma, 1990). *A. irregularis* was also found to be highly effective against *Aphelenchoides composticola.* Further studies revealed that *A. irregularis* did not compete with *Agaricus bisporus* mushroom and checked the nematode population significantly and increased the yield.

Grewal and Sohi (1988) mass produced *Arthrobotrys conoides* on wheat grain and, after incorporating it into compost at spawning, reduced a mixed nematode population of *A. composticola* and *Rhabditis* sp. by 42.5% in 25 days.

The bacterium, *Bacillus thuringiensis* is found to suppress the population of *Ditylenchus mycetophagus* (Cayrol, 1974).

(e) Host Resistance: *Agaricus edulis* has been reported to be resistant to *D. mycetophagus*. Thapa *et al.* (1987) reported that *Pleurotus sajor caju* and *Stropharia rugoso annulata* were found resistant to both *A. sacchari* and *D. mycetophagus*. K 30 strain of *A. bitorquis* showed resistance to *A. sacchari* only. Resistance in *Pleurotus sajor caju* against *A. sacchari, A. agarici* and *A. composticola* has been reported by several workers.

(f) Integrated Methods

- Cropping should be done in purposely built mushroom houses with proper ventilation.
- Cleanliness should be maintained inside and surroundings of mushroom farm so that flies, mites and other contaminants are avoided by spraying 2% formalin solution.
- Composting platform must be cemented to prevent the direct contact of compost with soil.
- No person or worker should be allowed to enter into the farm without proper disinfestation of their hands and feet.

- Wooden trays and handling tools should be disinfected before use by dipping them in boiling water or in 5% formalin for 2 min before use.
- Compost should be steamed at 60ºC for minimum of 2 hr before spawning.
- Casing material should be well sterilized before use. Formalin at 5% concentration can be used for this purpose.
- Irrigation water should be nematode-free.
- Proper cook-out of mushroom houses with steam at 70ºC for 5–6 hr or at 80ºC for 30–60 min is essential after the crop is over to avoid carrying of infestation to subsequent crops.
- The dry neem leaf powder may be mixed with compost at 2–5% concentration which not only check nematode multiplication in mushroom growing trays but also helps in proper mycelial growth and improved sporophore yield.

11 VEGETABLE CROPS UNDER PROTECTED CULTIVATION

With increased emphasis on commercialization of horticulture, cultivation of vegetables under polyhouse is gaining importance in India. There is a great potential to produce vegetables like tomato and capsicum under polyhouse for domestic and export market. Demand for vegetables during the off-season is high and prices are astronomical. Under these circumstances, many farmers and entrepreneurs are considering cultivation of vegetables under polyhouse or greenhouse conditions. Although it is highly remunerative, all vegetables cannot be grown under polyhouse conditions due to the high cost of infrastructure and maintenance, therefore, high value crops alone can be cultivated under polyhouse by adopting intensive cultivation practices.

The important crops grown under polyhouse conditions include capsicum, tomato, cucumber, melons and leafy vegetables. In India, protected cultivation of vegetables is at a nascent stage and only tomato and capsicum are being grown under polyhouse.

Plant protection holds the key for successful production of crops under polyhouse conditions, as these crops are highly vulnerable to pests and diseases. Prevalence of microclimate that is congenial for multiplication and spread of plant pathogens, high density of plant population, monocropping of susceptible genotypes and increased labour activities make polyhouse grown plants more prone to pest and disease threats. Qualitative damage is far more significant than quantitative loss in case of export oriented crops as slight distortion or discolouration makes them unfit for marketing. The important pest problems faced under protected cultivation include the tomato leaf curl virus transmitted by whitefly, *Bemisia tabaci*; thrips, *Scirtothrips dorsalis* and white mite, *Polyphagotarsonemus latus* on capsicum; and root-knot nematodes on both tomato and capsicum. The management strategies for protected cultivation should focus on effective, less persistent, non-phytotoxic pesticides along with safer alternatives like plant products, bioagents, cultural practices and mechanical methods.

11.1. TOMATO

11.1.1. Pests

11.1.1.1. Whitefly, *Bemisia tabaci*

(i) Damage: Whitefly infesting vegetables under greenhouse conditions is a vector of tomato leaf curl virus. Adults and nymphs suck the sap from leaves (Fig 11.1) resulting in curling of leaves. If the infestation is severe, sooty mould develops on the leaves due to excretion of honeydew by the whitefly, which reduces the photosynthetic activity of the plant and yield gets reduced drastically.

Figure 11.1. Whitefly infestation on tomato

(ii) Management

- Raise the seedlings in polyhouse using plastic seedling trays (portrays).
- Spray imidacloprid 200 SL (0.3 ml/l) or thiomethoxam 25 WP (0.3 g/l) in nursery at 15 days after sowing.
- Drench the base of the seedlings with imidacloprid 200 SL (0.3 ml/l) or thiomethoxam 25 WP (0.3 g/l) before transplanting.
- Remove the leaf curl infested plants as soon as disease symptoms are expressed. This helps in reducing source of inoculum of the disease.
- After transplanting give need-based sprays of imidacloprid 200 SL (0.3 ml/l) or thiomethoxam 25 WP (0.3 g/l) at 15 days after planting and do not repeat after fruiting stage as this may leave harmful residues in fruits.
- Install yellow sticky traps coated with adhesive or sticky glue at crop canopy level for monitoring adult whitefly population.
- If the traps indicate the whitefly activity, spray dimethoate 30 EC at 2 ml/l or neem seed kernel extract 4% (NSKE) or pongamia or neem oil (8–10 ml/l) or neem soap (10 g/l).

11.1.1.2. Red Spider Mite, *Tetranychus urticae*

(i) Damage: Red spider mites are highly polyphagous. They can cause serious damage by their fast multiplication rate in a very short time. They thrive under high temperature and dry weather. They are generally found on the lower surface of older leaves (Fig 11.2). However, when the infestation is very high they attack all parts of the plant and are observed in colonies covered by white-silky webs. Adults and nymphs lacerate the leaves causing yellowing and discoloration. Drying and dropping of attacked leaves is also observed when the infestation is severe.

Figure 11.2. Red spider mite damage in tomato

(ii) Management

- Remove the mite infested leaves and destroy before spraying.
- Spray need based application of acaricides like abamectin 1.9 EC at 0.5 ml/l or dicofol 18.5 EC at 2.5 ml/l or fenazaquin 10 EC at 1 ml/l in rotation with plant products like pongamia oil or neem oil (8–10 ml/l) or neem soap (10 g/l).

Note: Botanicals have very short residual toxicity, particularly when the temperature is high. In addition, these may cause phytotoxicity. Hence, sprays with botanicals should be done with caution and sprays to be given in the evenings when the temperature is low.

11.1.1.3. Serpentine Leaf Miner, *Liriomyza trifolii*

(i) Damage: Serpentine leaf miner is a tiny metallic fly with a characteristic yellow patch behind the compound eye. Damage starts in nursery itself. Initially the adult female punctures the developing leaf and feeds on the oozing sap. Larvae mine leaves in a serpentine manner (Fig 11.3) and on pupation they drop to soil. Severe damage leads to drying and dropping of leaves.

(ii) Management

- Apply neem cake at 250 kg/ha to beds while planting and repeat after 25 days to prevent pupal emergence from soil.
- Spray neem seed powder extract (4%) or neem soap 1%.

Figure 11.3. Leaf miner damage in tomato

- Frequent spraying of synthetic pesticides should be avoided as it may cause resurgence of the pest. At the most, one spray of deltamethrin 2.8 EC at 1 ml/l or cypermethrin 25 EC at 0.5 ml/l or triazophos 40 EC at 2 ml/l may be given if required.

11.1.1.4. Fruit Borer, *Helicoverpa armigera*

(i) Damage: This is a major problem in improperly maintained polyhouses. Eggs are laid on young leaves. Initially larvae feed on tender leaves and scrape the leaf surface, later they bore into fruits (Fig 11.4). Generally the posterior end of the body remains outside the hole. The damaged fruits become unmarketable.

Figure 11.4. Fruit borer damage to tomato

(ii) Management

- Spray *Ha* NPV at 250 LE/ha + 1% jaggery along with sticker (0.5 ml/litre) during evenings when the larvae are young.
- For grown up larvae spray indoxacarb 14.5 SC at 0.5 ml/l or thiodicarb 75 WP at 1 g/l.

11.1.1.5. Tobacco Caterpillar, *Spodoptera litura*

(i) Damage: This pest is also important in ill managed polyhouses. Eggs are laid in clusters on foliage. Young larvae feed gregariously on leaves. Mature larvae, migrate and cause extensive damage to leaves and fruits. They hide in soil and crop debris during day time.

(ii) Management

- Collection and destruction of egg masses and gregarious larvae.
- Spray *Spodoptera* NPV 250 LE/ha + 1% jaggery along with sticker (0.5 ml/litre) during evenings.
- Use poison baiting. Mix 10 kg of rice bran or wheat bran with 2 kg jaggery by adding a little water in the morning. In the evening add 250 g of methomyl or thiodicarb formulation and sprinkle over the bed. Caterpillars get attracted to fermenting jaggery, feed and get killed (Fig 11.5).

Figure 11.5. Dead tobacco caterpillar after baiting

Flow chart for insect pest management in tomato

Raise seedlings in portrays in polyhouse1:

5 DAP	• Spray the plants with imidacloprid or thiomethoxam.
One day before transplanting	• Drench the base of seedlings with imidacloprid or thiomethoxam.
At transplanting	• Apply neem cake at 250 kg/ha.
15 DAP	• Spray the seedlings with imidacloprid or thiomethoxam.

25 DAP	• Apply neem cake at 250 kg/ha.
Post flowering and fruiting stage	• Monitor for pests like fruit borer, tobacco caterpillar, leaf miner, whitefly and red spider mite. • Erect yellow sticky traps to monitor whitefly. • Spray NPV according to the pest. • Remove leaves severely infected with leaf miner/red spider mite. • Spray neem seed powder/neem soap for leaf miner. • Spray synthetic acaricide/botanical in rotation to control red spider mite. • Spray systemic insecticide/botanical to control whitefly.

11.1.2. Diseases

11.1.2.1. Early Blight, *Alternaria solani*

(i) Symptoms: The leaves show circular brown, concentric ring spots on the leaves. Yellow halo can be seen around the brown spots. When the number of leaf spots increase, leaves wither and drop. Affected fruits show brown to black depressed spots. Infected fruits fall to the ground. The pathogen also infects stem, petiole and calyx.

(ii) Management

- Regular rouging of older infected leaves reduces the inoculum load in the polyhouse.
- Foliar spray of 0.2% chlorothalanil or 0.2% mancozeb or 0.2% propane at fortnightly intervals checks the infection.

11.1.2.2. Powdery Mildew, *Leveillula taurica*

(i) Symptoms: The leaves show white powdery patches, on both sides of leaves. At later stages of infection, leaves turn yellowish brown and wither. If the infection is on lower side of the leaves, yellow patches are visible on the upper side.

(ii) Management

- At the onset of the disease, application of 0.075% hexaconazole or 0.1% dinocap or 0.3% wettable sulphur or 0.1% carbendazim reduces the disease.
- The ergosterol biosynthesis inhibitor fungicides like tridimefon/propiconazole/ triforine are more effective in reducing the disease intensity.

11.1.2.3. Septoria Leaf Spot, *Septoria lycopersici*

(i) Symptoms: The disease appears as circular spots with pale centers on leaves, stem and calyx and rarely on fruits. As the spots grow larger, they become more or less circular in outline and show definite brown coloured margins with grey centres. The disease on stem appears as small, slightly elongated, dark spots containing numerous black pycnidia. Small elongated specks appear on calyx, which are intermediate in size as those found on leaves and stems. Black circular spots may also be seen on young as well as mature fruits.

(ii) Management

- Destruction of infected plant material and use of clean seed is useful in checking the disease.
- Spraying of 0.1% carbendazim or 0.2% mancozeb or 0.3% copper oxychloride (COC) has been recommended.

11.1.2.4. Tomato Leaf Curl Virus (TLCV)

(i) Symptoms: TLCV is caused by Gemini virus which is transmitted by whitefly, *Bemisia tabaci*. Severe stunting of plants with downward cupping and crinkling of the leaves characterize the disease. The newly emerging leaves exhibit slight yellow colouration and later on show curling (Fig 11.6). Older leaves become leathery and brittle. The nodes and internodes are significantly reduced in size. The infected plant look pale and produce more lateral branches giving a bushy appearance. The fruits from infected plants are small and deformed. Losses due to the disease range from 30 to 92.5% depending on the stage of infection and cultivar.

Figure 11.6. Leaf curl virus on tomato

(ii) Management

- Raise tomato nursery under insect proof conditions.
- Handle the seedlings during transplanting properly.
- Maintain crop hygiene including removal of infected plants and weeds.
- Spray insecticides (0.1% dimethoate or 0.03% imidacloprid or 0.1% dichlorovas or 0.1% neem oil) regularly for the management of vectors.
- Grow resistant crop cultivars.

11.1.2.5. Tomato Spotted Wilt Virus (TSWV)

(i) Symptoms: The disease is characterized by bronzing of young leaves, chlorosis followed by necrotic ring spots. Necrotic spots appear under the growing tips and the entire twig dies back. In some cases this leads to severe stunting and cessation of growth. Early infected plants have deformed uneven ripe fruits while the fruits formed after late infection show concentric rings. A tospo virus (TSWV) causes the disease. TSWV is transmitted by several species of thrips and the main vector is *Frankliniella occidentalis*.

(ii) Management

- Raise seedlings under nylon nets (50-mesh) to prevent vectors (thrips).
- Apply carbofuran at 1 kg *a.i.*/ha to the nursery beds at the time of sowing.
- Spray the crop with 0.2% acephate or 0.03% imidachloprid or 0.1% fipronil.
- Rouge infected plants and weed hosts.
- Crop rotation with non-host crops of thrips.

11.2. BELL PEPPER

11.2.1. Pests

11.2.1.1. Thrips, *Scirtothrips dorsalis*

(i) Damage: Thrips are minute insects with fringed wings, serious during dry periods (high temperature). Both adult and nymphs suck the sap from young developing leaves. Affected leaves curl upwards along the margin and are reduced in size (Fig 11.7). They are also found feeding on floral parts and fruits. Fruit damage results in rough brown patches affecting their quality and drastically reducing the market value (Fig 11.7).

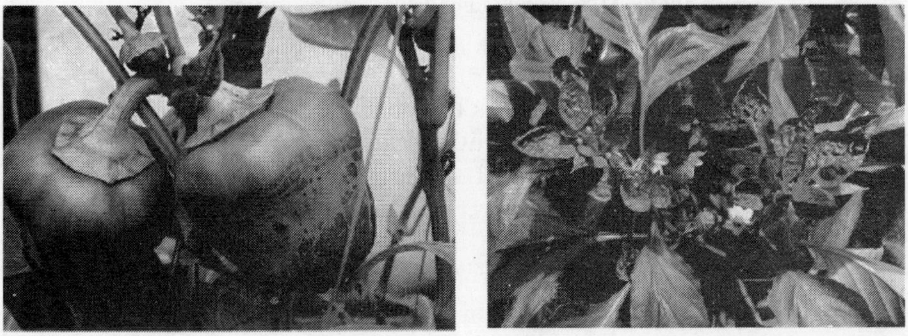

Figure 11.7. Thrips damage on capsicum fruits and leaves

(ii) Management

- Apply neem cake at 250 kg/ha to plant beds while planting and repeat after 30 days.
- Spray acephate 75 SP at 1.0 g/l or fipronil 5 SC at 1 ml/l or ethofenprox 10 EC at 1 ml/l in rotation.

- Spray acephate 75 SP (0.5 g/l) + pongamia oil (2 ml/l) + 1 ml sticker in one litre water after emulsifying (shaking thoroughly in a bottle).

11.2.1.2. White or Yellow Mite, *Polyphagotarsonemus latus*

(i) Damage: These are very minute mites and cannot be seen by naked eyes. Damage is more in hot and humid conditions. Adults and nymphs scrape terminal leaves and auxiliary shoots. As a result, leaves become narrow, twisted with elongated petiole. The damage is characterized by downward curling of leaves (Fig 11.8) and stunted growth of plants and dropping of flowers.

Figure 11.8. Yellow mite damage on capsicum

(ii) Management

- Need based application of acaricides like abamectin 1.9 EC at 0.5 ml/l or dicofol 18.5 EC at 2.5 ml/l or fenazaquin 10 EC at 1 ml/l in rotation with plant products like pongamia oil or neem oil (8–10 ml/l) or neem soap (10 g/l) when the leaves start curling down with all the precautions.

11.2.1.3. Aphids, *Aphis gossypii* and *Myzus persicae*

(i) Damage: These are small green insects and suck the sap from tender leaves and flower buds. Excretion of honeydew by aphids promotes sooty mould growth and thereby reducing photosynthesis and fruit quality. Aphids also act as vectors for transmitting mosaic virus diseases.

(ii) Management

- Spray acephate 75 SP at 1 g/l or dimethoate 30 EC at 2 ml/l in rotation when required and remove all the virus affected plants and destroy.

11.2.1.4. Tomato Fruit Borer, *H. armigera* and Tobacco Caterpillar, *Spodoptera litura*

(i) Damage: Sometimes tomato fruit borer and tobacco caterpillar may also attack capsicum under polyhouse (Fig 11.9).

Figure 11.9. Foliage damage by tobacco caterpillar in capsicum

(ii) Management

Follow the management practices of these pests given under tomato.

Flow chart for insect pest management in capsicum

Raise seedlings in portrays in polyhouse:

At transplanting	• Apply neem cake at 250 kg/ha.
15 DAP	• Spray acephate for thrips.
30 DAP	• Apply neem cake at 250 kg/ha.
Post flowering and fruiting stage	• Monitor for pests like fruit borer, tobacco caterpillar, thrips and yellow mite. • Spray NPV according to the pest. • Spray acephate/fipronil/ethofenprox/ acephate + pongamia oil in rotation for thrips control. • Spray synthetic acaricide/botanical in rotation to control yellow mite.

11.2.2. Diseases

11.2.1.1. Powdery Mildew, *Leveillula taurica*

(i) Symptoms: The symptoms appear as whitish floury patches on the leaves, stem and petiole and rarely on fruit. Infected leaves at later stages show yellowish patches. The affected plants yield poor quality fruits. The disease occurs at all stages of crop growth under protected cultivation. The delay in initiating plant protection measures increases severity of the disease.

(ii) Management

Since the spread of the disease is fast within the polyhouse, care should be taken to destroy the initial inoculum by application of fungicides (0.1% dinocap or 0.075% hexaconazole or 0.3% wettable sulphur) to prevent further spread of the inoculum.

11.2.1.2. Mosaic Virus

(i) Symptoms: The important symptoms of this disease are mottling, distortion and reduction in leaf lamina. Sometimes the infected plants show vein banding, yellowing and leaf curling symptoms. In case of early infection, stunted growth and reduction in fruit size and yield are seen. In later stages of infection, the fruits become malformed and unfit for consumption. Yellow and green streaks may be produced on the fruits.

(ii) Management

Effective management of thrips and mite population in the polyhouse reduces the disease incidence. Application of 0.15% acephate or 0.1% fipronil or 0.1% dicofol or 1% emulsified neem oil is recommended for insect vector management.

Flow chart for disease management in tomato and capsicum

15 DAP	• Spray acephate for thrips.
Raise seedlings in portrays in polyhouse:	
At transplanting	• Apply neem cake at 250 kg/ha. • Treat seed with captaf or thiram at 3 g/kg seed. • Enrich growing medium with *Trichoderma harzianum* at 1 kg/100 kg.
15 days after transplanting	• Drench the seedlings in portrays with 0.4% copper oxychloride (COC).
30 days after transplanting	• Spray 0.3% COC to seedling portrays.
Post flowering and fruiting stage	• Monitor for diseases and insect vectors. • Spray schedules may be undertaken as per the diseases noticed.

General Preventive Measures to Control Insect Pests, Diseases and Nematodes in Tomato and Capsicum

- Use of resistant cultivars.
- Destruction of infected plant debris to reduce the inoculum level.
- Soil fumigation prior to planting reduces the infective propagules of pathogens, nematodes, insects, etc.
- Seed treatment and drenching nursery with copper oxychloride.
- Spraying nursery with a fungicide and insecticide prior to transplantation.
- Spraying sides of polyhouse with fungicide and insecticide to destroy inoculum and to prevent their entry.
- Use of coats and changing the foot wear while entering the polyhouse by the workers to prevent the carrying of inoculum and pests inside the polyhouse.

12 PESTICIDE RESIDUES IN VEGETABLE CROPS

The increased use of pesticides for the management of horticultural crop pests (insects, mites, vertebrates, nematodes and diseases caused by fungi, bacteria, virus/mycoplasma/ phytoplasma) due to relatively heavy carry over of these pests, has created the problem of contamination of vegetables. Toxic pesticide residues on vegetables that are consumed fresh may create health hazards to the consumers due to inherent toxicity. Further, the biomagnifications of pesticide residues in human body over a period of time can also lead to long term health effects. While there is necessity to regulate the build up of pesticide residues in food chain for internal market, there is also economic need for such regulation in view of quality control on growing export of vegetables in global market. These factors suggest rationalizing the use of pesticides on vegetables and managing the resultant build up of toxic residues by working out safe waiting periods. Further, the persistence of pesticide residues can also be dislodged with suitable decontamination processes.

12.1. CONTAMINATION OF VEGETABLES WITH PESTICIDE RESIDUES

The indiscriminate use of pesticides in vegetable crops many times close to harvest as well as post-harvest applications during transit and vegetable yards is resulting in contamination of vegetable produce with the pesticide residues. The periodic monitoring of market basket samples of various vegetables carried out across the country showed 40–60% samples contaminated with pesticide residues (Handa, 1992). Many a times, the extent of residues are above prescribed maximum residue limits (MRL) in India as against only 18–19% samples found contaminated in developed countries. Further, it is alarming to find that most of the contaminated samples were found to be loaded with hard to degrade and highly toxic pesticides like DDT and HCH (Srivastav and Patel, 1990), is of great concern from the point of possible health hazards to consumers. The contamination of vegetables with DDT and HCH appears to be from lateral unauthorized use of these pesticides after harvest. However, there has been wide variation in level of contamination in different vegetables *viz.*, potato-45.6%, tomato-51.5%, brinjal-51.7%, chilli-74.0%, cabbage-48.1%, cauliflower-61.0%, okra-58.0%, beans-78.1%, onion-33.3%, gourds-92.3%, leafy vegetables-96.4%, were found contaminated across the country (Awasthi, 1998).

Increased use of newer pesticides under high-tech and off-season vegetable cultivation has simultaneously increased pesticide residue load on vegetable produce. Accordingly, the monitoring of farm gate samples of various vegetables showed the contamination level to the extent of 60% with varying levels of pesticide residues especially with insecticides such as monocrotophos, dimethoate, quinalphos, triazophos, endosulfan, lindane, deltamethrin, fenvalerate, cypermethrin; and fungicides like carbendazim and mancozeb (Awasthi and Ahuja, 1997; Ahuja *et al.*, 1998). The situation is considered to be still more serious where the consumption of raw vegetables like cucumber, tomato, bell pepper, radish, onion and cabbage as salad is popular. Further, dipping of vegetables in pesticide solutions by retailers, to keep the vegetables fresh looking, is the limit of ignorance from the point of possible health hazards.

12.2. PERSISTENCE OF TERMINAL PESTICIDE RESIDUES AND WAITING PERIODS FOR VEGETABLES

The pesticide residue studies on vegetables have been carried out in respect of the progressive and periodic evaluation of the persistence pattern of different insecticides, fungicides and nematicides under different agro-climatic conditions of recommended pesticide schedules for control of various pests in order to clear these schedules to be adopted. The information on persistence pattern of various pesticides on different vegetable crops in their fruiting stage along with the terminal residues at harvest was reviewed and suitable pre-harvest waiting periods, in terms of days to be allowed to lapse between pesticide application and harvest, were worked out to allow healthy residue-free harvest of vegetables (Bindra and Kalra, 1973; Agnihotrudu and Mithyantha, 1978; Awasthi, 1993a, b; Agnihotri, 1999).

The size, shape, growth pattern and stage of vegetables; chemical class of pesticide, formulation used; number of applications and climatic conditions were found to play an important role in persistence and dissipation of pesticide residues in or on vegetables.

12.3. IMPLICATIONS OF PESTICIDE RESIDUES AND SAFETY MEASURES

The toxicity due to pesticide residues, apart from acute effects, produces latent diseases and disorders such as cancer, heart diseases, brain, kidney and liver damage as well as sterility, spontaneous abortions and birth defects. The risk from acute to sub-acute toxicity is unlikely to be of concern in case of consumers as they seldom come in contact with pesticides directly. The consumer is mainly exposed to pesticides through consumption of contaminated food with toxic pesticide residues. The Government through legislation has enforced many safety factors to ensure that pesticide residues remain below the prescribed maximum residue limit (MRL) (Table 12.1) (Agnihotri, 1999). These include the rate of pesticide application, time of application, number of applications and duration between the last application and harvest, commonly called waiting periods.

Table 12.1. Maximum Residue Limit (MRL) values of pesticides on vegetables

Pesticide	Vegetable	Maximum Residue Limit (mg/kg)
Aldicarb	Potato	0.50
Aldrin, Dieldrin	Vegetables	0.10
Benomyl	Vegetables	0.50
Captafol	Tomato	5.00

Table. 12.1. (Contd.)

Pesticide	Vegetable	Maximum Residue Limit (mg/kg)
Carbaryl	Potato	0.20
Carbendazim	Vegetables	0.50
Carbofuran	Vegetables	0.10
Chlordane	Vegetables	0.20
Chlorfenvinphos	Vegetables	0.05
Chlorothalanil	Potato	0.10
Chlorpyriphos	Potato; Onion	0.01
	Other vegetables	0.20
Copper oxychloride	Potato	1.00
	Other vegetables	20.00
Cypermethrin	Brinjal	0.20
	Cabbage	2.00
	Okra	0.20
DDT	Vegetables	3.50
Decamethrin, Deltamethrin	Brassica leaf vegetables	0.05
	Bulbous vegetables	0.10
	Leafy vegetables	0.20
Diazinon	Cauliflower	0.50
Dichlorvos	Vegetables	0.15
Dicofol	Vegetables	5.00
Dimethoate	Vegetables	2.00
Dithiocarbamates	Potato	0.10
	Tomato	3.00
Endosulfan	Vegetables	2.00
Ethion	Cucumber; Squash	0.50
	Other vegetables	1.00
Fenitrothion	Vegetables	0.30
Fenthion	Onion; Beans	0.10
	Potato	0.05
	Peas; Tomato	0.50
	Muskmelon	2.00
	Other vegetables	1.00
Fenvalerate	Cauliflower; Brinjal; Okra	2.00
Formothion	Vegetables	2.00
Heptachlor	Vegetables	0.055

Table. 12.1. (Contd.)

Pesticide	Vegetable	Maximum Residue Limit (mg/kg)
Hexachlorocyclohexane (HCH)–Gamma isomer	Vegetables	3.00
Lindane	Vegetables	3.00
Malathion	Vegetables	4.00
Monocrotophos	Carrot; Turnip; Potato; Sugar beet	0.05
	Onion; Peas	0.10
	Other vegetables	0.20
Parathion	Vegetables	0.50
Parathion (methyl)	Vegetables	1.00
Permethrin	Cucumber	0.50
Phorate	Tomato	0.10
	Other vegetables	0.05
Phosalone	Potato	0.10
	Other vegetables	1.00
Phosphamidon	Vegetables	0.20
Pyrethrins	Vegetables	1.00
Quinalphos	Pigeonpea	0.01
Thiometon	Potato; Carrot; Sugar beet	0.05
	Other vegetables	0.50
Trichlorofon	Potato; Carrot; Sugar beet	0.05
	Other vegetables	0.50

12.4. MANAGEMENT OF EXCESS PESTICIDE RESIDUES

The periodic sprays, dustings, soil applications of pesticides for the management of various pest complexes make the final sink of their toxic residues in the vegetables, especially which are often consumed afresh. In order to avoid and minimize their adverse effects, the vegetables may be washed with water which would dislodge and reduce the residues to the extent of 20 to 25% (Awasthi, 1986). Treatment with common salt solution, dilute solutions of acid, alkali and non-toxic chemicals could bring down the residues to 20 to 40%. Washing vegetables with dilute detergent solution was found to reduce it by 55 to 60% from vegetable surface. Boiling of vegetables could reduce 40 to 100% of the residues. The processing of the treated vegetables into different products was found to reduce the residues to the extent of 30 to 80%.

Further, keeping in view the environmental pollution and the adverse effects due to excessive residues following repeated and indiscriminate use of pesticides, the strategy of integrated pest management (IPM) was mooted. IPM programmes depend on multi component suppression of pest population and restrict the regular prophylactic usage of broad spectrum pesticides. Therefore, the main component of IPM is not only pest management but also pesticide residue management by reducing pesticide load in order to protect the environment and life from possible hazards due to toxic pesticide residues.

12.5. PESTICIDES/PESTICIDE FORMULATIONS BANNED IN INDIA (AS ON 1ST JUNE 2005)

12.5.1. Pesticides Banned for Manufacture, Import and Use in India (25 Nos.)

1. Aldrin
2. Benzene hexachloride (BHC)
3. Calcium cyanide
4. Chlordane
5. Copper acetoarsenite
6. Dibromochloropropane (DBCP)
7. Endrin
8. Ethyl mercury chloride
9. Ethyl parathion
10. Heptachlor
11. Menazone
12. Nitrofen
13. Paraquat dimethyl sulphate
14. Pentachloro nitrobenzene (PCNB)
15. Pentachlorophenol (PCP)
16. Phenyl mercury acetate (PMA)
17. Sodium methane arsenate (MSMA)
18. Tetradifon
19. Toxafen
20. Aldicarb
21. Chloro benzilate
22. Dieldrin
23. Maleic hydrazide
24. Ethylene dibromide
25. Trichloro acetic acid (TCA)

12.5.2. Pesticides/Pesticide Formulations Banned for Use but their Manufacture is allowed for Export (2 Nos.)

1. Nicotine sulphate
2. Captafol 80% WP

12.5.3. Pesticide Formulations Banned for Import, Manufacture and Use (4 Nos.)

1. Methomyl 24% L
2. Methomyl 12.5% L

3. Phosphamidon 85% SL
4. Carbofuran 50% SP

12.5.4. Pesticides Refused Registration (as on 1st June 2005)

1. Calcium arsenate
2. EPM
3. Azinphos methyl
4. Lead arsenate
5. Mevinphos (Phosdrin)
6. 2, 4, 5-T
7. Carbophenothion
8. Vamidothion
9. Mephosfolan
10. Azinphos ethyl
11. Binapacryl
12. Dicrotophos
13. Thiodemeton/Disulfoton
14. Fentin acetate
15. Fentin hydroxide
16. Chinomethionate (Morestan)
17. Ammonium sulphamate
18. Leptophos (Phosvel)

12.5.5. Pesticides Withdrawn (7 Nos.)

1. Dalapon
2. Ferbam
3. Formothion
4. Nickel chloride
5. Paradichlorobenzena (PDCB)
6. Simazone
7. Warfarin

12.5.6. Pesticides for Restricted Use in India

1. Aluminium phosphide
2. DDT
3. Lindane
4. Methyl bromide
5. Methyl parathion
6. Sodium cyanide
7. Methoxy ethyl mercuric chloride (MEMC)
8. Monocrotophos (ban for use in vegetables)
9. Endosulfan (banned in Kerala)
10. Fenitrothion

13 ALLIED TOPICS IN PEST MANAGEMENT

13.1. USE OF POWER SPRAYERS

Power sprayers require less spray fluid per unit area as compared to high volume sprayers like gator, back pack, etc. While high volume spray require about 800–1000 litres of spray fluid per hectare, low volume spray with power sprayers require only 200–300 litres. Hence, while using power sprayers, spray solutions should be at least three times more concentrated than high volume sprays and spray solutions also should be reduced three times. For example, if 450 ml of monocrotophos is to be mixed with 300 litres of water for Gator sprayer, then mix 450 ml of monocrotophos in 100 litres of water for power sprayer.

13.2. USE OF STICKERS/SPREADERS

The addition of stickers/spreaders to the spray solution improves the retention, even spreading of the spray solution over sprayed surface.

Commonly used stickers and spreaders are as follows:

Sandovit	:	0.5 ml/l
Teepol	:	0.5 ml/l
Triton	:	0.5 ml/l
Citowet	:	0.5 ml/l

13.3. PREPARATION OF BORDEAUX MIXTURE AND BORDEAUX PASTE

13.3.1. Preparation of 1% Bordeaux Mixture

Dissolve one kg of copper sulphate (98% purity) in 50 litres of water in wooden barrel or plastic drum. Dissolve one kg of fresh hydrated lime in 50 litres of water separately. Both the solutions should be sieved. Pour slowly copper sulphate solution and lime water simultaneously by continuous stirring into the tank.

It should be ensured that there is no free copper in the mixture. To check this, a clean iron knife/blade should be held/dipped in the mixture for a few minutes, removed and observed for any reddish brown deposit of copper on it. If the copper deposit is noticed, add a little lime solution to the mixture to neutralize excess copper. The best way is to use litmus paper and check pH. The Bordeaux mixture should be neutral or slightly alkaline, otherwise it will injure the leaves/plants. Bordeaux mixture should always be prepared fresh and should not be stored for more than one day. However, mixture can be stabilized by adding 200 gm of jaggery per 100 litres of mixture.

13.3.2. Preparation of 10% Bordeaux Paste

- Dissolve 1 kg of copper sulphate in 5 litres of water.
- Dissolve 1 kg of quick lime in another 5 litres of water separately.
- Mix the above 2 solutions to get the Bordeaux paste.

13.3.3. Caution While Using Bordeaux Mixture

- Always spray freshly prepared mixture for effective management. Therefore, prepare only required quantity for the day.
- Use only wooden, earthen or plastic vassals, to avoid the corrosion of the metal vessals by the mixture.
- Do not use the Bordeaux mixture in combination with any other chemical or pesticide.
- To avoid choking of nozzle, it is advisable to strain the Bordeaux mixture through a cloth or a sieve before putting into the spray tank.
- The Bordeaux mixture tends to sediment easily. Therefore, its stirring while using is desirable.
- In exceptionally hot days, when the plants are showing signs of temporary wilting or when it is raining, the Bordeaux mixture should not be sprayed, particularly on newly emerged tender foliage. Exercise this caution particularly on nursery plants.
- After carrying out spray operations, the appliances should be thoroughly washed with plenty of water to remove any copper deposits.
- The left over Bordeaux mixture should not be dumped in the field as this may prove toxic to the subsequent sowings.

13.4. PREPARATION OF 4% NEEM SEED KERNEL EXTRACT

4 kg kernel of neem seed is required to prepare 100 litre of spray fluid. Grind 4 kg of kernel with 20 litres of water to get a fine paste. For grinding, a mixer or wet grinder may be used. This fine paste is kept for 12 hr and repeatedly filtered, washed and squeezed through a rough cloth and made up to 100 litres, which is used for spraying. About 800–1000 litres of solution is required to spray one hectare of vegetable crops.

13.5. SEED TREATMENT FOR STORAGE

Pests and diseases affect seeds in storage. Following seed treatments may be adopted for protecting seeds during storage (Table 13.1):

Table 13.1. Effect of different pesticides on storage of seeds

Causal organism	Pesticide	Dose
Insects	Carbaryl	4 g/kg seed
Diseases	Subeej/Derosal	2 g/kg seed
	Chlorothalonil/Captan/Iprodione	2 g/kg seed
	Thiride/Vitavax	2 g/kg seed

13.6. CONTROL OF RODENTS

There are three main methods to control rodents in the field. Trapping has been employed since the earliest times, and much skill has been developed in its use. Poisoning and environmental alteration have the biggest contribution to make to increasing crop yields.

13.6.1. Control in the Field

13.6.1.1. Poisons

There are three main kinds of poisons for rodent control, multi-dose, single-dose and gassing poisons. Warfarin, coumatetralyl, chlorophacinone, coumachlor, difenacoum and calciferol are multi-dose poisons. All except the last are anticoagulants, which have an effect on the blood system of the animal. These compounds will normally give up to 100% kill. The single dose poison is zinc phosphide. Generally both multi-dose and single-dose poisons have to be mixed with bait and have a lethal effect when eaten, but gassing poisons release poisonous gases which are breathed by the rodents. In most situations where there is a rodent problem in growing crops, the use of bait containing a multi-dose poison is the best method of control.

Detecting Live Burrows (where rodents are active): Locate all burrows in the field and close them with mud in the evenings. Next morning, check-up and find out the burrows which have opened. They are live burrows.

Baits: Rats and mice shall usually eat cereals (milled or whole grain) and often the cheapest locally available will be accepted as baits by them. Rodents tend to like sugar and a small amount of sugar is added to the bait to make it more attractive. Sometimes a mixture of cereals will give good bait.

Pre Baiting: To familiarize the rats with the food used for poisoning, pre-baiting is required. For this, prepare plain bait of a mixture of wheat, bajra, jowar or maize flour and groundnut chops mixed with vegetable oil for 2–3 days. When rats start accepting the bait, they can be poisoned.

Bait Mixing: It is very important that the correct amounts of poison may be mixed throughout the cereal so that it is all poisoned. The concentration of rodenticides has been calculated to produce the maximum kill of rats. The instructions on the label should be followed. Many of the anticoagulants such as warfarin are sold as a master-mix or concentrate which require to be mixed in the proportion of one part to 19 parts of bait. Mixing of zinc phosphide should be done just before use.

Suitable Mixtures: If a cereal flour or meal is chosen as bait, a satisfactory mix can be obtained by mixing the poison and the meal together for about one minute.

Quantities for Warfarin Bait

Fine cereal meal or flour	–	19 parts (1900 g)
Warfarin concentrate (0.5%)	–	1 part (100 g)

Quantities for Zinc Phosphide Bait

Fine cereal meal or flour	–	39 parts (975 g)
Zinc phosphide	–	1 part (25 g)

If whole grain bait is chosen, it is necessary to add a 'sticker' to ensure that the poison is evenly distributed. Corn oil or other vegetable oils are the best choice although a mineral oil is also acceptable.

Quantities for Warfarin Bait

Wheat or other grain	–	18 parts (1800 g)
Corn oil	–	1 part (100 ml)
Warfarin concentrate (0.5%)	–	1 part (100 g)

Quantities for Zinc Phosphide Bait

Wheat or other grain	–	36 parts (1800 g)
Corn oil	–	2 parts (50 ml)
Zinc phosphide	–	1 part (25 g)
or		
Zinc phosphide	–	3 parts
Flour	–	80 parts
Groundnut chops	–	16 parts
Edible oil	–	Sufficient to make a lump of the bait

Making Wax Blocks: A very useful way of making long lasting baits is by mixing the poison and the bait with melted paraffin wax. The wax sticks the bait and poison together and small blocks can be prepared by the following method:

Fine cereal meal or whole grain	–	10 parts (1000 g)
Sugar	–	1 part (100 g)
Warfarin concentrate (0.5%)	–	1 part (100 g)
Semi-refined paraffin wax (m.p. 60–65°C)	–	8 parts

To prepare the bait blocks, first heat the paraffin wax just enough to melt it. Next, stir the cereal, sugar and poison until all are thoroughly mixed. Then pour the mixture into flat tray and leave it to set. When set hard enough, cut into blocks each weighing about 100 g. The advantage of wax blocks over loose baits lies in their greater resistance to the weather. They remain palatable to rats even when exposed to rain and sun.

Prepare the bait as described above and keep some quantity in each live burrow and close it with mud. Check next day. If burrow remains closed, the rat is dead. If burrow is opened, rat is not dead. Hence, repeat the bating.

Gassing Poisons: Gassing powders act by releasing poisonous gases which are breathed in by the rodents. There are two main ones which have been used against rats; sodium cyanide and aluminium phosphide. They produce poisonous gases such as hydrogen cyanide and aluminium phosphine respectively, when they come in contact with the moisture.

Insert 6 grams (2 tablets) of cellophos (Aluminium phosphide) into each live burrow and close it with mud. Check next day. If burrow remains closed, the rat is dead. If burrow is opened, rat is not dead. Hence, repeat the fumigation.

13.6.1.2. Trapping

Even when most skillfully used, traps will rarely control rats as effectively as poisons. There use is probably confined to larger rodents being caught for food or to the few survivors of a poison treatment. Many different local traps will catch rats and those such as the break back trap which will kill the animal, solve the problem of having to kill rats that are caught in the traps. Any locally available food that the rats eat can be used for bait which should be placed in the centre of the treadle.

13.6.1.3. Weed and Bush Clearance

There is some evidence that weed control both within the crop and around the field edges will limit the numbers of rodents by reducing the sources of food and living spaces. Many species of rodents do not actually live in the crops they attack, but in the undergrowth, banks and grassland surrounding it. Bush clearance at intervals may well be less costly than poisoning. This may be a very useful additional benefit of weed control done primarily for improving crop yields directly.

13.6.2. Control in Storage

Making the store rodent-proof is probably the most satisfactory approach, but it is not always practicable and regular poisoning may be necessary to keep rodents in check. A very important and inexpensive method of minimizing a rodent problem is scrupulous attention to hygiene in and around the store. All potential rodent living areas—piles of unwanted building materials, wooden packing cases, waste paper and sweepings—should be removed completely and their further accumulation prevented. In addition, all undergrowth should be cut down and kept at a height which will enable any new rat holes and runs to be readily detected.

13.6.2.1. Poisons

The use of multi- or single-dose poisons outlined in the field can be used. Aluminium phosphide tablets will kill rodents in stores, as will other fumigating gases.

13.6.2.2. Trapping

Traps may be useful to get rid of the survivors of a poison treatment or to catch a few rodents getting into a store.

13.7. CONTROL OF TERMITES

- Locate termite colony, destroy mound, locate the queen and kill it.
- Most effective method of killing the mound is poisoning the mounds by fumigating with cellophos at 2 tablets/metre diameter mound and plaster all the openings with wet mud.
- Pour 60 ml Coroban 20 EC/Dursban 20 EC through a hole made in the centre of closed mounds or through open chimney in case of open mounds.

13.8. CONTROL OF ROOT GRUBS

Incorporate in the soil 20 kg phorate 10 G or 25 kg carbofuran 3G or 30 kg solvirex 5 G or 25 kg ekalux 5 G per hectare or 100 ml Dursban/Coroban in 10 litres of water (500 litres to cover 1 ha.) after first summer rains (April-May) for maximum kill of root grubs.

13.9. READY RECKONER FOR MAKING PESTICIDE SPRAY SOLUTIONS

Table 13.2. Ready reckoner for making pesticide spray solutions

% *a.i.* of pesticides in commercial products	Quantity of pesticide formulation to be added per litre of water in gram or ml to get the desired strength of solution (%)											
	0.01	0.02	0.03	0.04	0.05	0.06	0.07	0.08	0.09	0.1	0.2	0.3
2.5	4.00	8.00	12.00	16.00	20.00	24.00	28.00	32.00	36.00	40.00	80.00	120.00
5	2.00	4.00	6.00	8.00	10.00	12.00	14.00	16.00	18.00	20.00	40.00	60.00
10	1.00	2.00	3.00	4.00	5.00	6.00	7.00	8.00	9.00	10.00	20.00	30.00
15	0.67	1.30	2.00	2.60	3.30	4.00	4.66	5.20	6.00	6.66	13.30	20.00
20	0.50	1.00	1.50	2.00	2.50	3.00	3.50	4.00	4.50	5.00	10.00	15.00
25	0.40	0.80	1.20	1.60	2.00	2.40	2.80	3.20	3.60	4.00	8.00	12.00
30	0.33	0.66	1.00	1.30	1.65	1.98	2.31	2.64	2.97	3.33	6.60	9.90
35	0.29	0.58	0.87	1.16	1.45	1.74	2.03	2.32	2.61	2.85	5.70	8.55
40	0.25	0.50	0.75	1.00	1.25	1.50	1.75	2.00	2.25	2.50	5.00	7.50
50	0.20	0.40	0.60	0.80	1.00	1.20	1.40	1.60	1.80	2.00	4.00	6.00
75	0.13	0.27	0.40	0.53	0.66	0.79	0.92	1.07	1.20	1.33	2.66	4.00
80	0.12	0.25	0.37	0.50	0.63	0.75	0.88	1.00	1.13	1.25	2.50	3.75
85	0.12	0.25	0.36	0.48	0.59	0.70	0.82	0.96	1.06	1.18	2.35	3.54
100	0.10	0.20	0.30	0.40	0.50	0.60	0.70	0.80	0.90	1.00	2.00	3.00

Note: The pesticides should be measured in g or ml. *e.g.* for preparation of malathion 0.05% spray solution from 50% malathion, the value at the point of intersection of 10th row and 5th column should be taken *i.e.*, 1.0 ml should be dissolved in 1 litre of water.

Preparation of Spray Solution

The spray solution must be prepared to the exact concentrations as recommended. The quantity of the commercial pesticide formulation required for preparation of the spray solution is calculated by using the following formula:

Quantity of the pesticide formulation

$$= \frac{\text{Quantity of spray solution required} \times \text{Concentration of solution desired}}{\text{\% of } a.i. \text{ of pesticide in the formulation}}$$

Prepare spray solution first in small quantity and then increase the volume to desired level by adding water. In case of wettable powder, take required quantity of pesticide, add a little quantity of water, mix it thoroughly to prepare the paste and then add remaining quantity of water to this paste with constant stirring.

13.10. COMPATIBILITY CHART

Table 13.3. Compatibility Chart: Precautions to be taken while using insecticides, fungicides and other Agrochemicals

No.	Pesticides/Agrochemicals	1	2	3	4	5	6	7	8	9	10	11	12	13	14	15	16	17	18	19	20	21	22
1	Plant origin insecticides@	a	a	a	a	a	a	a	a	a	a	b	a	a	b	a	a	a	a	b	b	c	a
2	Carbaryl	a	a	a	b	a	c	a	d	a	a	a	d	a	d	a	d	a	a	a	d	a	a
3	DDT Group	a	a	a	a	a	a	a	a	a	a	a	d	a	a	a	b	a	a	a	b	a	a
4	Monocrotophos	a	b	a	a	a	a	a	a	a	a	a	b	a	a	a	a	a	a	a	b	a	a
5	Dimethoate	a	a	a	a	a	a	a	a	a	c	a	c	a	a	a	b	a	a	a	b	a	a
6	Endosulfan	c	a	a	a	a	a	a	d	a	a	a	b	a	c	a	b	a	a	a	c	a	c
7	Fenitrothion	a	a	a	a	a	a	a	a	a	a	a	b	d	a	a	b	a	a	a	b	a	a
8	Kerosene	a	d	a	a	a	d	a	a	a	a	a	b	b	a	b	b	b	b	b	b	a	c
9	Malathion, Parathion, Demeton	a	a	a	a	a	a	a	a	a	a	a	b	d	a	a	a	d	a	a	a	a	a
10	Oxy-demeton methyl	a	a	a	a	c	a	a	a	a	a	a	a	c	a	a	a	c	c	a	a	c	c
11	Phoslane	d	a	a	a	a	a	a	a	a	a	a	a	d	a	a	a	a	a	b	b	b	b
12	Bordeaux mixture	b	b	d	b	b	b	b	b	b	b	d	a	b	d	a	b	b	a	a	a	b	a
13	Captan, Difolatan	a	a	a	a	a	a	d	b	d	a	a	b	a	d	a	b	a	d	a	b	a	a
14	Copper oxychloride	a	b	a	a	a	c	a	a	a	c	a	a	d	d	d	b	b	a	b	a	a	a
15	Dithiocarbamate group#	a	a	a	a	a	a	b	a	a	a	a	a	d	a	b	d	a	a	a	a	a	—
16	Lime sulphur	b	b	b	b	b	b	b	b	b	b	d	b	b	b	b	a	b	a	a	a	a	b
17	Karathane	a	a	a	a	a	a	a	b	a	a	a	b	e	b	a	b	a	b	c	c	b	c
18	Wettable sulphur	a	a	a	a	a	a	a	b	a	a	a	a	d	a	a	a	a	a	a	a	a	a
19	2,4-D	b	—	a	a	a	c	a	b	a	c	c	a	a	b	a	a	c	a	a	a	c	a
20	Lime	b	b	b	b	b	c	b	b	b	c	a	b	a	a	a	a	a	a	a	a	a	a
21	Streptomycin	c	a	a	a	a	a	a	c	a	c	c	a	a	a	a	a	a	a	c	c	a	c
22	Urea	b	a	a	a	a	c	a	a	a	c	c	a	a	a	b	a	a	a	a	b	c	a
23	Carbendazim	a	a	a	a	—	a	—	—	a	—	—	a	a	—	a	—	—	a	a	—	—	a

@ = Plant origin insecticides: Rotenene, Pyrethrum. # = Dithiocarbamates: Maneb, Zineb, Nabam, Thiram, Ziram, Mancozeb.

a = Can be mixed. b = Not to be mixed. c = Information not available. d = Use immediately.

e = Usually not mixed.

13.11. WAITING PERIODS FOR DIFFERENT PESTICIDES ON VEGETABLE CROPS

Table 13.4. Waiting periods for different pesticides on vegetable crops

Pesticides * Dose/Crop	Waiting Period in Days														
	Sevin 0.1%	Furadan 1 kg *a.i.*/ha	Dursban/ Coroban 0.05%	Cymbush/ Cypermethrin 0.01%	Nuvan 0.1%	Decis 0.002 %	Endosulfan 0.07%	Folithion/ Sumithion	Sumicidin 0.01%	Lebacid 0.05%	Methyl Parathion 0.05%	Monocrotophos 0.05%	Thimet 1 kg *a.i.*/ha	Ekalux 0.05 %	Phosphamidon 0.05%
Okra	·Nil	50*		3		Nil	2	5	7	3		3		12	Nil
Cabbage	Nil	60*	7	7		1	5	3	11	2	7	2	50*	5	
Cauliflower	10			8			3		9				70*	20	
Brinjal	Nil	Nil		2		Nil	Nil	21	6			10	**	21	
Tomato	Nil			5		Nil	7	5	10			10	**		Nil
French bean	5	70*		3		8	2		3			6	70*	8	
Onion	20	90*		Nil									60*		
Pea	Nil	70*		3		8	3	7	3			6	70*	8	
Radish	60	60*							10		5				
Amaranthus							8		15						
Bitter gourd		60*		1					2			10		7	11
Musk melon		60*		1					2	3		10		7	
Mushroom					8										

*—Soil application while sowing. **—Safe at second harvest.

13.12. PEST AND DISEASE MANAGEMENT IN NURSERIES

Among numerous diseases which cause severe damage, damping-off is the most important disease of tomato, chilli, cabbage and cauliflower. Besides this, seed rotting, bacterial blight, downy mildew, black rot, termite, leaf miner, flea beetle, whitefly and nematodes are other diseases and pests of vegetables in nursery. By the time tender radical and plemule are put forth in the nursery bed, they are killed by several seed and soil-borne fungi and bacteria. The pre-emergence damping-off and seed rotting are the usual causes for seedling stand and cause hidden losses. The post-emergence damping-off is a more serious problem. Seedlings just after emergence get infected near the soil surface (collar region) and give wire stem symptoms. The tissue near collar region become soft, water-soaked, constricted and weak causing the seedlings to collapse and later on die. Damping-off is caused by several species of soil fungi such as *Rhizoctonia, Pythium, Fusarium, Phytophthora, Phomopsis, Sclerotium, Colletotrichum, Phoma* and *Alternaria*. Heavy soil, excessive rain, high soil moisture, dense planting, low light intensity and high dose of nitrogen favour damping-off. Leaf miner is a problem in tomato hybrids and off-season nursery. Flea beetle and bacterial leaf spots are associated together in kharif tomato. Bacterial blight is the second most severe problem in tomato and chilli nursery in kharif season causing severe blighting. Following practices have been recommended for the management of pests and diseases in vegetable nurseries (Pandey, 2000).

13.12.1. Soil Solarization

Soil solarization by the use of white transparent polythene film during hot summer is one of the most effective approaches for the management of soil-borne pests and diseases particularly in nursery beds. The main objective is to eliminate pathogens, weeds, toxic metabolites, insects, etc. The dominant soil pathogens like *Macrophomina phaseolina, Fusarium oxysporum* f. sp. *pisi, Meloidogyne* spp., *Pseudomonas* spp., *Streptomyces scabies, Verticillium dahliae, Phytophthora* spp. and *Pythium* spp. which are responsible for nursery diseases were effectively reduced by soil solarization (Katan, 1981) (Table 13.5). Organic amendments of cruciferous plants when combined with soil solarization gave effective control of cabbage yellows caused by *Pythium* spp. and *Sclerotium rolfsii*. Excellent control of bacterial blight of tomato seedlings to the tune of 55–60% was recorded in solarized beds (Pandey and Pandey, 2002a).

Table 13.5. Effect of soil solarization on the management of soil-borne pathogens of vegetable crops and their yield

Crop	Soil-borne pathogen controlled	% Increase in yield over control
Potato	*Verticillium dahliae,Pratylenchus thornei*	35
Onion	*Pyrochaeta terrestris*	60–125
Tomato	*Stemphylium botryosum*	245
	Pyrochaeta lycopersici	140–350
Brinjal	*Verticillium dahliae*	215

13.12.2. Cultural Methods

Nursery beds should be properly drained and pulverized. The soil should be light and made porous by adding equal quantity of sand and compost. Sowing of seeds should be thin enough to provide proper aeration and light to check damping-off infection. Germinating seeds and juvenile seedlings should be protected from heavy rain, scorching sunlight through agronet, muslin cloth, shading net or polythene portable tunnel arch.

13.12.3. Organic Amendments

Decomposition of organic matter in amended soil has a major impact on disease suppression. Reduction in disease is largely induced by high soil temperatures during decomposition of organic amendments. Many types of micro-organisms (*Trichoderma* spp., *Streptomyces* spp., *Pseudomonas* spp., *Bacillus* spp., *Enterobacter* spp., *Flaviobacterium* spp.) present in compost function as biocontrol agents and their population is substantially increased after peak heating. Organic amendments are known to reduce the population of nematodes like *Meloidogyne* spp., *Helicotylenchus* spp. and *Pratylenchus* spp. (Hoitinik and Peter, 1986). Composted municipal sludge when incorporated in soil at 10% by weight reduced *Aphanomyces* root rot of pea, *Rhizoctonia* root rot of bean and radish, *Fusarium wilt* of cucumber, and *Phytophthora* crown rot of pepper.

.12.4. Biological Control

Biological control includes practices that create an environment favourable to antagonists and mass introduction of antagonists, non-pathogenic strains of other beneficial organisms. Three antagonistic mechanisms *viz*., competition, antibiosis and mycoparasitism are involved in biological control of soil-borne pathogens. Competition mechanism works in the control of *Fusarium oxysporum* f. sp. *melonis* by *Trichoderma harzianum*. Antibiosis occurs when the biocontrol agent produce antibiotics which have direct effect on the pathogen. The antagonists like *T. virens* produce gliotoxin, viridin and trichodermin by *T. viride* and *T. polysporum* which are effective against *Pythium ultimum* and *Macrophamina phaseolina*. Soil-borne pathogens causing root rot, damping-off, collar rot and wilt (*Rhizoctonia solani, Pythium* spp., *Fusarium* spp., *Phytophthora* spp., *Macrophamina phaseolina, Sclerotinia sclerotiorum, Phoma* spp. and *Aspergillus* spp., *Botrytis* spp., *Armillaria* spp., *Verticillium* spp. and *Erysiphe* spp.) in different vegetable crops are the major targets of the biocontrol agents. Many foliar diseases like blight (*Alternaria solani*) and die-back/ anthracnose (*Colletotrichum capsici*) of chilli are also suppressed by the *Aspergillus niger* V isolate (Pandey and Pandey, 2001).

13.12.5. Chemical Control

The easiest control measure for damping-off is to treat soil and seed. Five litres of 0.3% captan should be uniformly applied/m^2 area of nursery bed. Seed treatment with 0.25% carbendazim should be preferred over captan before sowing, considering the internal seed-borne inoculum of the pathogen. In general, seeds of tomato and chilli should be treated with 0.25% imidachloprid for protection against sucking pests in nursery. After complete germination of all the seeds, spray 0.3% copper oxychloride followed by 150 ppm streptocycline in such a way that seedlings and their roots are covered by the chemicals in tomato and chilli.

13.13. SOIL SOLARIZATION

Soil solarization is a method of killing soil-borne plant pathogens (fungi, bacteria, nematodes), soil insects and weed seeds by heat. This technique consists of covering moist soil with a plastic film during periods of intense sunshine and heat, thereby capturing radiant heat energy from the sun, causing physical, chemical and biological changes in the soil. The soil temperature is increased (by 5–15°C) to levels lethal to many soil-borne fungi, bacteria, nematodes and weeds. The short wave incident solar radiation penetrates the polyethylene sheet but the long wave radiation is prevented from soil, thus trapping the heat resulting in thermal inactivation and production of heat shock proteins and irreversible heat injury. Solarization also improves plant nutrition by increasing the availability of N and other essential elements. This practice can be profitably used on a small scale and is especially useful for nursery beds.

Tarping with Linear Low Density Poly Ethylene (LLDPE) 100 gauge film during summer for 15 days (in between April 15th to June 15th) in tomato nursery at Anand, Gujarat gave 66 and 92% reduction of root-knot nematodes and weeds, respectively. This practice produced 615% more seedlings over control giving net profit of Rs. 7,661 from 1000 m^2 area with cost: benefit ratio of 1: 5.65.

Soil solarization using thin transparent polyethylene mulches for six weeks caused as much as 72.6 and 88.5% decrease in the population densities of *Tylenchorhynchus vulgaris* and *Hoplolaimus indicus*, respectively in nursery beds. Application of mahua cake before solarization usually caused slightly greater reduction in nematode population (Kamra and Gaur, 1998).

13.14. MASS PRODUCTION OF BIOCONTROL AGENTS

13.14.1. *Pseudomonas fluorescens*

A loopful of *P. fluorescens* should be transferred to 100 ml of King's B broth in a 250 ml conical flask under aseptic conditions. The flask can be incubated at room temperature for 48 hr. The population of *P. fluorescens* in the broth can be assessed (a minimum of 9×10^9 cfu/ml is necessary).

The pH of the substrate (talc powder) can be adjusted to 7.0 by addition of 150 g calcium carbonate/kg of talc powder and sterilized at 1.1 kg/cm^2 for 30 min for 2 successive days. One kg of the sterilized substrate (talc powder) and 5 g of carboxy methyl cellulose should be placed in a polythene bag or any sterile container under aseptic conditions, 400 ml of *P. fluorescens* suspension to be added and mixed thoroughly. The formulation can be stored in a polythene bags for one month. The product should contain a minimum of 2.5×10^8 cfu/g.

The product should have the following quality parameters:

- Fresh product should contain 2.5×10^8 cfu/g.
- Three months after storage at room temperature, the population should be $8–9 \times 10^7$ cfu/g.
- Storage period is 3–4 months.
- Minimum population load for use is 1×10^8 cfu/g.
- The product should be packed in white polythene bags.
- Moisture content of the final product should not be more than 20%.
- Population in broth should be $9 \pm 2 \times 10^8$ cfu/ml.

The composition of King's B medium is as follows:

Peptone	—	20.00 g
Dipotassium hydrogen phosphate	—	1.50 g
Magnesium sulphate	—	1.50 g
Glycerol	—	10.00 ml
Agar	—	15.00 g
Distilled water	—	1000 ml
Cycloheximide	—	100 ppm
Ampicillin	—	50 ppm
Chloramphenicol	—	12.50 ppm

(King's B broth is prepared without the agar and antibiotics)

Liquid Formulation Technology to Extend the Shelf Life of *P. fluorescens*

Poly Vinyl Pyrrolidone (PVP) (2%) was tested as an effective preservation agent in protecting the cells of *P. fluorescens* for more than a year without affecting the natural properties of bacteria. The antagonistic efficacy of preserved cells was also confirmed *in vitro* and *in vivo*.

13.14.2. *Trichoderma* spp.

Among the various fungal biological control agents which are effective against nematode and fungal diseases of vegetable crops, *Trichoderma viride* and *T. harzianum* have a good market. They can be easily mass produced on several organic substrates.

13.14.2.1. Small Scale

The mycelial discs grown in potato dextrose agar medium were transferred to 250 ml conical flasks containing sterilized molasses yeast medium (Yeast extract –5 g, Molasses –30 g, Distilled water –1 litre sterilized at 1.42 kg/cm^2 for 20 min) at 100 ml /flask and incubated for 10 days. The mycelial mat along with the broth was homogenized and mixed with talc powder (500-mesh) at 1: 2 ratio. Carboxy methyl cellulose (CMC) was added at 5 g/kg of the product as sticker. The product was shade dried until it attained 12% moisture level. The clumps were broken, homogenized and packed in polythene bags of 100 g capacity. The required population level at the time of packing should be

28×10^7 cfu/g of the product. The shelf life of the product was 4 months. At the end of expiry, the cfu should not be less than 20×10^6 cfu/g of the product.

13.14.2.2. Large Scale

Add 70 ml of molasses yeast medium in each 250 ml conical flasks and plug with non-absorbent cotton wool. Autoclave at 1.42 kg/cm^2 for 1 hr. Transfer 8 ml mycelial disc of *T. viride*/*T. harzianum* to each flask and incubate for 10 days. This mother culture is used for inoculating the medium in fermentor.

Add 50 litres of molasses yeast medium in 50 litre fermentor, sterilize at 1.42 kg/cm^2 for 20 min and cool to room temperature. Add 2.5 litres of inoculum (from mother culture) into the sterilized medium and incubate for 10 days. Provide aeration in the fermentor for 4–8 hr per day by agitation. Presence of greenish fungal growth indicates *T. viride*/*T. harzianum*. This broth culture contains fungal mat, spores and antibiotics. Draw samples 10 days after incubation and assess the spore load in the broth which should be at least 10^8 spores/ml. Mix one litre of broth culture with 2 kg talc powder, add CMC at 5 g/kg of the product, air dry (12% moisture content) and sieve. The population should not be less than 20×10^6 cfu/g of the product. It is packed in opaque polythene bags and stored in a cool dry place before use. The product can be safely stored for 4 months without losing its viability.

13.14.3. Arbuscular Mycorrhizal Fungi (*Glomus mosseae, G. fasciculatum*)

13.14.3.1. In Glasshouse

Cement pots or tanks are filled with sterilized sand: soil (1: 1 by volume) or soilrite: peat (1: 1 by volume) mixture. A standard starter culture of AMF obtained from any authentic source is placed as a band 3 cm below the soil surface. The seeds of Rhoades grass/finger millet/maize are sown on them and covered with soil. Pots are irrigated normally. When the plants are about 10–12 weeks old, the top shoots are cut off and the root system colonized by the fungus along with the soil containing chlamydospores is used as the inoculum.

13.14.3.2. On-farm Production

A suitable land (even waste land) is selected near the field where the fungus inoculum has to be used. The land is brought to the fine tilth by ploughing, harrowing and cultivator operations. A raised bed (20 cm height) of soil: sand (1: 1 by volume) mixture is prepared on an even land surface. The raised bed is sterilized by soil solarization for 2 to 3 weeks using polythene sheet. The starter culture of AMF is applied at a depth of 3–4 cm from the top by removing the top soil. The seeds of Rhoades grass/finger millet/maize are sown on them and covered with soil. Irrigation is done normally. When the crop attains a height of 10–15 cm, 12 weeks after sowing, the shoot portion is cut off at ground level. The soil is dug up to a depth of 20 cm along with root system and is used as the inoculum.

13.14.3.3. Method of Inoculation

Vegetable crops can be inoculated with AMF in the nursery beds. The AMF inoculum consisting of chlamydospores, soil and infected root bits is either spread uniformly in the nursery beds or placed in rows over which seeds of vegetable crops (tomato, chilli, capsicum, brinjal) are sown. The seedlings get colonized by the fungus in about 4–5 weeks after which they are transplanted in the main field. Inoculated plants show better growth and improved nutrient content.

13.14.4. NPV of *Spodoptera litura*

13.14.4.1. Production and Utilization

- Collect 7–9 days old larvae (4th instar) and starve them for 8 hr.
- Prepare *Sl* NPV suspension (10^8 POBs-Polyhedral Occlusion Bodies/ml) in 250 ml of water with teapot in a bottle.
- Dip clean castor leaves in *Sl* NPV suspension, shade dry for 15–20 min and feed the larvae for 2 days.
- Provide fresh non-infected leaves for the larvae thereafter.
- Alternatively, expose larvae to semi-synthetic diet treated with *Sl* NPV suspension.
- Larvae die after 7 days and their under surface turns pink and later white.
- Collect dead larvae and allow it to putrefy in water for a week.
- Grind dead larvae and filter through muslin cloth.
- Remove the supernatant and suspend the polyhedra in water.
- Centrifuge at 500 rpm for 5–10 min and discard the supernatant retaining the POBs.
- Centrifuge further at 2500 rpm for 5–10 min and suspend the white POBs in water.
- Count using Neubauer haemocytometer and express on larval basis or per unit of larval weight.
- Dry the POBs over calcium chloride or precipitate by acetone.
- Formulate by adding permitted spreaders/wetting agents.

13.14.4.2. Nursery and Field Sprays

- Spray *Sl* NPV at 250 larval equivalent (LE) mixed in 125 litres of water added with crude sugar at 1 ml/l of water and teepol at 1 ml/10 litres of water in the evening hour for one ha of nursery.
- Spray thrice at fortnightly intervals from the day neonate larvae are sighted.
- Continue spraying up to three weeks after planting.
- Spray when eggs start hatching in castor at weekly intervals.
- The dosage varies from 250 LE to 450 LE/ha.
- Larvae feeding on sprayed plants turn pinkish white, move to top of the plant, hang upside down and die (Fig 13.1).

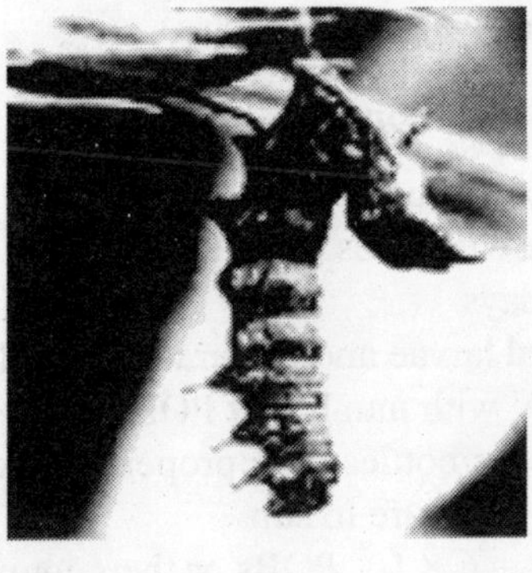

Figure 13.1. Infected larva hanging upside down (Tree top disease)

13.14.4.3. *Spodoptera litura* Culture

- Allow a pair of newly emerged moths in a well ventilated plastic container (20 × 15 cm) lined (inner surface) with paper to obtain eggs (Fig 13.2).

Figure 13.2. *Spodoptera* rearing unit

- Cover the bottom sponge lining with blotting paper.
- Provide 50% honey solution and water in small plastic cups.
- Cut out the egg batches on the paper.
- Dip eggs in 10% formalin for 30 min, wash in running water for 30 min and dry using blotting paper.
- Hatch in sterilized glass vials (20 g capacity).
- Transfer 3 day old larvae in bouquets of sterilized castor leaves for feeding.
- Collect the pupae, sterilize with sodium hypochlorite, wash in water and dry with blotting paper.
- Identify the sex with the help of ovipositor and last abdominal segment and keep over a wet sponge in an adult emergence cage (22 × 15 cm).
- Adults emerge after 10 days.

13.14.5. NPV of *Helicoverpa armigera*

13.14.5.1. Production and Utilization

- Prepare the viral solution (virus at 18×10^6 POBs/ml of distilled water).
- Feed 5–7 day old larvae with virus infected diet.
- Incubate at 26°C for 7 days.
- Collect the dead infected larvae and macerate in distilled water.
- Standardize the *Ha* NPV with number of POBs/ml in terms of LC_{50}.
- Store POBs in plastic cans/bottles with proper instructions on the containers even up to 6 months without exposure to sun.
- 1 larval equivalent (LE) = 6×10^9 POBs or their equivalent in activity.

13.14.5.2. Field Spray

- *Ha* NPV is sprayed to control *H. armigera* on chick pea, pigeonpea, field bean, and tomato.
- Spray 3–4 times of *Ha* NPV 250 LE/ha with crude sugar at 5 ml/l of water, groundnut oil cake at 1 ml/100 ml of water and chick pea flour at 1 ml/100 ml of water.
- After 3–5 days, dead larvae will be visible hanging upside down.

13.14.5.3. *Helicoverpa armigera* Culture

- Collect the adult insects in light traps or from endemic crop areas.
- Place the adults in oviposition cages (cylindrical iron frame-50 cm height × 30 cm diameter and black or white cloth cover) at 100 pairs/cage.
- Hang 2 cotton swabs dipped in 50% honey and water over a circular plastic mesh, 5 cm from frame base.
- Place the frame on aluminium tray (40 × 40 × 5 cm) with a 3 cm thick sponge at the bottom soaked in water.
- Remove eggs daily from the inner side of the cloth cover.
- Sterilize the eggs with 10% formalin, wash in distilled water for 10 min and dry in laminar air flow chamber.
- Place the eggs in a moist cotton inside a flask and wrap with aluminium foil.
- Larvae emerges after 3–4 days and place them in multi cellular trays.
- Larvae are fed with chick pea based nutritious semi-synthetic diet.
- Harvest pupae after 18–20 days, sterilize, wash and dry.
- Separate the male and female pupae and place in adult emergence cages.
- Adults emerge after 7–8 days.

13.14.6. Trichogrammatid Egg Parasitoids

13.14.6.1. Database

- Trichogrammatid egg parasitoids are multiplied on eggs of *Corcyra cephalonica*, a factitious host.

- One cc of *Corcyra* eggs contains 20,000 eggs. This is supplied in a card.
- Eight releases of Trichogrammatids and each release with 1,40,000 are recommended to contain first and second generation eggs/ha.
- Releases are spaced at weekly intervals.
- 56 cc of *Corcyra* eggs are to be produced/ha (7 cards/release × 8 releases).
- 56,000 cc of *Corcyra* are required/1000 ha. *i.e.* 56,000 cards.
- 150 g of sorghum grains are required to produce 1 cc of *Corcyra* eggs.
- 8,400 kg of sorghum grains are required as feed to produce 56,000 cc.
- Approximately 10 tonnes of sorghum grains are required as feed to produce natural enemies to cover 1000 ha.
- The parasitized *Trichogramma* egg card should be handed over to the growers within three days after despatch.
- It cannot be stored in dealers/retailers shop like antagonastic fungi/biofertilizers/ pesticides/fertilizers.

13.14.6.2. Production: *Corcyra cephalonica*

Ideal Conditions for Rearing

- **Temperature:** 27 ± 1°C.
- **RH:** 60 ± 5%.
- **Light:** 200–250 lux and avoid direct sun light.

Rearing

- In each box, put 2.5 kg broken cereal or millets, 10 tablets of dry yeast and 0.5 g streptomycin sulphate.
- Keep 0.5 cc *Corcyra* eggs and secure the lid of the charge box for 40 days.
- On 40th day, moths start to emerge and continue to emerge for 2 months with a peak emergence from 45th to 75th day.
- Emergence rate is 5–15 moths/box/day in the beginning and it reaches 75–100 moths/ box/day during the peak.
- Each box produces 0.5 cc eggs/day during the peak emergence.
- Collect the moths daily using mechanical moth collection devices and transfer them to special oviposition cages.
- Discontinue the use of box and burn the contents properly after 80 days of infestation when less than 10 moths/box emerges.
- Pass the collected eggs through 15, 30 and 40-mesh sieves and run over a slope of paper or by automatic egg cleaner to eliminate the dust particles.
- Treat the eggs with UV rays (30 W UV tube for 45 min at a distance of 60 cm) to prevent eggs from hatching, if they are meant for *Trichogramma* production.

Note: Eggs meant for multiplication of *Corcyra* should not be treated with UV rays.

13.14.6.3. Production and Utilization

- Collect cleaned and UV treated *Corcyra* eggs.
- Glue the eggs to "tricho" cards measuring 15 × 10 cm, which are prepunched into 16 pieces of 4 × 3 cm leaving some space for stapling.
- Expose the eggs to adult female *Trichogramma* in the ratio of 30 : 1 for 24 hr and allow it to parasitize (Fig 13.3).

Figure 13.3. *Tricho* adult oviposit on *Corcyra* egg, *Tricho* card

- Expose females to the eggs in the ratio of 40 : 1 on cards inside polythene bags, where female oviposits till they die.
- Transfer 3–5 day old parasitized egg cards in safe polythene bags and place dried honey for emerging parasitoids.

13.14.6.4. Field Release

- Cut "tricho cards" into small bits.
- Release cards 8 days after parasitization for emergence of adults.
- Staple the bits to lower side of the leaves at 100 spots/ha in the evening hours at weekly intervals (Fig 13.3).
- Release is done based on the appearance of eggs of host.

13.14.7. Chrysopids

13.14.7.1. Production and Utilization

Ideal Conditions for Rearing

- **Temperature:** 27 ± 1°C.
- **RH:** 70%.
- **Light:** Normal illumination by a fluorescent tube.
- **Room area:** 6 × 6 m.

Rearing

- Rear the adults with artificial honey based diet and larvae on *Corcyra* eggs separately.
- Keep 200 adults in the oviposition cage measuring 75 × 30 × 30 cm with a sliding top cover.
- Cover the sides of the cage with smooth nylon lining mesh of '0' size which is unsuitable for egg laying.
- Fit a black cloth on the top cover for laying of eggs.
- Replace the top cover with eggs on alternate days starting from 5th day.
- One batch is kept for 30–35 day and dead adults are removed every alternate day.
- Wait for 24 hr till the eggs harden and gently rub with a soft sponge.
- Expose *Corcyra* eggs to UV rays 30 W tube light for 45 min at 60 cm distance and inactivate them.
- Mix 3 day old 120 chrysopid eggs with 1 cc *Corcyra* eggs in a plastic container (27 × 18 × 6 cm).
- On hatching, *Chrysoperla* larvae feeds on *Corcyra* eggs.
- Remove the larvae on the 3rd day and transfer them into individual 2.5 cm cubical cells of plastic louvers.
- Each louver can hold 192 larvae.
- Sprinkle *Corcyra* eggs in all cells.
- Feed the larvae in 2 doses, first with 1.5 cc and second with 2.5 cc for 100 larvae with a gap of 3–4 days.
- Total quantity of *Corcyru* eggs required for rearing 100 *Chrysopid* larvae is 5 cc.
- Secure each louver on one side by brown paper and after transfer of larvae cover it with acrylic sheet. Brown paper would facilitate clear visibility of *Corcyra* eggs and also formation of cocoon by Chrysopid larvae.
- Collect cocoons after 24 hr of formation by removing paper from one side.
- Place them in adult oviposition cages for emergence.
- One set of louver remains in use for 13–15 days, then they are completely cleaned before reuse.

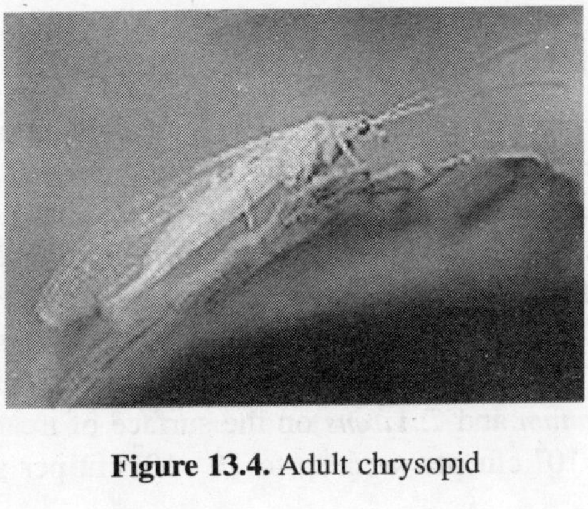

Figure 13.4. Adult chrysopid

13.14.7.2. Field Release

- Mix viable *Chrysoperla* eggs with *Corcyra* eggs. Add saw dust and paper strips. This is done before shipment. Once larvae emerge, they are released in the field.
- Release in the early morning by broadcasting larvae on the plants along with saw dust and paper strips.
- 10,000–20,000 larvae/ha is required in 2 releases which varies with crop, pest predator load.

13.15. ENRICHMENT OF FYM WITH BIOAGENTS (*TRICHODERMA HARZIANUM, PAECILOMYCES LILACINUS*)

Commercial formulation of the bioagent (*T. harzianum/P. lilacinus*) (1 kg/ton) was mixed with moist FYM along with 50 kg of neem/pongamia cake. Mixture was spread 15 cm high heap under shade, covered with polyethylene sheet and incubated for 3 weeks at 25 to 30°C and sprayed with water regularly to maintain moisture.

Colonized FYM can be used at 6 g/kg soil in pot culture experiments (12 g/2 kg soil in 15 cm diameter pots). It can also be used for main field treatment at 5 tonnes/ha before sowing of seeds or transplanting of seedlings. It can be applied at 2 kg/plant of perennial crops along with 400 g of neem/pongamia cake once in every 6 months after planting for the management of nematodes and other soil-borne pathogens.

T. harzianum colonized compost significantly enhanced germination and growth of both tomato and okra as compared to non-colonized FYM. Population of *T. harzianum* in colonized compost was several folds higher (48.6×10^{12}) than non-colonized FYM (3×10^4).

Colonization of FYM by *T. harzianum* helps to improve the quality of compost. It significantly enhanced total and water soluble humic matter. A number of macro and micro-nutrients like P, K, S, Zn, Cu and Fe (both total and water soluble content) were significantly higher in *T. harzianum* colonized compost. Six times increase in water soluble humic content due to colonization of *T. harzianum* seems quite important as it is reported to enhance the plant growth.

13.16. SEED BIOPRIMING/SOLID MATRIX PRIMING

Seed biopriming (treatment of seeds with a slurry of biocontrol agents mixed with an organic carrier and incubated for 4 days under warm and moist conditions with relative humidity of 80–100% until just prior to radical emergence) has potential advantages over simple coating of seeds as it results in rapid and uniform seedling emergence. *Trichoderma* required less moisture for growth. *Trichoderma* conidia germinate on the seed surface and form a layer around bioprimed seeds. Such seeds tolerate adverse soil conditions better. Biopriming could also reduce the amount of biocontrol agents that is applied to the seed. Population of *T. harzianum* and *T. virens* on the surface of treated seeds of tomato and brinjal increases from 10^4 cfu per seed up to 10^6–10^7 cfu per seed after biopriming.

13.17. FORMULATION OF ALGINATED PELLETS OF BIOCONTROL AGENTS

Alginate is a water soluble polysaccharide gum that has excellent gel-forming properties. Living organisms that are capable of biologically controlling a variety of nematodes can easily be entrapped in a calcium or sodium alginate matrix in aqueous form at room temperature. A pellet formulation of biocontrol agents was prepared by mixing sodium or calcium alginate, clay and streptomycin sulfate. The mixture was homogenized in a blender and then mixed with bioagent propagules. The gel beads, granules or seed coatings that are produced by the process are biodegradable. The encapsulation in Ca or Na alginate maintained the viability of the bioagents for 12 months at 7°C and 5 months at room temperature. Antagonistic fungi (species of *Paecilomyces, Trichoderma, Gliocladium, Talaromyces* and *Pencillium*) and bacteria (species of *Pseudomonas* and *Bacillus*) that can control soil-borne pathogens can be formulated in Ca or Na alginate (Connick, 1988). Alginated pellets can be used for soil application at 400 to 800 pellets/sq.m.

13.18. PRODUCTION OF BIOAGENT COLONIZED SEEDLINGS

The horticultural crops are invariably attacked by soil borne pathogens (nematodes, fungi, bacteria) at the seedling stage itself. Securing healthy seedlings of horticultural crops is essential to ensure freedom from soil-borne pathogens, optimum plant stand, good growth and higher yields. The following methods may be adopted to produce bioagent colonized seedlings of horticultural crops:

13.18.1. Seed Treatment

Seeds have to be treated with talc based formulations of *Pseudomonas fluorescens, Trichoderma harzianum* and *T. viride* at 10 g/kg seed before sowing.

13.18.2. Nursery Bed Treatment

The nursery beds may be treated with *P. fluorescens/T. harzianum/T. viride* at 50 g/m^2 along with neem/pongamia cake at 200 g/m^2 before sowing.

13.18.3. Treatment of Coco Peat or Any Other Substrate

The coco peat or any other substrate used for producing the nursery seedlings in portrays is treated with *P. fluorescens/T. harzianum/T. viride* at 1 kg/tonne before filling with the substrate.

13.18.4. Treatment of Soil Mixture

The soil mixture used for producing the nursery seedlings in polythene bags in a conventional manner is treated with *P. fluorescens/T. harzianum/T. viride* at 2 kg/tonne along with neem/pongamia cake at 50 kg/tonne before sowing.

13.19. TRADITIONAL PEST CONTROL

See table as given on next page.

Product	Preparation	Pests controlled
Neem leaf extract	Chop 6 kg fresh neem leaves and make it into a paste. Soak the paste in 6 litres of water overnight. On the following morning boil the suspension and filter by pouring over a coarse cloth. Add 150 g of detergent soap powder to the filtrate. Make up the volume to 60 litres. This is ready to spray.	Most of the leaf feeding caterpillar pests.
Vitex leaf extract (Nochi)	Grind 6 kg fresh leaves into a paste. Soak the paste overnight in 3 litres of cow's urine. On the following morning, stir the solution and boil for 3 hours. Allow it to cool. Pour it over a coarse cloth and squeeze. To the filtrate, add 150 g of detergent soap powder. The filtrate is then made up to 60 litres by adding water. This is ready to spray.	Castor semi looper.
Calotropis leaf extract	Chop 1 kg calotropis leaves and make it into a fine paste. Mix the paste with 50 litres water and filter the suspension and spray.	Caterpillar pests.
Jatropha leaf extract	Grind 5 kg of leaves and stem of Jatropa into fine paste. Soak the paste in 5 litres of cow's urine overnight. On the following morning, add 2 litres of water and boil for 3 hours. The solution is then concentrated to 5 litres and filtered; this is ready to spray.	All pests.
Tobacco leaf extract	Cut 5 kg of tobacco waste into pieces and soak in 5 litres of cow's urine for 1 week. Filter the suspension and mix 150 g of detergent soap powder and 250 ml of water. Spray at 50 ml/l water.	Aphids, thrips, beetles, lepidopteran caterpillars.

Table. (Contd.)

Product	Preparation	Pests controlled
Agave flesh extract	Grind chopped agave leaves and make it to a paste by thoroughly mixing it Soak it in water for 24 hr. Mix the suspension and filter it with a coarse cloth. Add 10 g of detergent soap powder and 500 ml of cow's urine to every 100 ml of extract and spray.	Leaf folder.
Garlic clove extract	Chop garlic cloves and make it into a fine paste. Soak it in equal volume of kerosene over night Mix and filter it. Add 150 g of detergent soap powder.	Brown leaf hopper, green leaf hopper, thrips, diamond back moth.
Mahua oil spray	Mix 30 ml of mahua oil and 5 g of detergent soap powder with 1litre of water and spray.	
Nerium seed extract	Seeds of nerium are dried, powdered, sieved and soaked in water over night. Filter with coarse cloth and spray the filtrate.	Rhinocerous beetle.
Cow dung extract	Mix 100 g cow dung in 100 ml water. Filter it and spray.	
Ipomea leaf extract	Clip 5 kg of fresh Ipomea leaves and grind to fine paste. Mix paste with equal volume of water and boil. Add a little of cow's urine to the boiling mixture. When the mixture is ¼ its volume stop boiling and allow it to cool overnight. Filter the solution. Repeat the washing with cow's urine to make a volume of 4.5 litres. Add 250 g of detergent soap powder in 500 ml water. Mix the extract with the soap solution. Use 1.25 litres of solution in 25 litres water.	All pests.

14 SOURCES OF CRITICAL INPUTS FOR THE MANAGEMENT OF PESTS AND DISEASES

14.1. INSECTICIDES

Table 14.1. Common and trade names and sources of availability of insecticides

Group	Common Name	Trade Name/s (Source)
Chlorinated Hydrocarbons	Endosulfan	Agrosulfan (16), Dawn (42), Devigor (9), Endocel (14), Endocin (23), Endogaurd (15), Endodhan (13), Endotaf (37), Endonil (11), Endoin (20), Endosul (46), Endosol (28), Hexasulfan (32), Hildan (17), Hysulfan (18), Lusulfan (26), Nagusulphan (29), Parrysulfan (18), Sedostar (41), Speed (31), Thiodan (4), Tagsulfan (50), Thiokill (52), Topsulfan (49).
	Gamma BHC	Lindane
Organophos-phates	Acephate	Acatin (23), Acehero (39), Acekill (5), Acemil 75 WP (28), Acet (19), Acetop (49), Acifete (26), Aimthene (1), Artin (32), Asataf (37), Dhanraj (11), Hilphate (17), Hythene 75, SP (18), Lancer (52), Laser (20), Orthane (9), Pace (31), Specthane (43), Starthane (41), Tamaron Gold (4), Tarpedo 75 SP (40), Tremor (55), Volace (54).
	Chlorfenvinphos	Chlorfenvinphos
	Chlorpyriphos	Agro Chlore (16), Blaze (21), Chorobon (52), Chlorgaurd (15), Classic (26), Coroban (7), Corocin (23), Cyphos (19), Dhanwan 20 (11), Doomer (5), Durmet 20 EC (8), Force (31), Hexaban (32), Hilban (17), Hyban (18), Nuklor 20 EC (12), Pyrivol (54), Radar (40), Robust (39), Ruban (9), Scout (1), Specphos20 (42), Starban (41), Strike (55), Sulban (46), Tafaban (37), Tagban 20 EC (50), Tarkash (3).
	Diazinon	Basudin (47), Tik-20 (37).

Table. 14.1. (Contd.)

Group	Common Name	Trade Name/s (Source)
	Dichlorvos (DDVP)	Agro-DDVP (16), Amidos (1), Dicotop (49), Divap 100 (35), Doom (52), Hilfol (17), Luvan (26), Marvex Super (32), Nuvan (47), Specvos (43), Suchlor (45), Vantaf (37), Vapona (3).
	Dimethoate	Hygro (18), Rogar (37), Rogrin (20), Rogar (24), Tara-909.
	Ethion	Acarin (18), Crush (20), Dhanumit (11), Fosmite (35), Indothion (21), Laser (26), MIT-505 (41), Phosmil (21), Tafethion (37), Volthion (54).
	Fenitrothion	Folithion (4), Hexafen (32), Sumithion (37).
	Fenthion	Baytex (4), Lebaycid (4).
	Formathion	Formathion
	Malathion	Agromala (16), Hilmala (17), Luthion (26), Malamar (32), Malataf (37), Cythion (8), Malatop (49), Mlamal (20), Maltox (1), Specmal (43), Sulmalathion (46).
	Methyl parathion	Devithion (32), Dhanumar (11), Luthion (26), Metacid (4), Milion (28), Prataf (49), Paratox (1), Profit (20), Tagpar (16).
	Monocrotophos	Agro Monark (16), Atom (21), Azodrin (8), Balwan (37), Bilphos (4), Corophos (7), Croton (24), Gaurdian (15), Hilcron (17), Hycrophos (18), Kadett (35), Luphos (26), Macrophos (50), Milphos (20), Moncar (5), Monochem (30), Monocil (9), Monocin (23), Monocron (31), Monodhan (11), Monophos (32), Monosect (2), Monostar (41), Monovol (54), Nuvacron (47), Parryphos (13), Phoskill (52), Specron (42), Sufos (45), Topcil (49).
	Oxy-demeton methyl	Dhanussystox (11), Hexasystox (32), Metasystox (4), Mode (2).
	Phenthoate	Aimsan (1), Amaze (55), Delsan (7), Elsan (52), Phendal (32), Penthomet (11), Tagsan (50).

Table. 14.1. Common and trade names and sources of availability of insecticides

Group	Common Name	Trade Name/s (Source)
	Phorate	Agro-phorate (16), Dhan 100 (11), Foratox (35), Forcin (23), Granutox (1), Helmet (50), Hilphorate (17), Luphate (26), Milate (28),Specphor (43), Starphor (41), Thimet (8), Umet (52), Volphor (53), Warrant (40).
	Phoslane	Voltas Phosalone (54), Zolone (38).
	Profenophos	Ajanta (33), Banjo (22), Bolero (24), Carina (35), Curacron (47), Proven (27), Prabal (37).
	Quinalphos	Agroquin (16), Basquin (5), Bayruil (4), Chemlux (30), Dhanulux (11), Ekalux (47), Ekatop (49), Flash (21), Hilquin (17), Hyguin (18), Kinalux (52), Knock (42), Krush (55), Miluax (20), Quick (31), Quinal (46), Quinaltaf (37), Quinatox (1), Quingaurd (15), Shakti (13), Smash (40), Starlux (41), Suquin (45), Tagquin (50), Vazara (26), Volquin (54).
	Triazophos	Hostothion (4), Josh (52), Titan (20), Trelka, Truzo (50), Trihero (39).
Carbamates	Carbaryl	Agroryl (16), Dhanuvin (11), Hexavin (32), Kildiryl (25), Killex Cabaryl (34), Parry Win 50 WP (13), Sevidol (38), Sevin 50 WP (38), Sevin D (38), Sevin Flo 42 (38).
	Carbofuran	Carbocil (9), Caardon (52), Diafuran (33), Furadon (37), Furacarb (1), Furatox (1), Fury (31), Hexafuran (32), Sumo (20).
	Carbosulfan	Marshall (37).
	Indoxacarb	Avant (12), Dash (21), Daksha (37), Dhava (11), Fego (52).
	Methomyl	Astra (13), Dunnet (11), Lannata (12).
	Thiodicarb	Larvin (2).
Pyrethroides	Alphacypermethrin	Forward 10 EC (24), Sherpa Alpha 10 EC (4).
	B-Cyfluthrin	Bulldock (4).
	Cypermethrin	Agro-Cyper (16), Angel (18), Ankush (3), Auzar (55), Bilcyp (4), Challenger (50), Cilcord (9), Colt-25 (35), Cybil 25 EC (4), Cymbush (19), Cyper Top (49), Cypercin (23), Cypergaurd (15),

Table 14.1. (Contd.)

Group	Common Name	Trade Name/s (Source)
		Cypermar (32), Cypermil (28), Cypersul (46), Cyproid (1), Cyrux 25 (52), Hilcyperin (17), Hi-Power (46), Hycyper (18), Indothrin (21), Jawa (12), Laser 10 EC (24), Piothrin (5), Polytrin (47), Ralothrin (37), Ripcord (8), Shakti (26), Sherpa Cyper 25 EC (4), Simper (19), Spec (43), Spec-Cyperin (42), Starcip (41), Superkiller (11), Suraksha (31), Trofy 25 EC (24), Trofy 25 EC (40), Ustad (52), Volcyper (54), White Gold (30).
	Decamethrin	Decakill (18), Decasyn (1), Decis (2), Easytab 25 EC (2), Faulad Rukrin (8).
	Ethofenfrox	Punakaso.
	Fenvalerate	Agrofen (16), Anchor (19), Bhaskarfen (5), Bhasma (55), Bilfen (4), Fen Fen (32), Fenchem (30), Fencron (47), Fenkill (52), Fennock 20 (9), Fenny (31), Fenval (40), Flame (20), Hilfen (17), Lufen (26), Milfen (28), Newfen (15), Starfen (41), Sumitox (1), Trfen (50), Tatafen (37), Topfen (49), Trumpcard (11), Viper (43).
	Lambda-Cyhalothrin	Agent (21), Bravo (20), Cliva (6), Karate (19), Reeva 5 (37).
	Permethrin	Ambush (19), Hilthrin (17), Permasect (32).
	Pyrethrum	
Insect growth regulators	Diflubenzuron	Dimilin 18 EC, 25 WP (15).
	Diafenthuran	Polo (4), Pegasus (47).
	Flufenoxuron	Cascade 10 EC (3).
	Lufenuron	Match 5 EC (47).
	Novaluron	Rimon 10 EC (21), Rim.
	Tiflubenzuron	Nomolt (9).
Combi products	B-Cyfluthrin 12.5 g + Chlorpyriphos 250 g	Bulldock 262.5 EC (4).
	Imidachloprid 50 g + B-Cyfluthrin 50 g	Confider Ultra 100 EC (4).
	Alphamethrin 1% + Chlorpyriphos 16%	Ducord 17 EC (4), Alfatoon, Lethal Plus (20).
	Fenvalerate 3% + Acephate 25%	Koranda 28 EC (29).

Table 14.1. (Contd.)

Group	Common Name	Trade Name/s (Source)
	Cypermethrin 5% + Ethion 40%	Nagata 45 EC (37), Rimjim (20).
	Cypermethrin 5% + Chlorpyriphos 50%	Noorani 55 EC (37), Nurello D-505, Lrthal Super 50 (20), Synergy 505 (9).
	Cypermethrin 4% + Profenphos 40%	Polytrin 44 EC (36).
	Cypermethrin 3% + Quinalphos 20%	Prachanda 23 EC, Viraat 23 EC (52).
	Deltamethrin 8% + Endosulfan 32%	Sevit 40 EC
	Deltamethrin 1% + Trizophos 35%	Spark 36 EC (4).
New molecules	Acetamaprid	Lift (21), Mnik (37), Pride (3), Sharp (20).
	Chlorfenapyr	Intrepid (3).
	Emamectin benzoate	Proclaim (47), Prabhav (37).
	Fipronil	Reagent (4).
	Imidachloprid	Admit (24), Atom (21), Confidar (4), Chemida (6), Imida Gold (52), Sabmida (39), Tata Mida (48), Victor.
	Spinosad	Tracer (9), Spintar (4).
	Thiomethoxam	Actra (4), Ananth 25 WP (37), Click (21), Crusier (47), Renova (52), Reno 70 WP.
Acaricides	Abamectin	Vertimec (47).
	Diafenthiuron	Polo, Pegasus.
	Dicofol	Kelthane (9), Colonel-S (21).
	Ethion	Acarin (18), Crush (20), Dhanumit (11), Fosmite (35), Indothion (21), Laser (26), MIT-505 (41), Phosmil (21), Tafethion (37), Volthion (54).
	Fenazaquin	Magister (12)
	Fenpropathrin + Pyriproxifen	Prempt 20 EC.
	Sulphur	Dhanusulf (11), Farmsulf (41), Hexasul (32), Insulf (52), Microsulf (32), Sulcol, Wet-Sulf (14), Sulfin M-20, (16), Sulphosan (1), Sultaf (37), Thiovit (47).
	Propagate	Omite (11).

14.2. FUNGICIDES

Table 14.2. Common and trade names and sources of availability of fungicides

Group	Common Name	Trade Name/s (Source)
Copper fungicides	Copper hydroxide	Kocide 101 (12).
	Copper oxychloride	Blitox (37), Blue Copper (47), Captox (1), Cupramar (32), Cuprastar (41), Devicopper (10), Dhanucop (11), Fytolan (19), Hexamar Copper (32), Hil Copper (17), Blue Vit (25), Trucop (21).
	Cuprous oxide	Cuprous oxide
Di-carboxymides	Iprodione	Rovral (38).
	Vinclozolin	
Heterocyclic nitrogenous fungicides	Captafol	Foltaf.
	Captan	Captra (21), Caftap (37), Deltan (7), Dhanutan (11), Hexacap (32).
	Folpet	
Mercury fungicides	Methoxy ethyl mercuric chloride	Bagalol (52),
	Phenyl mercury acetate	
Sulphur fungicides (Inorganic)	Wettable Sulphur	Dhanusulf (11), Farmsulf (41), Hexasul (32), Insulf (52), Microsulf (32), Sulcol, Wet-Sulf (14), Sulfin M-20 (16), Sulphosan (1), Sultaf (37), Thiovit (47).
Sulphur fungicides (Organic– Carbamates)	Ferbam	Hexferb (32).
	Maneb/Mancozeb	Abie (47), Agromanco (16), Aimcozeb (1), Dhanuka (11), Dithane M-45 (4), Hilthane (17), Indofil (21), Luzen (26), Leo M-45 (20), Zebthane M-45, Flowin-HT, Luzim (26), Saviour (9), Shield (35), Sparsh (55), Spic Mancozeb (44), Uthane M-45 (52), Zeb (31), Zebthane (37).
	Nabam	
	Thiram	Thiride (20), Hexathir (32), Vegfru Thiram (35).
	Vapam	
	Zineb	Discon-Z (1), Devizeb (10), Hexathane (32), Indofil Z-78 (21).

Table 14.2. (Contd.)

Group	Common Name	Trade Name/s (Source)
	Ziram	Cumin L (47), Hexazir (32), Tagziron (50), Ziride (20), Vegfru Zitox (35), Dhanuka Z-27 (11).
Quinine fungicides	Chloronil	
	Dichlone	
Systemic fungicides Benzimidazoles	Benomyl	Benlate (32), Benofit (13).
	Carbendazim	Agni (21), Agrozim (16), Aimocozin (1), Arrest (24), Bavistin (3), Benfin (21), Bengard (9), Calzin (26), Benzin (5), Carzim (26), Derosal (2), Dhamusten (11), Diafuran (33), Hycarb (18), Nirmool (41), Sahara (20), Stare (32), Volzim (54), Zen (31), Zoom (52).
	Thiabendazole	
Thiophanates	Thiophanate	Thiophanate.
	Thiophanate Methyl	Cercobin, Roko (55), Prism (20).
Oxathin compounds	Carboxin	Carboxin.
	Oxy-Carboxin	Morestein.
Phenylamides	Oxadixyl	
	Metalaxyl	Apron (24), Matco, Himilgard.
Triazoles	Bitertanol (Biloxazol)	Baycor (4).
	Dichlobutrazol	Vigil.
	Difenconazole	Score (47).
	Flusilazol	
	Flutriafol	
	Hexaconazole	Avone (20), Contaf (37), Conquer (52), Danzole(6), Manage (24), Sitara (21).
	Myclobutanil	Systhane (9), Boon (21).
	Penconazole	Topas (47).
	Propiconazole	Radar (37), Tilt (47), Swing, Sona (20), Dhan (21).
	Triadimefon	Bayleton (4).
	Triadimenol	
	Tricyclazole	Beam (21), Force-11 (20).
Pyrimidines	Fenarimol	
Morpholines	Dodemorph	Dodemorph.

Table 14.2. (Contd.)

Group	Common Name	Trade Name/s (Source)
	Tridemorph	Calixin (3).
Organophosphorus compounds	Fosetyl –AL	Aliette (2).
	Kitazine & Kitazine-P	
	Pyrazophos	Apagon.
Acetamides	Cymoxynil	
Carbamates	Propamocarb	
	Propineb	Antracol (4).
Di-caroxymides	Procymidone	
Imidazoles	Fenapanil	
Piperazine	Triforine	
Phenol derivatives	Chloroneb	
Miscellaneous systemic fungicides	Antibiotics	Aureofungin, Cycloheximide, Griseofulvin, Kasugamycin, Validamycin, Streptocycline.
	Etridiazol	
	Pyroxychlor	
Miscellaneous fungicides	Chinomethionat	
	Clorothalanil	Bravo, Kavach (47).
	Clorothaniz	
	Dinocap	Karathane (4).
	Dodine	Sylil (4), Syllet (28), Noor (21).
	Edifenphos	Hinosan (4).
	Glyodin	
	Morocide	Morocide
	Prochloraz	
	Quintozine	
Combi products	Metalaxyl 8% + Mancozeb 64%	Ridomil MZ (47), Master (48), Himil (20), Matco-8-64 (21).
	Iprodione 25% + Carbendazim 25%	Quintal (4).
	Carbendazim 12% + Mancozeb 63%	Care (20), Companion (21), Top-20 (18), Sixer 75 WP (11), Saaf (52).
	Cymoxanil 8% + Mancozeb 64%	Curzate 72 WP (12).
	Propioconazole + Difenoconazole 30%	Armure 30 EC.

14.3. RODENTICIDES

Table 14.3. Common and trade names and sources of availability of rodenticides

Common name	Trade name/s (Source)
Aluminium phosphide	Synfume, Phosfume, Celphos (14), Quickphos (52), Phostoxin, Delacea (32), Alphos (1), Fumino (52).
Bromadiolone	Lainirat (47), Moosh-Moosh (37).
Coumatetralyl	Racumin (4).
Magnesium phosphide	Magnophos.
Warfarin	Rodafarin, Ratafin, Fumarin, Warfotax (1).
Zinc phosphide	Synphos, Rasol (52), Zincfox (1), Rasox.
D-D 1: 1 mixture (Dichloro propene – Dichloro propane)	Telone, Vidden-D.
EDCT 3: 1 mixture (Ethylene dichloride carban tetrachloride)	Kiloptera, Fumex, Dichromek.

Source of Pesticides

1. AIMCO
2. Aventis
3. BASF
4. Bayer
5. Bhaskar Agro Chemicals
6. Cheminova
7. Coromandal Indag
8. Cynamid Agro
9. De Nocil
10. Devi Dayal
11. Dhanuka
12. Du Pont
13. EID Parry
14. Excel
15. Gardha Chemicals
16. Gujarat Agro
17. Hindusthan Insecticides
18. Hyderabad Chemicals
19. ICI Zeneca
20. IIL
21. Indofil
22. Insecticides India
23. IOCL
24. IS Agro
25. Kilpest
26. Lupin
27. Meghamani Organics Ltd.
28. Montari
29. Nagarjuna Chemicals
30. New Chemi
31. NFCL
32. Parry Chemicals
33. Parry India Ltd.
34. Paushak
35. Pesticides India
36. Polychem
37. Rallis India
38. Rhone-Poulenc
39. Sabero
40. Searle India
41. Shaw Wallace
42. Southern Pesticides
43. SPEC
44. Spic

45. Sudarshan Chemicals
46. Sulphur Mills
47. Syngenta
48. Tata
49. Thakar Chemicals
50. Tropical Agro
51. T. Stanes
52. United Phosphorus Ltd.
53. Volrho
54. Voltas
55. Wockardt

14.4. NEMATICIDES

Table 14.4. Globally important nematicides and their manufacturers

Active substance	Chemical group	Trade name	Manufacturer
1, 3-Dichloropropene	Halogenated hydrocarbon	Telone	Bayer Crop Science
Dazomet	Methyl isothiocyanate liberator	Basamid	BASF Corporation
Metham sodium	Methyl isothiocyanate liberator	Vapam	Amvac Chemical Corpn.
Ethoprophos	Organophosphorus	Mocap	Bayer Crop Science
Phenamiphos	Organophosphorus	Nemacur	Bayer Crop Science
Fosthiazate	Organophosphorus	Nemathorin	Syngenta
Cadusafos	Organophosphorus	Rugby	FMC Corporation
Aldicarb	Oxime carbamate	Temik	Bayer Crop Science
Oxamyl	Oxime carbamate	Vydate	Du Pont
Carbofuran	Carbamate	Furadan	FMC Corporation

14.5. PREDATORS

Table 14.5. Sources of availability of predators

Predators	State	Sources of availability
Amblyscius sp.	Tamil Nadu	Rajendra Foundn for Agri Res, Chittar
Apertochrysa sp.	Karnataka	Project Directorate of Biol Control, Bangalore
Blastostethus pallescens	Karnataka	Project Directorate of Biol Control, Bangalore
Brumoides suturalis	Karnataka	Project Directorate of Biol Control, Bangalore. Central IPM Centre, Bangalore
	New Delhi	Entomol Divn, Indian Agri Res Inst, New Delhi
Cardiastethus exiguous	Karnataka	Project Directorate of Biol Control, Bangalore
Cheilomenes sexmaculata	Gujarat	Entomol Dept, Anand Agri Univ, Anand
	Karnataka	Project Directorate of Biol Control, Bangalore. Central IPM Centre, Bangalore
	New Delhi	Entomol Divn, Indian Agri Res Inst, New Delhi
Chilocorus nigrita	Himachal Pradesh	Central IPM Centre, Solan. Ent Dept, Y S Paramar Univ of Hort & Forestry, Solan

Table. 14.5 (Contd.)

Predators	State	Sources of availability
	Jammu & Kashmir	Central IPM Centre, Jammu; Central IPM Centre, Srinagar; Entomol Dept, SK Univ Agri Sci & Technol, Srinagar
	Karnataka	Project Directorate of Biol Control, Bangalore. Bio Control Res Lab, PCI Pvt Ltd., Bangalore
	Tamil Nadu	Rajendra Foundn for Agri Res, Chittar
	West Bengal	Central IPM Centre, Burdwan
Chrysoperla carnea	Andhra Pradesh	Central Tobacco Res Inst, Rajahmundry. Ent Dept, Acharya N G Ranga Agri Univ, Hyderabad. Pest Control (India) Pvt Ltd, Hyderabad. SOM-IPM Systems (I) Ltd, Hyderabad
	Assam	Pest Control (India) Pvt Ltd, Guwahati
	Bihar & Jarkhand	Pest Control (India) Pvt Ltd, Jamshedpur
	Gujarat	Entomol Dept, Anand Agri Univ, Anand. Pest Control (India) Pvt Ltd, Vadodara
	Haryana	Pest Control (India) Pvt Ltd, Chandigarh
	Karnataka	Project Directorate of Biol Control, Bangalore. Central IPM Centre, Bangalore. Univ of Agri Sci, Entomol Dept, Dharwad. Kadur Agro Pvt. Ltd., Bangalore. Pest Control (India) Pvt Ltd, Bangalore
	Kerala	Pest Control (India) Pvt Ltd, Cochin
	Madhya Pradesh	Indore Biotech Inputs & Res (P) Ltd., Indore. Pest Control (India) Pvt Ltd, Bhopal
	Maharashtra	Entomol Dept, Mahatma Phule Agri Univ, Rahuri. Nathkrupa Biocontrol Lab, Nagpur. Pest Control (India) Pvt Ltd Mumbai-400 063. Pest Control (India) Pvt Ltd, Mumbai-400 059. Pest Control (India) Pvt Ltd, Thane-400 601. Pest Control (India) Pvt Ltd, New Mumbai. Pest Control (India) Pvt Ltd, Nashik-422 009. Pest Control (India) Pvt Ltd, Pune-411 033. Pest Control (India) Pvt Ltd, Pune-411 030
	New Delhi	Biotech Intern Ltd., New Delhi. Pest Control (India) Pvt Ltd, New Delhi
	Orissa	Pest Control (India) Pvt Ltd, Bhubaneswar
	Punjab	Pest Control (India) Pvt Ltd, Chandigarh

Table. 14.5 (Contd.)

Predators	State	Sources of availability
	Rajasthan	Pest Control (India) Pvt Ltd, Jaipur
	Tamil Nadu	Entomol Dept, TNAU, Madurai.
		Basarass Biocontrol Res, Vallavar.
		Greentech Agro Prod P Ltd, Coimbatore.
		Pest Control (India) Pvt Ltd, Chennai.
		Rajendra Foundn for Agri Res, Chittar.
		TN Co-op Sugar Fedn Ltd, Chengalpattu
	Uttarakhand	Pest Control (India) Pvt Ltd, Dehradun
	Uttar Pradesh	Pest Control (India) Pvt Ltd, Bareilly.
		Pest Control (India) Pvt Ltd, Lucknow.
		Pest Control (India) Pvt Ltd, Kanpur
	West Bengal	Pest Control (India) Pvt Ltd, Kolkata
Coccinella septumpunctata	Tamil Nadu	Jeypee Bio Techs, Virudhunagar
Cryptolaemus montrouzieri	Andhra Pradesh	Acharya N G Ranga Agri Univ, Hyderabad.
		Pest Control (India) Pvt Ltd, Hyderabad
	Assam	Pest Control (India) Pvt Ltd, Guwahati
	Bihar & Jarkhand	Pest Control (India) Pvt Ltd, Jamshedpur
	Gujarat	Pest Control (India) Pvt Ltd, Vadodara
	Haryana	Pest Control (India) Pvt Ltd, Chandigarh
	Karnataka	Project Directorate of Biol Control, Bangalore.
		Agri Technol Infmn Centre, IIHR, Bangalore.
		Kadur Agro Pvt. Ltd, Bangalore.
		Pest Control (India) Pvt Ltd, Bangalore
	Kerala	Pest Control (India) Pvt Ltd, Cochin
	Madhya Pradesh	Central IPM Centre, Indore.
		Indore Biotech Inputs & Res (P) Ltd, Indore.
		Pest Control (India) Pvt Ltd, Bhopal
	Maharashtra	Pest Control (India) Pvt Ltd, Mumbai-400 063.
		Pest Control (India) Pvt Ltd, Mumbai-400 005.
		Pest Control (India) Pvt Ltd, Thane.
		Pest Control (India) Pvt Ltd, New Mumbai.
		Pest Control (India) Pvt Ltd, Nashik.
		Pest Control (India) Pvt Ltd, Pune-411 033.
		Pest Control (India) Pvt Ltd, Pune-411 030
	New Delhi	Entomol Divn, Indian Agri Res Inst, New Delhi.
		Pest Control (India) Pvt Ltd, New Delhi
	Orissa	Pest Control (India) Pvt Ltd, Bhubaneswar
	Punjab	Pest Control (India) Pvt Ltd, Chandigarh
	Tamil Nadu	Basarass Biocontrol Res, Vallavar.
		Greentech Agro Prod P Ltd, Coimbatore.

Table. 14.5 (Contd.)

Predators	State	Sources of availability
		Pest Control (India) Pvt Ltd.
		Rajendra Foundn for Agri Res, Chittar.
		Shri Durga Agro Services, Coimbatore
	Uttarakhand	Pest Control (India) Pvt Ltd, Dehradun
	Uttar Pradesh	Pest Control (India) Pvt Ltd, Bareilly.
		Pest Control (India) Pvt Ltd, Lucknow.
		Pest Control (India) Pvt Ltd, Kanpur
	West Bengal	Pest Control (India) Pvt Ltd, Kolkata
Curinus coeruleus	Karnataka	Project Directorate of Biol Control, Bangalore
Mallada spp.	Gujarat	Entomol Dept, Anand Agri Univ, Anand
	Karnataka	Project Directorate of Biol Control, Bangalore
	New Delhi	Entomol Divn, Indian Agri Res Inst, New Delhi
	Punjab	Entomol Dept, Punjab Agri Univ, Ludhiana
M. astur	Karnataka	Project Directorate of Biol Control, Bangalore
M. boninensis	Karnataka	Project Directorate of Biol Control, Bangalore
Orius tantillus	Karnataka	Project Directorate of Biol Control, Bangalore
Parena nigrolineata	Karnataka	Project Directorate of Biol Control, Bangalore
Pharoscymnus horni	Karnataka	Project Directorate of Biol Control, Bangalore
Scymnus coccivora	Karnataka	Project Directorate of Biol Control, Bangalore

14.6. PARASITOIDS

Table 14.6. Sources of availability of parasitoids

Parasitoids	State	Sources of availability
Apanteles angaleti	Karnataka	Project Directorate of Biol Control, Bangalore
	Maharashtra	Entomol Dept, Agri College, Pune
	New Delhi	Entomol Divn, Indian Agri Res Inst, New Delhi
Botryoideclava bharatiya	Karnataka	Project Directorate of Biol Control, Bangalore
Brachymeria nosatoi	Andhra Pradesh	Ent Dept, Acharya N G Ranga Agri Univ, Hyderabad
	Karnataka	Central IPM Centre, Bangalore
	Kerala	CPCRI Regional Station, Kayangulam
B. nephantidis	Andhra Pradesh	Ent Dept, Acharya N G Ranga Agri Univ, Hyderabad
	Karnataka	Project Directorate of Biol Control, Bangalore
		Central IPM Centre, Bangalore
	Kerala	CPCRI Regional Station, Kayangulam
Bracon brevicornis	Karnataka	Project Directorate of Biol Control, Bangalore

Table. 14.6 (Contd.)

Parasitoids	State	Sources of availability
	New Delhi	Entomol Divn, Indian Agri Res Inst, New Delhi. Pest Control (India) Pvt Ltd, New Delhi
	Tamil Nadu	Central IPM Centre, Tiruchirapalli. Basarass Biocontrol Res, Vallavar. Rajendra Foundn for Agri Res, Chittar. TN Coconut Parasite Breeding Centre, Coimbatore. TN Co-op Sugar Fedn Ltd, Chengalpattu
B. hebetor	Andhra Pradesh	Ent Dept, Acharya N G Ranga Agri Univ, Hyderabad
	Karnataka	Project Directorate of Biol Control, Bangalore. Central IPM Centre, Bangalore
	Kerala	CPCRI Regional Station, Kayangulam
	Maharashtra	Entomol Dept, Agri College, Pune. Nathkrupa Biocontrol Lab, Nagpur
	New Delhi	Entomol Divn, Indian Agri Res Inst, New Delhi
	Tamil Nadu	Jeypee Bio Techs, Virudhunagar Rajendra Foundn for Agri Res, Chittar. Sakthi Biocontrol & Biofert Centre, Kallipatti Gobi
B. kirkpatricki	Karnataka	Project Directorate of Biol Control, Bangalore
	Kerala	CPCRI Regional Station, Kayangulam
Campoletis chloridae	Karnataka	Project Directorate of Biol Control, Bangalore
Cephalonomia stephanoderis	Gujarat	Entomol Dept, Anand Agri Univ, Anand
	Karnataka	Coffee Research Sub Station, Chethalli
	Kerala	Entomol Dept, Kerala Agri Univ, Trichur
	Maharashtra	Central Inst of Cotton Res, Nagpur
	Tamil Nadu	Regional Coffee Res Stn, Thandigudi. Centre for Pl Prot Studies, TNAU, Coimbatore
Chelonus blackburni	Karnataka	Project Directorate of Biol Control, Bangalore
Coccidoxenoides peregrinus	Karnataka	Project Directorate of Biol Control, Bangalore
Copidosoma koehleri	Bihar & Jarkhand	Central IPM Centre, Patna
	Karnataka	Project Directorate of Biol Control, Bangalore
	Maharashtra	Entomol Dept, Mahtma Phule Agri Univ, Rahuri
	New Delhi	Entomol Divn, Indian Agri Res Inst, New Delhi
Cotesia flaviceps	Karnataka	Project Directorate of Biol Control, Bangalore
	Orissa	Central IPM Centre, Bhubaneswar

Table. 14.6 (Contd.)

Parasitoids	State	Sources of availability
	Punjab	Entomol Dept, Punjab Agri Univ, Ludhiana
	Tamil Nadu	Divn of Pl Prot, SBI, Coimbatore
	Uttar Pradesh	Entomol Divn, IISR, Lucknow
C. plutellae	Karnataka	Project Directorate of Biol Control, Bangalore
C. africanus	Andhra Pradesh	Central Tobacco Res Inst, Rajahmundry
	Karnataka	Project Directorate of Biol Control, Bangalore
	Maharashtra	Central IPM Centre, Nagpur
Dipha aphidivora	Karnataka	Agri Res Stn, UAS, Bheemarayanagudi
Elasmus nephantidis	Andhra Pradesh	Ent Dept, Acharya N G Ranga Agri Univ, Hyderabad
	Karnataka	Project Directorate of Biol Control, Bangalore Agri Technol Infmn Centre, IIHR, Bangalore
	Kerala	CPCRI Regional Station, Kayangulam
	Tamil Nadu	Basarass Biocontrol Res, Vallavar
Encarsia spp.	Himachal Pradesh	Central IPM Centre, Solan
	Jammu & Kashmir	Entomol Dept, SK Univ Agri & Technol, Srinagar
	Karnataka	Central IPM Centre, Bangalore
Encarsia guadeloupae	Karnataka	Project Directorate of Biol Control, Bangalore. Agri Technol Infmn Centre, IIHR, Bangalore
Epiricania melanoleuca	Maharashtra	Vasanthdada Sugar Institute, Pune
	Tamil Nadu	Central IPM Centre, Tiruchirapalli. Rajendra Foundn for Agri Res, Chittar
	Uttar Pradesh	Entomol Divn, IISR, Lucknow. Central IPM Centre, Gorakhpur
Eriborus argenteopilosus	Karnataka	Project Directorate of Biol Control, Bangalore
Goniozus nephantidis	Andhra Pradesh	Ent Dept, Acharya N G Ranga Agri Univ, Hyderabad
	Karnataka	Project Directorate of Biol Control, Bangalore Pest Control (India) Pvt Ltd, Bangalore
	Kerala	CPCRI Regional Station, Kayangulam
	Tamil Nadu	Central IPM Centre, Tiruchirapalli Greentech Agro Prod P Ltd, Coimbatore Rajendra Foundn for Agri Res, Chittar TN Coconut Parasite Breeding Centre, Coimbatore
Leptomastix dactylopii	Karnataka	Project Directorate of Biol Control, Bangalore Agri Technol Infmn Centre, IIHR, Bangalore Entomol Divn, Central Coffee Res Inst, Chikmagalur

Table. 14.6 (Contd.)

Parasitoids	State	Sources of availability
	Kerala	Entomol Dept, Kerala Agri Univ, Trichur
	Tamil Nadu	Greentech Agro Prod P Ltd, Coimbatore
Micromus igarotus	Karnataka	Regional Res Stn, UAS, Raichur
Oomyzus sokolowskii	Karnataka	Project Directorate of Biol Control, Bangalore
Sturmiopsis inferens	Karnataka	Agri Technol Infmn Centre, IIHR, Bangalore
Telenomus spp.	Karnataka	Project Directorate of Biol Control, Bangalore
T. remus	Andhra Pradesh	Central Tobacco Res Inst, Rajahmundry
	Karnataka	Project Directorate of Biol Control, Bangalore
	Madhya Pradesh	Central IPM Centre, Indore
Tetrastichus howardi	Karnataka	Project Directorate of Biol Control, Bangalore
T. israeli	Gujarat	Entomol Dept, Anand Agri Univ, Anand
	Karnataka	Project Directorate of Biol Control, Bangalore. Central IPM Centre, Bangalore
	Kerala	CPCRI Regional Station, Kayangulam
	Tamil Nadu	Jeypee Bio Techs, Virudhunagar. Rajendra Foundn for Agri Res, Chittar. TN Coconut Parasite Breeding Centre, Coimbatore
Trichogramma achaeae	Karnataka	Project Directorate of Biol Control, Bangalore. Agri Technol Infmn Centre, IIHR, Bangalore. Central IPM Centre, Bangalore
	Punjab	Entomol Dept, Punjab Agri Univ, Ludhiana
	Tamil Nadu	Rajendra Foundn for Agri Res, Chittar. Shri Durga Agro Services, Coimbatore. Sun Agro Ind India, Chennai
T. brasiliensis	Karnataka	Project Directorate of Biol Control, Bangalore. Agri Technol Infmn Centre, IIHR, Bangalore
	Tamil Nadu	Khatau Agro Tech, Mumbai. National Biocontrol Laboratory, Nagpur
T. chilonis, *T. japonicum*	Andhra Pradesh	Central Tobacco Res Inst, Rajahmundry. Ent Dept, Acharya N G Ranga Agri Univ, Hyderabad. Pest Control (India) Pvt Ltd. SOM-IPM Systems (I) Ltd., Hyderabad
	Assam	State Biocontrol Laboratory, Guwahati. Pest Control (India) Pvt Ltd, Guwahati
	Bihar & Jarkhand	Pest Control (India) Pvt Ltd, Jamshedpur
	Goa	Central IPM Centre, Madgaon
	Gujarat	Pest Control (India) Pvt Ltd, Vadodara

Table. 14.6 (Contd.)

Parasitoids	State	Sources of availability
	Haryana	Entomol Dept, CCS Haryana Agri Univ, Hisar
		Regional Agri Centre, CCSHAU, Karnal
		State Biocontrol Laboratory, Sirsa
		Biocontrol Laboratory, Yamuna Nagar
	Karnataka	Project Directorate of Biol Control, Bangalore
		Agri Technol Infmn Centre, IIHR, Bangalore.
		Central IPM Centre, Bangalore.
		Entomol. Dept, Univ of Agri Sci, Dharwad.
		Inst of Pulses & Oilseeds Res, Gulbarga.
		Regional Res Stn, UAS, Raichur.
		AAO, Parasite Laboratory, Bailhongal.
		AO, Parasite Laboratory, Gangavathi.
		AAO, Parasite Laboratory, Mandya.
		Bio Solujan, Bangalore.
		Hospet Sugar Mills, Hospet.
		Kadur Agro Pvt. Ltd., Bangalore.
		Karnataka Agri Res Foundn, Sameerwadi.
		Pest Control (India) Pvt Ltd, Bangalore
	Kerala	Pest Control (India) Pvt Ltd, Cochin
	Madhya Pradesh	Indore Biotech Inputs & Res (Pvt.) Ltd., Indore.
		Pest Control (India) Pvt Ltd, Bhopal
	Maharashtra	Central Inst of Cotton Res, Nagpur.
		Entomol Dept, Mahtma Phule Agri Univ, Rahuri.
		Entomol Dept, Marathwada Agri Univ, Parbhani.
		Entomol Dept, Konkan Agri Univ, Dapoli.
		KNS Biotech, Nanded.
		Nathkrupa Biocontrol Lab, Nagpur.
		Pest Control (India) Pvt Ltd, Mumbai-400 063.
		Pest Control (India) Pvt Ltd, Mumbai-400 059.
		Pest Control (India) Pvt Ltd, Thane-400 601.
		Pest Control (India) Pvt Ltd, New Mumbai.
		Pest Control (India) Pvt Ltd, Nashik-422 009.
		Pest Control (India) Pvt Ltd, Pune-411 030.
		Pest Control (India) Pvt Ltd, Pune-411 033.
		R.S. Kherdekar, Talpathri.
		Sandoz (I) Ltd., Mumbai
	New Delhi	Entomol Divn, Indian Agri Res Inst, New Delhi.
		Biotech Intern Ltd., New Delhi.
		Navdanya (NGO), New Delhi.
		Pest Control (India) Pvt Ltd, New Delhi.

Table. 14.6 (Contd.)

Parasitoids	State	Sources of availability
	Orissa	Pest Control (India) Pvt Ltd, Bhubaneswar
	Punjab	Pest Control (India) Pvt Ltd, Chandigarh
	Rajasthan	Biocontrol Lab, ARS, RAU, Sri Ganganagar.
		JDA (Pl Path), Biocontrol Lab, Durgapura.
		DDA (Agro), IPM Lab, Ajmeer.
		DDA (Agro), IPM Lab, Malikpur.
		DDA (Agro), IPM Lab, Rampura.
		DDA (Agro), IPM Lab, Cjjatrapura.
		DDA (Agro), IPM Lab, Chittorgarh.
		DDA (Agro), IPM Lab, Hanumangarh.
		DDA (Pl Path), IPM Lab, Banaswara.
		Adarsh Biocontrol Laboratories, Alwar.
		Pest Control (India) Pvt Ltd, Jaipur
	Tamil Nadu	Divn of Pl Prot, SBI, Coimbatore.
		Regional Coffee Res Stn, Thandigudi.
		Bio-control Lab., Papparapatty.
		Bio-control Lab, Coimbatore.
		Bio-control Lab., Salem.
		Bio-control Lab, Villupuram.
		Bio-control Lab., Trichy.
		Bio-control Lab., Panjupettai.
		Bio-control Lab., Kuruppanaichenpalyam.
		Bio-control Lab., Vinayagapuram.
		Bio-control Lab., Tirunelveli.
		Bio-control Lab., Tanjavur.
		Bio-control Lab., Namakkal.
		Agro Res Lab, Coimbatore.
		Basarass Biocontrol Res, Vallavar.
		Biocontrol Res Centre, Kootham Poondi.
		Greentech Agro Prod P Ltd, Coimbatore.
		Jeypee Bio Techs, Virudhunagar.
		Main Biocontrol Res Lab, Chengalpattu.
		Pest Control (India) Pvt Ltd, Chennai.
		Rajendra Foundn for Agri Res, Chittar.
		Sakthi Biocontrol & Biofert Centre, Kallipatti, Gobi.
		Shri Durga Agro Services, Coimbatore.
		Sinagro Industries India, Chennai.
		Sun Agro Ind India, Chennai.
		Vigneswara Biocontrol, Pollachi

Table. 14.6 (Contd.)

Parasitoids	State	Sources of availability
	Uttarakhand	Pest Control (India) Pvt Ltd, Dehradun
	Uttar Pradesh	Entomol Divn, IISR, Lucknow.
		Biocontrol Lab, SBBP Agri Univ, Meerut.
		Bioved Res Soc, Allahabad.
		Crop Health Bioprod Res Centre, Ghaziabad.
		Pest Control (India) Pvt Ltd, Bareilly.
		Pest Control (India) Pvt Ltd, Lucknow.
		Pest Control (India) Pvt Ltd, Kanpur
	West Bengal	Pest Control (India) Pvt Ltd, Kolkata
T. dendrolimi	Karnataka	Project Directorate of Biol Control, Bangalore.
		Agri Technol Infmn Centre, IIHR, Bangalore
	Tamil Nadu	Rajendra Foundn for Agri Res, Chittar
T. embryophagum	Himachal Pradesh	Central IPM Centre, Solan.
		Ent Dept, Y S Paramar Univ of Hort & Forestry, Solan
	Jammu & Kashmir	Central IPM Centre, Jammu.
		Central IPM Centre, Srinagar.
		Entomol Dept, SK Univ Agri Sci & Technol, Srinagar
	Karnataka	Project Directorate of Biol Control, Bangalore
T. evanescens	Himachal Pradesh	Central IPM Centre, Solan.
		Y S Paramar Univ of Hort & Forestry, Solan
	Jammu & Kashmir	Central IPM Centre, Jammu.
		Central IPM Centre, Srinagar.
		Entomol Dept, SK Univ Agri & Technol, Srinagar
	Karnataka	Project Directorate of Biol Control, Bangalore
T. maidis	Karnataka	Project Directorate of Biol Control, Bangalore
T. pretiosum	Himachal Pradesh	Central IPM Centre, Solan.
		Ent Dept, Y S Paramar Univ of Hort & Forestry, Solan
	Jammu & Kashmir	Central IPM Centre, Jammu.
		Central IPM Centre, Srinagar.
		Entomol Dept, SK Univ Agri Sci & Technol, Srinagar
	Karnataka	Project Directorate of Biol Control, Bangalore
Trichogrammatoidea armigera	Karnataka	Project Directorate of Biol Control, Bangalore
T. bactrae	Karnataka	Project Directorate of Biol Control, Bangalore

Table. 14.6 (Contd.)

Parasitoids	State	Sources of availability
Trichospilus pupivorus	Karnataka	Central IPM Centre, Bangalore
	Kerala	CPCRI Regional Station, Kayangulam
Xanthopimpla punctata	Karnataka	Project Directorate of Biol Control, Bangalore. Central IPM Centre, Bangalore
	Kerala	CPCRI Regional Station, Kayangulam

14.7. PATHOGENS

14.7.1. Arbuscular Mycorrhizal Fungi

Table 14.7. Trade names and source of availability of Arbuscular Mycorrhizal Fungi

Arbuscular Mycorrhizal Fungi	Trade Name	State	Source of Availability
Glomus mosseae	Symbion VAM	Karnataka	Agri Technol Infmn Centre, IIHR, B'lore
		Maharashtra	Bharatiya Agro Industries Foundation, Pune
		Tamil Nadu	T Stanes & Co, Coimbatore
G. fasciculatum	Symbion VAM	Karnataka	Agri Technol Infmn Centre, IIHR, B'lore
		Maharashtra	Bharatiya Agro Industries Foundation, Pune
		Tamil Nadu	T Stanes & Co, Coimbatore
G. aggregatum		Maharashtra	Bharatiya Agro Industries Foundation, Pune
Gigaspora margarita		Maharashtra	Bharatiya Agro Industries Foundation, Pune
Arbuscular mycorrhizal fungi	Multiplex Trishul	Karnataka	Karnataka Agro Chemicals, B'lore

14.7.2. Antagonistic Fungi Against Plant Diseases and Nematodes

Table 14.8. Trade names and sources of availability of antagonistic fungi against Plant Diseases and nematodes

Antagonistic Fungi/ Trade Name	State	Source of Availability
Ampelomyces quisqualis (Bio-Dewcon)	Tamil Nadu	T. Stanes & Co, Coimbatore
Arthrobotrys cladodes var. *macroides*	Karnataka	Project Directorate of Biol Control, Bangalore
Arthrobotrys oligospora	Karnataka	Project Directorate of Biol Control, Bangalore
Aspergillus nizer (Kalisena SD, Kalisena SL, Sanjeevni)	Gujarat	Cadila Pharmaceuticals Ltd., Ahmedabad

Table 14.8. (Contd.)

Antagonistic Fungi/ Trade Name	State	Source of Availability
	New Delhi	Plant Path Divn, Indian Agri Res Inst, New Delhi
Dactylella brochophaga	Karnataka	Project Directorate of Biol Control, Bangalore
Fusarium oxysporum	Karnataka	Project Directorate of Biol Control, Bangalore
Fusarium sporotrichoides	Karnataka	Project Directorate of Biol Control, Bangalore
Gliocladium catenulatum	Karnataka	Project Directorate of Biol Control, Bangalore
G. deliquescens	Karnataka	Project Directorate of Biol Control, Bangalore
G. roseum	Karnataka	Project Directorate of Biol Control, Bangalore
Phoma glomerata	Karnataka	Project Directorate of Biol Control, Bangalore
Pochonia chlamydosporia (Bionema)	Karnataka	Project Directorate of Biol Control, Bangalore. Agri Technol Infmn Centre, IIHR, Bangalore. Central Sericultural Res &Trng Inst, Mysore
Paecilomyces lilacinus Bio-Nemator, Biocon, Niyantran, Multiplex Sparsha, Abtec Paecilomyces)	Karnataka	Project Directorate of Biol Control, Bangalore. Agri Technol Infmn Centre, IIHR, Bangalore
T. aureoviride	Karnataka	Project Directorate of Biol Control, Bangalore
T. hamatum	Karnataka	Project Directorate of Biol Control, Bangalore
T. harzianum (Binab-T, Ecoderma, F-Stop, Supravit, Tricodex, Trichoderma)	Andhra Pradesh	ADA (BCL), P.M. Palem. ADA (BCL), Kakinada. ADA (BCL), Nidadavole. ADA (BCL), Ibrahimpatnam. ADA (BCL), Ongole. ADA (BCL), Nellore. DDA (FTC), Nandyal. ADA (BCL), Anantapur. DDA (FTC), Rajendranagar. ADA (Oilseeds) (BCL), Mahabubanagar. ADA (BCL), Nalagonda. ADA (BCL), Mulugu Road. ADA (BCL), Karimnagar. ADA (BCL), Adilabad. Pest Control (India) Pvt Ltd, Hyderabad
	Assam	State Biocontrol Laboratory, Guwahati. Pest Control (India) Pvt Ltd, Guwahati
	Bihar & Jarkhand	Pest Control (India) Pvt Ltd, Jamshedpur
	Gujarat	Pest Control (India) Pvt Ltd, Vadodara
	Haryana	State Biocontrol Laboratory, Sirsa Pest Control (India) Pvt Ltd, Chandigarh

Table. 14.8 (Contd.)

Antagonistic Fungi/ Trade Name	State	Source of Availability
	Karnataka	Project Directorate of Biol Control, Bangalore.
		Agri Technol Infmn Centre, IIHR, Bangalore.
		Central IPM Centre, Bangalore.
		Central Sericultural Res & Trng Inst, Mysore.
		Coffee Research Sub Station, Chethalli.
		Central Coffee Res Inst, Entomol Divn, Chikmagalur.
		Univ of Agri Sci, Entomol Dept, Dharwad.
		Inst of Pulses & Oilseeds Res, Gulbarga.
		Regional Res Stn, UAS, Raichur.
		Agri Res Stn, UAS, Bijapur.
		Kadur Agro Pvt. Ltd., Bangalore.
		Pest Control (India) Pvt Ltd, Bangalore.
		P.J.Margo Pvt. Ltd., Bangalore
	Kerala	Travancore Organic Fertilizers, Kottayam.
		VRM Biotech Products, Changanacherry
	Madhya Pradesh	Pest Control (India) Pvt Ltd, Bhopal
	Maharashtra	Bio Era Technologies, Nagpur.
		Hoechst Schering Agr. Envt, Mumbai.
		KNS Biotech, Nanded.
		Nathkrupa Biocontrol Lab, Nagpur.
		Om Agro Organics, Yavatmal.
		Pest Control (India) Pvt Ltd, Mumbai-400 063.
		Pest Control (India) Pvt Ltd, Mumbai-400 059.
		Pest Control (India) Pvt Ltd, Thane.
		Pest Control (India) Pvt Ltd, New Mumbai.
		Pest Control (India) Pvt Ltd, Nashik.
		Pest Control (India) Pvt Ltd, Pune-411 033.
		Pest Control (India) Pvt Ltd, Pune-411 030.
		Sanvardhini Agro Pvt. Ltd., Satara.
		Soman Biofertilizers, Pune.
		Vidyas Biotech Lab, Nagpur.
		West Coast Herbochem P Ltd, Mumbai
	New Delhi	Biotech Intern Ltd., New Delhi.
		Pest Control (India) Pvt Ltd, New Delhi
	Orissa	Pest Control (India) Pvt Ltd, Bhubaneswar
	Punjab	Pest Control (India) Pvt Ltd, Chandigarh
	Rajasthan	Biocontrol Lab, ARS, RAU, Sri Ganganagar.
		JDA (Pl Path), Biocontrol Lab, Durgapura.
		DDA (Agro), IPM Lab, Ajmeer.

Table. 14.8 (Contd.)

Antagonistic Fungi/ Trade Name	State	Source of Availability
		DDA (Agro), IPM Lab, Malikpur.
		DDA (Agro), IPM Lab, Rampura.
		DDA (Agro), IPM Lab, Cjjatrapura.
		DDA (Agro), l IPM Lab, Chittorgarh.
		DDA (Agro), IPM Lab, Hanumangarh.
		DDA (Pl Path), IPM Lab, Banaswara.
		Pest Control (India) Pvt Ltd, Jaipur
	Tamil Nadu	Centre for Pl Prot Studies, TNAU, Coimbatore.
		Bio-control Lab, Papparapatty.
		Bio-control Lab, Coimbatore.
		Bio-control Lab, Salem.
		Bio-control Lab, Villupuram.
		Bio-control Lab, Trichy.
		Bio-control Lab, Panjupettai.
		Bio-control Lab, Kuruppanaichenpalyam.
		Bio-control Lab, Vinayagapuram.
		Bio-control Lab, Tirunelveli.
		Bio-control Lab, Tanjavur.
		Bio-control Lab, Namakkal.
		Basarass Biocontrol Res, Vallavar.
		Greentech Agro Prod P Ltd, Coimbatore.
		Jeypee Bio Techs, Virudhunagar.
		Pest Control (India) Pvt Ltd, Chennai.
		Rajendra Foundn for Agri Res, Chittar.
		Shri Durga Agro Services, Coimbatore.
		Shrishti Bioprod P Ltd, Coimbatore.
		Sun Agro Ind India, Chennai.
		TN Co-op Sugar Fedn Ltd, Chengalpattu.
		TAP Industries, Chennai
	Uttarakhand	Pl Path Dept, GB Pant Uni Agri & Technol, Pantnagar.
		Pest Control (India) Pvt Ltd, Dehradun
	Uttar Pradesh	Crop Health Bioprod Res Centre, Ghaziabad.
		Pest Control (India) Pvt Ltd, Bareilly.
		Pest Control (India) Pvt Ltd, Lucknow.
		Pest Control (India) Pvt Ltd, Kanpur
	West Bengal	Pest Control (India) Pvt Ltd, Kolkata

Table. 14.8 (Contd.)

Antagonistic Fungi/ Trade Name	State	Source of Availability
T. koningii	Karnataka	Project Directorate of Biol Control, Bangalore
T. longibactriatum	Karnataka	Project Directorate of Biol Control, Bangalore
T. polysporum	Karnataka	Project Directorate of Biol Control, Bangalore
T. pseudokoningii (Nursery Guard)	Karnataka	Project Directorate of Biol Control, Bangalore. Central Sericultural Res &Trng Inst, Mysore
T. virens (Gligard, WRC-GV)	Karnataka	Project Directorate of Biol Control, Bangalore
T. viride (Bio-Cure, Bio-Cure F, Agroderma, Antogon TV, Bioderma, Dermapak, Ecofit, Monitor, Multiplex Nisarga, Rakshak, Trichosan, Trichoft, Trichoderm, Trichonik, Tricho-X, Trichodermin)	Andhra Pradesh	ADA (BCL), P.M. Palem. ADA (BCL), Kakinada. ADA (BCL), Nidadavole. ADA (BCL), Ibrahimpatnam. ADA (BCL), Ongole. ADA (BCL), Nellore. DDA (FTC), Nandyal ADA (BCL), Anantapur. DDA (FTC), Rajendranagar. ADA (Oilseeds) (BCL), Mahabubanagar. ADA (BCL), Nalagonda. ADA (BCL), Mulugu Road. ADA (BCL), Karimnagar. ADA (BCL), Adilabad. Pest Control (India) Pvt Ltd, Hyderabad
	Assam	State Biocontrol Laboratory, Guwahati. Pest Control (India) Pvt Ltd, Guwahati
	Bihar & Jarkhand	Pest Control (India) Pvt Ltd, Jamshedpur
	Gujarat	Pest Control (India) Pvt Ltd, Vadodara
	Haryana	State Biocontrol Laboratory, Sirsa. Pest Control (India) Pvt Ltd, Chandigarh
	Karnataka	Project Directorate of Biol Control, Bangalore. Coffee Research Sub Station, Chethalli. Central Coffee Res Inst, Entomol Divn, Chikmagalur. Kadur Agro Pvt. Ltd., Bangalore. Pest Control (India) Pvt Ltd, Bangalore. P.J.Margo Pvt. Ltd., Bangalore. Travancore Org Fert Co P Ltd., Bangalore
	Kerala	Travancore Organic Fertilizers, Kottayam. VRM Biotech Products, Changanacherry

Table. 14.8 (Contd.)

Antagonistic Fungi/ Trade Name	State	Source of Availability
	Madhya Pradesh	Pest Control (India) Pvt Ltd, Bhopal
	Maharashtra	Bio Era Technologies, Nagpur.
		Hoechst Schering Agr. Envt, Mumbai.
		KNS Biotech, Nanded.
		Nathkrupa Biocontrol Lab, Nagpur.
		Om Agro Organics, Yavatmal.
		Pest Control (India) Pvt Ltd, Mumbai-400 063.
		Pest Control (India) Pvt Ltd, Mumbai-400 059.
		Pest Control (India) Pvt Ltd, Thane.
		Pest Control (India) Pvt Ltd, New Mumbai.
		Pest Control (India) Pvt Ltd, Nashik.
		Pest Control (India) Pvt Ltd, Pune-411 033.
		Pest Control (India) Pvt Ltd, Pune-411 030.
		Sanvardhini Agro Pvt. Ltd., Satara.
		Soman Biofertilizers, Pune.
		Vidyas Biotech Lab, Nagpur.
		West Coast Herbochem P Ltd, Mumbai
	New Delhi	Biotech Intern Ltd., New Delhi.
		Pest Control (India) Pvt Ltd, New Delhi
	Orissa	Pest Control (India) Pvt Ltd, Bhubaneswar
	Punjab	Pest Control (India) Pvt Ltd, Chandigarh
	Rajasthan	Biocontrol Lab, ARS, RAU, Sri Ganganagar.
		JDA (Pl Path), Biocontrol Lab, Durgapura.
		DDA (Agro), IPM Lab, Ajmeer.
		DDA (Agro), IPM Lab, Malikpur.
		DDA (Agro), IPM Lab, Rampura.
		DDA (Agro), IPM Lab, Cjjatrapura.
		DDA (Agro), l IPM Lab, Chittorgarh.
		DDA (Agro), IPM Lab, Hanumangarh.
		DDA (Pl Path), IPM Lab, Banaswara.
		Pest Control (India) Pvt Ltd, Jaipur
	Tamil Nadu	Centre for Pl Prot Studies, TNAU, Coimbatore.
		Bio-control Lab, Papparapatty.
		Bio-control Lab, Coimbatore.
		Bio-control Lab, Salem.
		Bio-control Lab, Villupuram.
		Bio-control Lab, Trichy.
		Bio-control Lab, Panjupettai.

Table. 14.8 (Contd.)

Antagonistic Fungi/ Trade Name	State	Source of Availability
		Bio-control Lab, Kuruppanaichenpalyam.
		Bio-control Lab, Vinayagapuram.
		Bio-control Lab, Tirunelveli.
		Bio-control Lab, Tanjavur.
		Bio-control Lab, Namakkal.
		Basarass Biocontrol Res, Vallavar.
		Greentech Agro Prod P Ltd, Coimbatore.
		Jeypee Bio Techs, Virudhunagar.
		Pest Control (India) Pvt Ltd, Chennai.
		Rajendra Foundn for Agri Res, Chittar.
		Shri Durga Agro Services, Coimbatore.
		Shrishti Bioprod P Ltd, Coimbatore.
		Sun Agro Ind India, Chennai.
		TN Co-op Sugar Fedn Ltd, Chengalpattu.
		TAP Industries, Chennai.
		T. Stanes & Co, Coimbatore
	Uttarakhand	Pl Path Dept, GB Pant Univ Agri & Technol, Pantnagar.
		Pest Control (India) Pvt Ltd, Dehradun
	Uttar Pradesh	Crop Health Bioprod Res Centre, Ghaziabad.
		Pest Control (India) Pvt Ltd, Bareilly.
		Pest Control (India) Pvt Ltd, Lucknow.
		Pest Control (India) Pvt Ltd, Kanpur
	West Bengal	Pest Control (India) Pvt Ltd, Kolkata
Verticillium lecanii	Karnataka	Project Directorate of Biol Control, Bangalore.
		Pest Control (India) Pvt Ltd, Bangalore
	Maharashtra	Sio Agro Research Lab, Mumbai
	Tamil Nadu	Sakthi Biocontrol & Biofert Centre, Kallipatti Gobi
V. suchlosporium	Karnataka	Project Directorate of Biol Control, Bangalore
Arthrobotrys sp. + *Plilacinus* + *Verticillium* sp. (Nemastin)	Maharashtra	Kumar Krishimitra Bioproducts, Pune
T. harzianum + *T. viride* (Ecoderma, NIPROT)	Karnataka	Bio Control Res Lab, PCI Pvt. Ltd., Bangalore.
		P.J. Margo Pvt. Ltd., Bangalore
T. harzianum + *T. viride* + *T. virens* (Combat)	Maharashtra	Kumar Krishimitra Bioproducts, Pune

14.7.3. Entomopathogenic Fungi

Table 14.9. Trade names and sources of availability of entomopathogenic fungi against insect pests

Entomopathogenic Fungi	State	Source of Availability
Beauveria bassiana (Bio-Power, Tracobea, Dispel, Larvocel, Don Muscardin, BABA, Abtech Beauveria)	Karnataka	Project Directorate of Biol Control, Bangalore. Entomol Dept, Univ of Agri Sci, Dharwad. Travancore Org Fert Co P Ltd., Bangalore
	Maharashtra	Vidyas Biotech Lab, Nagpur
	New Delhi	Defence Inst of Physiol, New Delhi. Navdanya (NGO), New Delhi
	Rajasthan	Pesticides India Ltd., Udaipur.
	Tamil Nadu	Divn of Pl Prot, SBI, Coimbatore. Basarass Biocontrol Res, Vallavar. Greentech Agro Prod P Ltd, Coimbatore. Jeypee Bio Techs, Virudhunagar. Rajshree Sugars & Chem Ltd, Varadaraj Nagar. Shri Durga Agro Services, Coimbatore. T. Stanes & Co, Coimbatore
Botryodiplodia theobromae	Karnataka	Project Directorate of Biol Control, Bangalore
Fusarium cocophilum	Karnataka	Project Directorate of Biol Control, Bangalore
F. moniliformae	Karnataka	Project Directorate of Biol Control, Bangalore
F. roseum	Karnataka	Project Directorate of Biol Control, Bangalore
Metarhizium anisopliae (Bio-Magic)	Gujarat	Entomol Dept, Anand Agri Univ, Anand
	Karnataka	Project Directorate of Biol Control, Bangalore. Entomol Dept, Univ of Agri Sci, Dharwad
	Kerala	CPCRI Regional Station, Kayangulam
	Maharashtra	Bio Era Technologies, Nagpur. National Biocontrol Laboratory, Nagpur
	New Delhi	Biotech Intern Ltd., New Delhi. Navdanya (NGO), New Delhi
	Tamil Nadu	Divn of Pl Prot, SBI, Coimbatore. Basarass Biocontrol Res, Vallavar. Biocontrol Res Centre, Kootham Poondi. Rajendra Foundn for Agri Res, Chittar. Shri Durga Agro Services, Coimbatore.

Table. 14.9. (Contd.)

Entomopathogenic Fungi	State	Source of Availability
Nomuraea rileyi	Andhra Pradesh	Directorate of Oil Seeds Res, Hyderabad
	Karnataka	Project Directorate of Biol Control, Bangalore.
		Agri Technol Infmn Centre, IIHR, Bangalore.
		Entomol Dept, Univ of Agri Sci, Dharwad
Paecilomyces fumosoroseus (Priority)	Tamil Nadu	Jeypee Bio Techs, Virudhunagar.
		T. Stanes & Co, Coimbatore
Scopulariopsis brevicaulis	Karnataka	Project Directorate of Biol Control, Bangalore
Verticillium lecanii (Bio-Catch, Tracovert, Verticel, Varsha, Abtech Verticillium)	Karnataka	Project Directorate of Biol Control, Bangalore.
		Entomol Dept, Univ of Agri Sci, Dharwad.
		Agri Res Stn, UAS, Bijapur.
		Travancore Org Fert Co P Ltd., Bangalore
	Maharashtra	Wockhardt Ltd., Mumbai
	Tamil Nadu	T. Stanes & Co, Coimbatore

14.7.4. Bacterial Biopesticides

Table 14.10. Trade names and sources of availability of bacterial biopesticides

Bacterial Biopesticides/ Trade Name/s	State	Source of Availability
Botryodiplodia theobromae	Karnataka	Project Directorate of Biol Control, Bangalore
B. sphaericus	New Delhi	Biotech Intern Ltd, New Delhi
	Tamil Nadu	Sinagro Industries India, Chennai
B. subtilis (Kill Dew DP Epic, Kodiac)	Maharashtra	Kumar Krishimitra Bioproducts, Pune
B. thuringiensis var. *kurstaki* (Biolep, Biosap, Delfin, Dipel, Bibot, Spicturin, Halt)	Andhra Pradesh	Indian Inst of Chem Technol, Hyderabad
	Madhya Pradesh	Indore Biotech Inputs & Res (Pvt.) Ltd., Indore
	Maharashtra	Lupin Laboratories (I) P Ltd., Mumbai.
		Sandoz (I) Ltd., Mumbai.
		Wockhardt Ltd., Mumbai
	New Delhi	Biotech Intern Ltd., New Delhi.
		Hindustan Insecticides, New Delhi
	Tamil Nadu	Tuticorn Alkali Chem & Fert Ltd, Chennai
Pseudomonas fluorescens (Bio-Cure B, Bioshield, Tracomonas, Sudocel, Dagger-G)	Assam	State Biocontrol Laboratory, Guwahati

Table. 14.10. (Contd.)

Bacterial Biopesticides/ Trade Name/s	State	Source of Availability
	Haryana	State Biocontrol Laboratory, Sirsa
	Karnataka	Agri Technol Infmn Centre, IIHR, Bangalore.
		Travancore Org Fert Co P Ltd., Bangalore
	Kerala	Plantrich Chem & Fert Ltd, Kottayam
	New Delhi	Biotech Intern Ltd., New Delhi
	Tamil Nadu	Bio-control Lab, Papparapatty.
		Bio-control Lab, Coimbatore.
		Bio-control Lab, Salem.
		Bio-control Lab, Villupuram.
		Bio-control Lab, Trichy.
		Bio-control Lab, Panjupettai.
		Bio-control Lab, Kuruppanaichenpalyam.
		Bio-control Lab, Vinayagapuram.
		Bio-control Lab, Tirunelveli.
		Bio-control Lab, Tanjavur.
		Bio-control Lab, Namakkal.
		Basarass Biocontrol Res, Vallavar.
		Greentech Agro Prod P Ltd, Coimbatore.
		Jeypee Bio Techs, Virudhunagar.
		Rajendra Foundn for Agri Res, Chittar.
		Sun Agro Ind India, Chennai.
		T. Stanes & Co, Coimbatore

14.7.5. Combination Products of Biocontrol Agents

Table 14.11. Source of availability of combination products of biocontrol agents

Combination Products of Biocontrol Agents/ Trade Name/s	State	Source of Availability
B. pumilis + *P. chlamydosporia* (Bacillus Plus-P)	Karnataka	Agri Technol Infmn Centre, IIHR, Bangalore
B. subtilis + *P. chlamydosporia* (Bacillus Plus-S)	Karnataka	Agri Technol Infmn Centre, IIHR, Bangalore
P. fluorescens + *P. chlamydosporia* (Pseudomonas Plus)	Karnataka	Agri Technol Infmn Centre, IIHR, Bangalore
Arthrobotrys sp. + *P.lilacinus* + *Verticillium* sp. (Nemastin)	Maharashtra	Kumar Krishimitra Bioproducts, Pune

Table. 14.11 (Contd.)

Combination Products of Biocontrol Agents/ Trade Name/s	State	Source of Availability
Aspergillus niger + *T. viride* + *P. lilacinus* (Multiplex Shock)	Karnataka	Karnataka Agro Chemicals, B'lore
P. chlamydosporia + *B. pumilis* (Bacillus Plus-P)	Karnataka	Agri Technol Infmn Centre, IIHR, Bangalore
P. chlamydosporia + *B. subtilis* (Bacillus Plus-S)	Karnataka	Agri Technol Infmn Centre, IIHR, Bangalore
P. chlamydosporia + *P.lilacinus* (Biovert Plus)	Karnataka	Agri Technol Infmn Centre, IIHR, Bangalore
P. chlamydosporia + *P.fluorescens* (Pseudomonas Plus)	Karnataka	Agri Technol Infmn Centre, IIHR, Bangalore
P. chlamydosporia + *T. harzianum* (Trichorich)	Karnataka	Agri Technol Infmn Centre, IIHR, Bangalore
P. chlamydosporia + *T. viride* (Trichovert)	Karnataka	Agri Technol Infmn Centre, IIHR, Bangalore
T. harzianum + *T. viride* (Ecoderma, NIPROT)	Karnataka	Bio Control Res Lab, PCI Pvt Ltd, Bangalore. Margo Pvt Ltd, Bangalore
T. harzianum + *T. viride* + *T. virens* (Combat)	Maharashtra	Kumar Krishimitra Bioproducts, Pune

14.7.6. Entomopathogenic Viruses

Table 14.12. Trade names and sources of availability of entomopathogenic viruses

Entomopathogenic Viruses/Trade Name/s	State	Source of Availability
Ha NPV of *Helicoverpa armigera* (Helivax, Heliokill, Helicide, Helinash, Heliocel)	Andhra Pradesh	ADA (BCL), P.M. Palem. ADA (BCL), Kakinada. ADA (BCL), Nidadavole. ADA (BCL), Ibrahimpatnam. ADA (BCL), Ongole. ADA (BCL), Nellore. DDA (FTC), Nandyal. ADA (BCL), Anantapur. DDA (FTC), Rajendranagar.

Table. 14.12 (Contd.)

Entomopathogenic Viruses/Trade Name/s	State	Source of Availability
		ADA (Oilseeds) (BCL), Mahabubanagar.
		ADA (BCL), Nalagonda.
		ADA (BCL), Mulugu Road.
		ADA (BCL), Karimnagar.
		ADA (BCL), Adilabad.
		Kamal Herbal & Pheromone Exp (P) Ltd., Hyderabad.
		Pest Control (India) Pvt Ltd, Hyderabad.
		Sneha Biotech, Vijayawada
	Assam	State Biocontrol Laboratory, Guwahati.
		Pest Control (India) Pvt Ltd, Guwahati
	Bihar & Jarkhand	Pest Control (India) Pvt Ltd, Jamshedpur
	Gujarat	Pest Control (India) Pvt Ltd, Vadodara
	Karnataka	Project Directorate of Biol Control, Bangalore.
		Agri Technol Infmn Centre, IIHR, Bangalore.
		Entomol Dept, Univ of Agri Sci, Dharwad.
		Inst of Pulses & Oilseeds Res, Gulbarga.
		Regional Res Stn, UAS, Raichur.
		Agri Res Stn, UAS, Bheemarayanagudi.
		Bio-Pest Management P Ltd, Bangalore.
		Cee Kay Associates, Bangalore.
		Innovative Pest Control Lab, Bangalore.
		Kadur Agro Pvt. Ltd., Bangalore.
		Pest Control (India) Pvt Ltd, Bangalore.
		P.J. Margo Pvt. Ltd., Bangalore
	Kerala	Pest Control (India) Pvt Ltd, Cochin
	Madhya Pradesh	Indore Biotech Inputs & Res (Pvt.) Ltd., Indore.
		Pest Control (India) Pvt Ltd, Bhopal.
		Samriddi Bioculture Pvt. Ltd., Indore
	Maharashtra	Central Inst of Cotton Res, Nagpur.
		Entomol Dept, Mahtma Phule Agri Univ, Rahuri.
		Entomol Dept, Marathwada Agri Univ, Parbhani.
		Entomol Dept, Punjabrao Desmukh Agri Univ, Akola.
		Entomol Dept, Konkan Agri Univ, Dapoli.
		Ajay Biotech Lab (P) Ltd, Pune.
		Anand Biocontrol Lab, Jalgaon.
		Ankur Seeds Pvt. Ltd., Ratnagiri.
		Arag Biotech, Nagpur.
		Bio Era Technologies, Nagpur.

Table. 14.12 (Contd.)

Entomopathogenic Viruses/Trade Name/s	State	Source of Availability
		Chintawar Biocontrol Lab, Yeotmal.
		Harit Biocontrol Laboratories, Yeotmal.
		JR Biocontrol Lab, Yeotmal.
		KNS Biotech, Nanded.
		Nathkrupa Biocontrol Lab, Nagpur.
		National Biocontrol Laboratory, Nagpur.
		Pest Control (India) Pvt Ltd, Mumbai-400 063.
		Pest Control (India) Pvt Ltd, Mumbai-400 059.
		Pest Control (India) Pvt Ltd, Thane.
		Pest Control (India) Pvt Ltd, New Mumbai.
		Pest Control (India) Pvt Ltd, Nashik.
		Pest Control (India) Pvt Ltd, Pune-411 033.
		Pest Control (India) Pvt Ltd, Pune-411 030.
		Sanvardhini Agro Pvt. Ltd., Satara.
		Tirupati Agro Inputs, Latur.
		Universal Biotech, Nagpur
	New Delhi	Biotech Intern Ltd., New Delhi.
		JK Traders, Delhi.
		Pest Control (India) Pvt Ltd, New Delhi
	Orissa	Pest Control (India) Pvt Ltd, Chandigarh
	Punjab	Pest Control (India) Pvt Ltd, Chandigarh
	Rajasthan	Biocontrol Lab, ARS, RAU, Sri Ganganagar.
		Adarsh Biocontrol Laboratories, Alwar.
		Pest Control (India) Pvt Ltd, Jaipur
	Tamil Nadu	Centre for Pl Prot Studies, TNAU, Coimbatore.
		Entomol Dept, TNAU, Madurai.
		Bio-control Lab, Papparapatty.
		Bio-control Lab, Coimbatore.
		Bio-control Lab, Salem.
		Bio-control Lab, Villupuram.
		Bio-control Lab, Trichy.
		Bio-control Lab, Panjupettai.
		Bio-control Lab, Kuruppanaichenpalyam.
		Bio-control Lab, Vinayagapuram.
		Bio-control Lab, Tirunelveli.
		Bio-control Lab, Tanjavur.
		Bio-control Lab, Namakkal.

Table. 14.12 (Contd.)

Entomopathogenic Viruses/Trade Name/s	State	Source of Availability
		Basarass Biocontrol Res, Vallavar.
		Greentech Agro Prod P Ltd, Coimbatore.
		Jeypee Bio Techs, Virudhunagar.
		Pest Control (India) Pvt Ltd, Chennai.
		Rajendra Foundn for Agri Res, Chittar.
		Shri Durga Agro Services, Coimbatore.
		Spinco Biotech Pvt. Ltd, Chennai.
		Sun Agro Ind India, Chennai.
		T. Stanes & Co, Coimbatore
	Uttarakhand	Pest Control (India) Pvt Ltd, Dehradun
	Uttar Pradesh	Biocontrol Lab, SBBP Agri Univ, Meerut.
		Crop Health Bioprod Res Centre, Ghaziabad.
		Pest Control (India) Pvt Ltd, Bareilly.
		Pest Control (India) Pvt Ltd, Lucknow.
		Pest Control (India) Pvt Ltd, Kanpur
	West Bengal	Pest Control (India) Pvt Ltd, Kolkata
Sl NPV of *Spodoptera litura* (Spodovax, Spodokill, Spodonash)	Andhra Pradesh	Central Tobacco Res Inst, Rajahmundry
		ADA (BCL), P.M. Palem.
		ADA (BCL), Kakinada.
		ADA (BCL), Nidadavole.
		ADA (BCL), Ibrahimpatnam.
		ADA (BCL), Ongole.
		ADA (BCL), Nellore.
		DDA (FTC), Nandyal.
		ADA (BCL), Anantapur.
		DDA (FTC), Rajendranagar.
		ADA (Oilseeds) (BCL), Mahabubanagar.
		ADA (BCL), Nalagonda.
		ADA (BCL), Mulugu Road.
		ADA (BCL), Karimnagar.
		ADA (BCL), Adilabad.
		Kamal Herbal & Pheromone Exp (P) Ltd., Hyderabad.
		Pest Control (India) Pvt Ltd, Hyderabad
	Assam	Pest Control (India) Pvt Ltd, Guwahati
	Bihar & Jarkhand	Pest Control (India) Pvt Ltd, Jamshedpur
	Gujarat	Pest Control (India) Pvt Ltd, Vadodara

Table. 14.12 (Contd.)

Entomopathogenic Viruses/Trade Name/s	State	Source of Availability
	Karnataka	Project Directorate of Biol Control, Bangalore.
		Agri Technol Infmn Centre, IIHR, Bangalore.
		Entomol Dept, Univ of Agri Sci, Dharwad.
		Regional Res Stn, UAS, Raichur.
		Bio-Pest Management Pvt. Ltd., Bangalore.
		Pest Control (India) Pvt Ltd, Bangalore.
	Kerala	Pest Control (India) Pvt Ltd, Cochin
	Madhya Pradesh	Pest Control (India) Pvt Ltd, Bhopal
	Maharashtra	National Biocontrol Laboratory, Nagpur.
		Pest Control (India) Pvt Ltd, Mumbai-400 063.
		Pest Control (India) Pvt Ltd, Mumbai-400 059.
		Pest Control (India) Pvt Ltd, Thane-400 601.
		Pest Control (India) Pvt Ltd, New Mumbai.
		Pest Control (India) Pvt Ltd, Nashik-422 009.
		Pest Control (India) Pvt Ltd, Pune-411 030.
		Pest Control (India) Pvt Ltd, Pune-411 033
	New Delhi	Biotech Intern Ltd., New Delhi.
		Pest Control (India) Pvt Ltd, New Delhi
	Orissa	Pest Control (India) Pvt Ltd, Chandigarh
	Punjab	Pest Control (India) Pvt Ltd, Chandigarh
	Rajasthan	Biocontrol Lab, ARS, RAU, Sri Ganganagar.
		Adarsh Biocontrol Laboratories, Alwar.
		Pest Control (India) Pvt Ltd, Jaipur
	Tamil Nadu	Centre for Pl Prot Studies, TNAU, Coimbatore.
		Entomol Dept, TNAU, Madurai.
		Basarass Biocontrol Res, Vallavar.
		Jeypee Bio Techs, Virudhunagar.
		Pest Control (India) Pvt Ltd, Chennai.
		Shri Durga Agro Services, Coimbatore.
		Sun Agro Ind India, Chennai
	Uttarakhand	Pest Control (India) Pvt Ltd, Dehradun
	Uttar Pradesh	Pest Control (India) Pvt Ltd, Bareilly.
		Pest Control (India) Pvt Ltd, Lucknow.
		Pest Control (India) Pvt Ltd, Kanpur
	West Bengal	Pest Control (India) Pvt Ltd, Kolkata

14.7.7. Entomopathogenic Nematodes

Table 14.13. Trade names and sources of availability of entomopathogenic nematodes

Entomopathogenic Nematodes/Trade Name/s	State	Source of Availability
Heterorhabditis spp. (Green Cammandos)	Karnataka	Project Directorate of Biol Control, Bangalore
	Maharashtra	Bio Sense Crop Protection, Mumbai
Steinernema spp. (Soil Commandos)	Karnataka	Project Directorate of Biol Control, Bangalore
	Maharashtra	Bio Sense Crop Protection, Mumbai

14.8. PHEROMONES

Table 14.14. Source of availability of pheromones of fruit crop pests

Pheromone	State	Source of Availability
Oriental fruit moth	Maharashtra	Agri Sense BCS, Mumbai. Bio Sense Crop Protection, Mumbai
Gypsy moth	Maharashtra	Bio Sense Crop Protection, Mumbai
Codling moth	Maharashtra	Agri Sense BCS, Mumbai. Bio Sense Crop Protection, Mumbai
San Jose scale	Maharashtra	Agri Sense BCS, Mumbai
Apple leaf roller	Maharashtra	Agri Sense BCS, Mumbai
Apple fruit miner	Maharashtra	Agri Sense BCS, Mumbai
Red banded leaf roller	Maharashtra	Agri Sense BCS, Mumbai
Pear leaf roller	Maharashtra	Agri Sense BCS, Mumbai
Peach tree borer	Maharashtra	Agri Sense BCS, Mumbai
Peach twig borer	Maharashtra	Agri Sense BCS, Mumbai
Grape torteix	Maharashtra	Agri Sense BCS, Mumbai
Mediterranean fruit fly	Maharashtra	Agri Sense BCS, Mumbai
California red scale	Maharashtra	Agri Sense BCS, Mumbai

14.9. NEEM-BASED AND OTHER BOTANICALS

Table 14.15. Source of availability of neem–based and other botanicals

Neem-Based & other Botanicals	State	Source of Availability
Achook	Maharashtra	Bahar Agrochemicals, Ratnagiri Mahagrapes, Pune
Aphidin	Maharashtra	Eco-Max Agrosystems Ltd, Mumbai
Avana	Tamil Nadu	EID Parry (I) Ltd., Bioprod Divn, Chennai

Table. 14.15 (Contd.)

Neem-Based & other Botanicals	State	Source of Availability
Aza	Andhra Pradesh	SOM-IPM Systems (I) Ltd., Hyderabad
Azadit	Rajasthan	Pesticides India Ltd., Udaipur
Biogold	Tamil Nadu	SPIC, Bio-Tech Divn, Chennai
Bioneem	Maharashtra	Ajay Biotech Lab (P) Ltd, Pune
Bionema	Uttar Pradesh	Bioved Res Soc, Allahabad
Biopest	Uttar Pradesh	Bioved Res Soc, Allahabad
Biosol	Tamil Nadu	A.V. Thomas Pvt. Ltd., Chennai
Eco-Neem	Karnataka	Murkumbi Bio Agro P Ltd., Belgaum. Pest Control (India) Pvt Ltd, Bangalore
Eco-Neem Plus	Karnataka	Murkumbi Bio Agro P Ltd., Belgaum. Pest Control (India) Pvt Ltd, Bangalore
Ecotin	Karnataka	Monarch Biofert & Res Centre, Bangalore
Field Marshal	Gujarat	Ganesh Biocontrol System, Gondal
Fortuna Aza	Andhra Pradesh	Fortune Bio-tech Limited, Secunderabad
Jai Neem	Haryana	Jai Chemicals, Faridabad
Jawan	Maharashtra	Mc DA Agro Pvt. Ltd., Mumbai
Jeevan Crop Protector	Maharashtra	Mc DA Agro Pvt. Ltd., Mumbai
Juerken	Tamil Nadu	Madurai Chem & Agro Ind Ltd., Madurai
Limonool	Karnataka	Bio-Multi Tech Pvt. Ltd., Bangalore
Margocide-CK	Karnataka	Monofix Agro Products Ltd., Hubli
Margosan-O	Karnataka	Monofix Agro Products Ltd., Hubli
Neemactin	Maharashtra	Wockhardt Ltd., Mumbai
Neemak	Karnataka Maharashtra	West Coast Herbochem P Ltd, Bangalore West Coast Herbochem P Ltd, Mumbai
Neemarin	New Delhi	Biotech Intern Ltd., New Delhi
Neemark	Karnataka	West Coast Herbochem P Ltd, Bangalore
Neemasol	Tamil Nadu	EID Parry (I) Ltd., Pesticides Divn, Chennai
Neemax	Maharashtra	Eco-Max Agrosystems Ltd, Mumbai
Neemazal-F	Tamil Nadu	EID Parry (I) Ltd., Bioprod Divn, Chennai
Neemazal-T/S	Tamil Nadu	EID Parry (I) Ltd., Bioprod Divn, Chennai
Neemocide	West Bengal	Vinayak Fats & Proteins P Ltd, Kolkata
Neemol	Andhra Pradesh Tamil Nadu	Ramson Laboratories, Vijayawada Agro Links, Trichy

Table. 14.15 (Contd.)

Neem-Based & other Botanicals	State	Source of Availability
Neemolin	Maharashtra	Khatau Agro Tech, Mumbai
Neemosan	Tamil Nadu	Agronule Industries, Trichy
Neemox	Maharashtra	Prakash Farm Chemicals, Mumbai
Neempourn	Tamil Nadu	Prabhakar Oil Mills, Trichy
Neemta	Gujarat	A.J. Chemicals, Ahmedabad
Neethrin	Uttar Pradesh	Amitul Agro Chem Pvt. Ltd., Gorakhpur
Neem Base	New Delhi	Agrochem Divn, Indian Agri Res Inst, New Delhi
Neem Based pesticides	Maharashtra	Bahar Agrochemicals, Ratnagiri. KNS Biotech, Nanded. Om Agro Organics, Yavatmal
	Tamil Nadu	Ecobiocides & Botanicals Pvt. Ltd., Theni. Jeypee Bio Techs, Virudhunagar
Neem Cake	Uttarakhand	Sri Ram Solvent Extraction Plant, Jaspur
Neem Gold	Tamil Nadu	EID Parry (I) Ltd., Bioprod Divn, Chennai. SPIC, Bio-Tech Divn, Chennai
Neem Guard	Maharashtra	Gharda Chemicals Pvt. Ltd., Mumbai
Neem Hit	Maharashtra	Skylark Agro Chemicals, Thane
Neem Oil	Karnataka	Agro Extracts Ltd., Bangalore
	Maharashtra	Mc DA Agro Pvt. Ltd., Mumbai. National Organic Chem Ind Ltd., Mumbai. Sio Agro Research Lab, Mumbai
	Uttarakhand	Sri Ram Solvent Extraction Plant, Jaspur
Neem Plus	Maharashtra	BD Keathen & Co., Mumbai
Neem Rich	Maharashtra	National Chem Lab, Pune
Neem seed	Maharashtra	Cheminova Agro-Tech, Nandubar
Neem Soap	Karnataka	Agri Technol Infmn Centre, IIHR, Bangalore
Neem Top	Tamil Nadu	Sri Krishna Co, Coimbatore
Nim-76	New Delhi	Defence Inst of Physiol, New Delhi
Nimba	New Delhi	Entomol Divn, Indian Agri Res Inst, New Delhi
Nimbasal	New Delhi	Nimba Foods & Agrochem, New Delhi. Pesto-Chem India Ltd, New Delhi
Nimbicidine	Tamil Nadu	T. Stanes & Co, Coimbatore
Nimbitor	Maharashtra	Zandu Pharmaceutical Works, Mumbai
Nimbosol	Tamil Nadu	Victoria Laboratories, Salem

Table. 14.15 (Contd.)

Neem-Based & other Botanicals	State	Source of Availability
Nimitox	Karnataka	Rallis India Ltd., Bangalore
Nimlin	Maharashtra	Sunline Agro Chem P Ltd, Dhulia
Nimbo Bas	Tamil Nadu	Basarass Biocontrol Res, Vallavar
Ozoneem	Haryana	Ozone Biotech, Faridabad
Phytowin	Tamil Nadu	Phyto Products Ltd., Thiruthuripoordi
Pongamia oil-based biopesticide	Tamil Nadu	Jeypee Bio Techs, Virudhunagar
Pongamia Soap	Karnataka	Agri Technol Infmn Centre, IIHR, Bangalore
Rakshak	Karnataka	Murkumbi Bio Agro P Ltd., Belgaum. P.J. Margo Pvt. Ltd., Bangalore
RD-9 Repellin	Andhra Pradesh	ITC Ltd., Hyderabad
Replin 555	Tamil Nadu	Micro Chemicals Ltd., Madurai
Shaktiman	Gujarat	Khetiwadi Corner, Vadodara. Krishna Biotech Pvt. Ltd., Veravel. Oceon Agro India Ltd., Baroda
Sukrina	Tamil Nadu	Canster Chemicals Ltd., Chennai
Suneem	Maharashtra	Sunida Exports Ltd., Mumbai
Swaticure	Maharashtra	Swathi Industries Ltd., Mumbai
Tric	Uttar Pradesh	Amitul Agro Chem Pvt. Ltd., Gorakhpur
Tricure	Karnataka	Murkumbi Bio Agro P Ltd., Belgaum. P.J. Margo Pvt. Ltd., Bangalore
Vapacide	Andhra Pradesh	Indian Inst of Chem Technol, Hyderabad

REFERENCES

Abuzar, S. and Haseeb, A. 2006. Efficacy of carbofuran, bavistin, neem, *Trichoderma harzianum* and *Aspergillus niger* against *Meloidogyne incognita* and *Fusarium oxysporum* f. sp. *vasinfectum* disease complex on okra. *Indian J. Nematol.* **36**: 299–300.

Adams, S. 1997. Seeing red colored mulch starves nematodes. *Agri. Res.*, October, p. 18.

Ahlawat, Y.S. and Chenulu, V.V. 1982. Studies on transmission of radish mosaic virus by the aphid. *Indian Phytopath.* **35**: 633–38.

Ahlawat, Y.S. and Chenulu, V.V. 1984. Radish mosaic–A new disease caused by turnip mosaic virus in India. *Trop. Agric.* (Trinidad) **61**: 188–92.

Ahmed, J.A. and Choudhury, B.N. 2004. Management of root-knot nematode, *Meloidogyne incognita* using organic amendments in French bean. *Indian J. Nematol.* **24**: 137–39.

Ahuja, A.K., Mohapatra, S., Sharma, D. and Awasthi, M.D. 1998. Monitoring of vegetables for pesticide contamination at harvest. *Proc. Nat. Symp. Pest Mangmt. Hort. Crops*, Bangalore, pp. 243–46.

Ahuja, S. 1983. Studies on the variability in infection of root-knot nematode *Meloidogyne incognita* on potato. *Third Nematol. Symp.,* Y.S. Paramar Uni. Hort. and Forestry, Solan, Himachal Pradesh, p. 4.

Ahuja, S. and Mukhopadhyaya, M.C. 1984. Efficacy of some agricultural residues as compared to nematicides/insecticides in the control of root-knot nematode (*Meloidogyne incognita*) infestation in okra (*Abelmoschus esculentus*). *Proc. Symp. Utilization Agril. Wastes*, pp. 131–35.

Agnihotri, N.P. 1999. Monitoring of pesticide residues. *In: Pesticide Safety Evaluation and Monitoring* (N.P. Agnihotri, *ed.*), pp. 94–118. All India Co-ordinated Res. Project on Pesticide Residues, New Delhi.

Agnihotrudu, V. and Mithyantha, M.S. 1978. *Pesticide residues–A review of Indian Work.* Rallis India Ltd., Bangalore.

Alam, M.M., Khan, A.M. and Saxena, S.K. 1973. Some new host records of root-knot nematode, *Meloidogyne incognita* (Kofoid & White, 1919) Chitwood, 1949. *Indian J. Nematol.* **3**: 159–60.

Alam, M.M., Khan, A.M. and Saxena, S.K. 1977. Influence of different cropping sequences on soil population of plant parasitic nematodes. *Nematol. Medit.* **5**: 71–78.

Alam, M.M., Saxena, S.K. and Khan, A.M. 1977. Influence of interculture of marigold and margosa with some vegetable crops on plant growth and nematode population. *Acta Bot. Indica* **5**: 33–39.

Amalraj, S.F.A. and Rao, G.R. 1992. Potato cyst nematode pathogens in the southern hills of India. *Indian J. Nematol.* **22**: 82–85.

Anita, B. and Subramanian, S. 1998. Management of the reniform nematode *Rotylenchulus reniformis* in tomato. *In: Nematology–Challenges and Opportunities in 21st[st] Century* (Usha K. Mehta, *ed.*), pp. 249–50. Sugarcane Breeding Inst., Coimbatore.

Anjaneyulu, A. and Ramakrishnan, K. 1972. Chemotherapy of egg plant little leaf disease. *Mysore J. Agric. Sci.* **6**: 135.

Anon., 1985. *Kufri Swarna, a High Yielding Late Blight and Cyst Nematode Resistant Potato Cultivar Suitable for Cultivation in Tamil Nadu Hills.* No. 21(E), Central Potato Res. Inst., Shimla, 4 pp.

Anon., 1989a. *Biennial Report (1987–89) of AICRP on Plant Parasitic Nematodes with Integrated Approach for their Control*, Indian Agri. Res. Inst., New Delhi, 80 pp.

Anon., 1990b. *Annual Report of the Central Tuber Crops Research Institute*, Trivandrum.

Anon., 1992. *Research Highlights of AICRP on Agricultural Ornithology*, 1992. Andhra Pradesh Agri. Univ., Hyderabad, 16 pp.

Anon., 1993b. *Biennial Report (1991–93) of AICRP on Plant Parasitic Nematodes with Integrated Approach for their Control*, Dept. of Nematology, Haryana Agri. Univ., Hisar, 80 pp.

Anon., 2000. *Bean Fly Information System. Database on Bean Flies*. Extn. Folder No. 73, Indian Inst. of Hort. Res., Bangalore, 6 pp.

AVRDC. 1996. *AVRDC 1995 Progress Report*. Asian Vegetable Res. and Devp. Centre, Shanhua, Taiwan, 187 pp.

Awasthi, M.D. 1986. Effect of chemical treatments on the decontamination of synthetic pyrethroids residues from brinjal fruits. *Pestic. Sci.* **17**: 89–90.

Awasthi, M.D. 1993. Pesticide residues in vegetable crops. *In: Advances in Horticulture–Vegetable Crops* (K.L. Chadha and G. Kalloo, *eds*.), pp. 921–76. Malhotra Publishing House, New Delhi.

Awasthi, M.D. 1998. Pesticide residues in food chain in India–Fruits and vegetables. *In: Pesticides, Man and Biosphere* (O.P. Shukla, Omkar and A.K. Kulshrestha, *eds*.), pp. 121–60. APH Publishing Corp., New Delhi.

Awasthi, M.D. 2001. Pesticide residues in fruits and vegetables. *In: Integrated Pest Management in Horticultural Ecosystems* (P. Parvatha Reddy, A. Verghese and N.K. Krishna Kumar, *eds*.), pp. 263–78. Capital Publishing Co., New Delhi.

Ayyar, P.N.K. 1926. A preliminary note on the root nematode, *Heterodera radicicola* Mitter and its economic importance in south India. *Sci. Cult.* **22**: 391–93.

Bagyaraj, D.J., Manjunath, A. and Reddy, D.D.R. 1979. Interaction of vesicular arbuscular mycorrhiza with root-knot nematodes in tomato. *Plant and Soil* **51**: 397–403.

Bajaj, H.K. and Jain, R.K. 2001. CCS Haryana Agriculture University, Hisar, Haryana. *In: Indian Nematology–Problems and Perspectives* (S.C. Dhawan, H.S. Gaur, Pankaj, K.K. Kaushal, S. Ganguly, G. Chawla, R.V. Singh and Anil Sirohi, *eds*.), pp. 78–95. Divn. of Nematol., Indian Agri. Res. Inst., New Delhi.

Balasubramanian, P. and Ramakrishnan, C. 1983. *Nematol. Medit.* **11**: 203–04.

Bandopadhyay, S. and Chattopadhyay, S.B. 1986a. Survival of *Xanthomonas campestris* pv. *campestris* in the host tissue. *Indian J. Mycol. Res.* **24**: 57–63.

Bandopadhyay, S. and Chattopadhyay, S.B. 1986b. Study of incidence on black rot disease of cabbage and cauliflower under different months of planting following natural conditions of infection. *Indian J. Mycol. Res.* **24**: 125–33.

Bansal, R.K. and Verma, V.K. 2002. Antagonistic efficacy of *Azotobacter chroococcum* against *Meloidogyne javanica* infecting brinjal. *Indian J. Nematol.* **32:** 132–34.

Bath, D.S. and Singh, D. 1989. Studies on the economic threshold level of mustard aphid, *Lipaphis erysimi* (Kalt.) on the radish seed crop in India. *Trop. Pest Mangmt.* **35**: 154–56.

Bera, S.C. 1986. Control of black rot disease of cauliflower by some chemicals. *Pesticides* **20**: 51–52.

Bhagawati, B. and Goswami, B.K. 2000. Interaction of *Meloidogyne incognita* and *Fusarium oxysporum* f. sp. *lycopersici* on tomato. *Indian J. Nematol.* **30**: 93–94.

Bhagawati, B., Goswami, B.K. and Singh, C.S. 2000. Management of disease complex of tomato caused by *Meloidogyne incognita* and *Fusarium oxysporum* f. sp. *lycopersici* through bioagents. *Indian J. Nematol.* **30**: 16–22.

Bhagawati, B. and Phukan, P.N. 1990. Chemical control of *Meloidogyne incognita* on pea. *Indian J. Nematol.* **20**: 73–83.

Bhargava, K.S. and Joshi, R.D. 1960. A mosaic disease of vegetable marrow (*Cucurbita pepo* L.) in Uttar Pradesh. *Proc. 47th Indian Sci. Cong. Part III*, P. 314.

Bhatnagar, V.S., Lateef, S.S., Sithanathan, S., Pauler, C.S. and Reed, W. 1982. Research on *Heliothis* management at ICRISAT. *Proc. Int. Workshop on Heliothis Management*, Patancheru, Hyderabad, pp. 385–97.

Bhattacharyya, S.K., Shekhawat, G.S. and Singh, B.P. 1990. *Potato Late Blight*. Tech. Bull. No. 27, Central Potato Res. Inst., Shimla, India, 40 pp.

Bhatti, D.S. and Jain, R.K. 1977. Estimation of loss in okra, tomato and brinjal yield due to *Meloidogyne incognita*. *Indian J. Nematol.* **7**: 37–41.

Bhosle, B.B., Sehgal, M., Puri, S.N., Sardana, H.R. and Singh, D.K. 2006. Efficacy of organic amendment in management of root-knot nematode, *Meloidogyne incognita* on okra. *Indian J. Nematol.* **36**: 37–40.

Bijay Kumar. 2007. *Indian Horticulture Database 2006*. National Hort. Board, Gurgaon, Haryana, India, 324 pp.

Black, L.L., Bond, W.P., Story, R.N. and Batti, J.M. Jr. 1986. Tomato spotted wilt virus in Louisiana: Epidemiological aspects. *Proc. Workshop on Epidemiology of Plant Virus Diseases*, Orlando, Florida.

Borah, A. and Phukan, P.N. 2000. Effect of VAM fungus, *Glomus fasciculatum* and root-knot nematode, *Meloidogyne incognita* on brinjal. *J. Agri. Sci. Soc. North-Eastern India* **13**: 212–14.

Borah, A. and Phukan, P.N. 2004. Comparative efficacy of *Glomus fasciculatum* with neem cake and carbofuran for the management of *Meloidogyne incognita* on brinjal. *Indian J. Nematol.* **24**: 129–32.

Borkakaty, D. 1993. *Management of Root-knot Nematode,* Meloidogyne incognita *(Kofoid and White, 1919) Chitwood, 1949 on Brinjal by Integrated Approach.* M. Sc. (Agri.) Thesis, Assam Agri. Univ., Jorhat.

Butani, D.K. and Jotwani, M.G. 1984. *Insects in Vegetables.* Periodical Expert Book Agency, New Delhi, India, 356 pp.

Butler, E.J. 1906. The wilt disease of pigeon-pea and pepper. *Agric. J. India* **1**: 25–36.

Cannayane, I. and Sivakumar, C.V. 2001. Bare root dip treatment with *Paecilomyces lilacinus* against *Meloidogyne incognita* in tomato. *Ann. Pl. Prot. Sci.* **9**: 351–54.

Capoor, S.P. and Sharma, D.C. 1965. Purple leaf disease of beet. *Indian Phytopath.* **18**: 88–89.

Capoor, S.P. and Varma, P.M. 1948. Yellow mosaic of *Phaseolus lunatus* L. *Curr. Sci.* **17**: 152–53.

Cappel, E.L. 1892. *Report of Department of Land Records in Agriculture, Bombay,* 16p.

Castillo, M.B. 1985. Some studies on the use of organic soil amendments for nematode control. *Philippine Agriculturist* **68**: 76–93.

Chahal, P.P.K. and Chahal, V.P.S. 1988. Biological control of root-knot nematode with *Azotobacter chroococcum* on brinjal. *In: Advances in Plant Nematology*, (M.A. Maqbool, A. Morgan Golden, A. Gaffar and L.R. Krusberg, *eds.*), pp. 257–63. Natl. Nematol. Centre, Univ. of Karachi, Karachi, Pakistan.

Chahal, S.S., Sahoo, M.S. and Gill, K.S. 1979. Source of resistance against bacterial blight of guar (*Cyamopsis tetragonoloba*). *Indian J. Mycol. Pl. Path.* **9**: 103–04.

Chakravarti, B.P., Shekhawat, P.S. and Choudhary, S.L. 1978. Compatibility of bavistin with agrimycin in controlling bacterial blight of cowpea. *Indian J. Mycol. Pl. Path.* **8**: 28.

Chandramohan, N. 1994. Seasonal incidence of diamondback moth, *Plutella xylostella* L. and its parasitoids in Nilgiris. *J. Biocontrol* **8**: 77–80.

Chari, M.S. 1970. *The Bionomics and Control of Cluster Bean Pod Gall Midge, Asphodylia cyamopsii* N. Sp. (Cecidomyiidae: Diptera). Ph. D. Thesis, Sardar Patel Univ., Anand, 76 pp.

Dahiya, R.S., Bansal, R.K. and Paruthi, I.J. 1998. Efficacy of mustard cake and poultry manure for the management of *Meloidogyne javanica* in bottle gourd and okra. *Proc. of National Symposium on Rational Approaches in Nematode Management for Sustainable Agriculture*, Nematol. Soc. of India, New Delhi, pp. 100–02.

Dalal, M.R. and Vats, R. 1998. Estimation of loss in pea (*Pisum sativum* L.) due to *Rotylenchulus reniformis*. *In: Nematology–Challenges and Opportunities in 21st Century,* (Usha K. Mehta, *ed.*), pp. 8–9. Sugarcane Breeding Institute, Coimbatore.

Darekar, K.S. and Mahse, N.L. 1988. Assessment of yield losses due to root-knot nematode, *Meloidogyne incognita* Racc 3 in tomato, brinjal and bitter gourd. *Int. Nematol. Network Newsl.* **5** (4): 7–9.

Darling, H.M. 1959. *Plant Disease Reporter* **23**: 239–42.

Das, A.K. 1994. *Assessment of Yield Loss Due to Meloidogyne incognita on French Bean and its Management*. M. Sc. (Agri.) Thesis, Assam Agri. Univ., Jorhat.

Das, N. and Sinha, A.K. 2005. Integrated management of root-knot nematode (*Meloidogyne incognita*) on okra [*Abelmoschus esculentus* (L.) Moench]. *Indian J. Nematol.* **35**: 175–82.

Davide, R.G. and Zorilla, R.A. 1986. Evaluation of a fungus, *Paecilomyces lilacinus* for the biological control of root-knot nematodes *Meloidogyne incognita* on okra. *Int. Nematol. Network Newsl.* **3** (3): 32–33.

Devi, Geetanjali. 1993. *Pathogenicity, Crop Loss Assessment and Management of* Meloidogyne incognita *(Kofoid and White, 1919) Chitwood, 1949 on Carrot* (*Dacus carota* L.). M. Sc. (Agri.) Thesis, Assam Agri. Univ., Jorhat.

Dhawan, S.C. and Sethi, C.L. 1977. Inter-relationship between root-knot nematode, *Meloidogyne incognita* and little leaf of brinjal. *Indian Phytopath.* **30**: 55–63.

Egunjobi, O.A. 1985. Effect of cocoa pod husks soil amendment on cowpea infestation by *Meloidogyne* spp. *Pakistan J. Nematol.* **3**: 99–103.

Ekanayake, H.M.R.K. and Jayasundara, N.J. 1994. Effect of *Paecilomyces lilacinus* and *Beauvaria bassiana* in controlling *Meloidogyne incognita* on tomato in Sri Lanka. *Nematol. Medit.* **22**: 87–88.

Fassuliotis, G. and Sparrow, A.H. 1955. Preliminary report of *x*-ray studies on the golden nematode. *Pl. Dis. Reptr.* **39**: 572.

French, E.R. 1986. *ACIAR Bacterial Wilt Newsletter No.* **1**: 5–6.

Gadewar, A.V., Trivedi, T.P. and Shekhawat, G.S. 1991. *Potato in Karnataka*. Tech. Bull. No. 17, Central Potato Res. Inst., Shimla, India, 72 pp.

Gandhi, S.K. and Chand, J.N. 1985a. Effect of nitrogen and phosphorus on the bacterial blight of cluster bean. *Indian J. Agric. Sci.* **55**: 376–77.

Gandhi, S.K. and Chand, J.N. 1985b. Yield losses in guar due to bacterial blight caused by *Xanthomonas campestris* pv. *cyamopsidis*. *Indian Phytopath.* **38**: 516–17.

Gandhi, S.K. and Chand, J.N. 1987. Horizontal resistance of cluster bean to *Xanthomonas campestris* pv. *cyamopsidis*. *Indian J. Agric. Sci.* **57**: 755–57.

Gandhi, S.K. and Chand, J.N. 1988. Plant age, temperature and free moisture in relation to susceptibility of cluster bean to *Xanthomonas campestris* pv. *cyamopsidis*. *Proc. Indian Acad. Sci., Plant Sci.* **98**: 49–53.

Garabedian, S. and Van Gundy, S.D. 1983. Use of avermectins for the control of *Meloidogyne incognita* on tomatoes. *J. Nematol.* **15**: 503–10.

Gautam, A., Siddiqui, Z.A. and Mahmood, I. 1995. Integrated management of *Meloidogyne incognita* on tomato. *Nematol. Medit.* **23**: 245–47.

Gopalakrishnan, C. 1998. *Studies on Entomopathogenic Fungus Paecilomyces farinosus (Holmskiold) Brown and Smith for the Control of Cabbage Diamondback Moth, Plutella xylostella* L. Ph. D. Thesis, Bangalore Univ., Bangalore.

Gopalakrishnan, C. and Mohan, K.S. 1993. Epizootics of *Erynia neoaphidis* (Zygomycetes: Entomophthorales) in field population of *Brevicoryne brassicae* (Homoptera: Aphididae) on cabbage. *Entomon* **18**: 21–24.

Gopalakrishnan, C. and Mohan, K.S. 1996. Field efficacy of entomopathogenic fungus *Nomuraea rileyi* against Noctuid pests of tomato (Abstr.). *Natl. Symp. Hortil. Biotechnol.*, Bangaore, 65 pp.

Gopinatha, K.V., Nanje Gowda, D.N. and Nagesh, M. 2002. Management of root-knot nematode *Meloidogyne incognita* on tomato using bioagent *Verticillium chlamydosporium*, neem cake, marigold and carbofuran. *Indian J. Nematol.* **32**: 179–81.

Goswami, B.K. and Chenulu, V.V. 1974. Interaction of root-knot nematode, *Meloidogyne incognita* and tobacco mosaic virus in tomato. *Indian J. Nematol.* **4**: 69–80.

Goswami, B.K. and Singh, H. 2001. Biomanagement of disease complex caused by *Meloidogyne incognita* and *Fusarium oxysporum* on tomato. *National Conference on Centenary of Nematol. In: India–Appraisal and Future Plans*, Indian Agri. Res. Inst., New Delhi, 169 pp.

Goswami, B.K., Uma Rao and Singh, S. 1998. Potentiality of some fungal bioagents against root-knot nematode, *Meloidogyne incognita* infecting tomato. *Proceedings of First National Symposium on Pest Management in Horticultural Crops*, Bangalore, pp. 304–07.

Govindaswamy, C.V., Mariappan, V., Murugesan, S., Padmanabhan, C., Thangamani, G. and Janaki, I.P. 1970. Studies on a cowpea mosaic virus disease from Coimbatore. *Madras Agric. J.* **57**: 405–15.

Gour, H.N., Gupta, V.P. and Singh, R.P. 1981. Preliminary alterations in resistant and susceptible cowpea varieties inoculated with *Xanthomonas campestris* pv. *vignicola* (Abstr.). *Proc. 3rd Intn. Symp. Pl. Path.*, New Delhi, 120 pp.

Gowdar, S.B. and Kulkarni, S. 1999. Compatibility effect of antagonists and seed dressers against *Fusarium udum*–the causal agent of pigeonpea wilt. *Karnataka J. Agri. Sci.* **12**: 197–99.

Grewal, P.S. 1989. Effect of leaf-matter incorporation on *Aphelenchoides composticola* (Nematode), mycofloral composition, mushroom compost quality and yield of *Agaricus bisporus*. *Ann. Appl. Biol.* **115**: 299–312.

Grewal, P.S. and Grewal, S.K. 1988. Solar heat pasteurization of casing soil and chicken manure for mushroom cultivation. *Mushroom J. Trop.* **8**: 35–45.

Grewal, P.S. and Sohi, H.S. 1988. A new and cheaper technique for the rapid multiplication of *Arthrobotrys conoides* and its potential as a bio-nematicide in mushroom culture. *Curr. Sci.* **57**: 44–46.

Gupta, D.C. and Mukhopadhyaya, M.C. 1971. Effect of N, P and K on the root-knot nematode, *Meloidogyne javanica* (Treub.) *Chitwood. Sci. Cult.* **37**: 246–47.

Gupta, D.K. 1978. Yield losses in guar (*Cyamopsis tetragonoloba*) caused by bacterial blight (*Xanthomonas cyamopsidis*). *Indian J. Mycol. Pl. Path.* **8**: 27.

Gupta, J.C. and Atwal, A.S. 1972. Biology and ecology of *Hoplolaimus indicus* (Hoplolaiminae: Nematoda). 3. Influence of population density on tomato and citrus plants. *Proceedings of XIth International Symposium on Nematology,* Reading, U. K., 26 pp.

Gupta, P.K., Agarwal, R.K. and Sharma, S.L.1986. Efficacy of soil amendments in managing 'Stalk rot' (*Sclerotinia sclerotiorum*) of cauliflower. *Indian J. Pl. Path.* **4**: 129–31.

Gurfan, S.M. and Chakravarti, B.P. 1960. *Proc. Bihar Acad. Agri. Sci.* **8 & 9**: 126–27.

Hackney, R.W. and Dickerson, O.J. 1975. Marigold, castor bean and chrysanthemum as controls of *Meloidogyne incognita* and *Pratylenchus alleni. J. Nematol.* **7**: 84–90.

Handa. 1992. Monitoring of pesticide residues in Indian environment. *In: Pest Management and Pesticides–Indian Scenario* (B.V. David, *ed.*), pp. 221–24. Namratha Publn., Madras, India.

Handa, D.K. and Mathur, B.N. 1981. Control of root-knot of chillies with nematicides. *Second Nematology Symposium*, Tamil Nadu Agri. Univ., Coimbatore, 48 pp.

Haq, S., Singh, R.P. and Saxena, S.K. 1985. Control of the root-knot nematode *Meloidogyne incognita* on tomato with soft coal fly ash. *Int. Nematol. Network Newsl.* **2** (3): 4–6.

Hare, W.W. 1951. *Phytopathology* **41**: 16.

Hasan, N. and Jain, R.K. 1992. Field application of *Paecilomyces lilacinus* in combination with certain organic matters for controlling *Meloidogyne incognita* infecting cowpea followed by lucerne. *First Afro-Asian Nematol. Symp.*, Aligarh Muslim Univ., Aligarh, pp. 10–11.

Hasan, N. and Jain, R.K. 1998. Nematological research in Uttar Pradesh: An overview. *In: Phytonematology in India* (P.C. Trivedi, *ed.*), pp. 150–69. C.B.S. Publishers and Distributors, New Delhi.

Haseeb, A., Kumar, V., Shukla, P.K. and Ahmad, A. 2006. Effect of different bioinoculants, organic amendments and pesticides on the management of *Meloidogyne incognita–Fusarium solani* disease complex on tomato *cv.* K-25. *Indian J. Nematol.* **36**: 65–69.

Haseeb, A., Shukla, P.K., Kumar, V. and Khan, R.U. 2004. Comparative efficacy of bio-control agents, neem seed powder and pesticides against *Meloidogyne incognita* and *Fusarium oxysporum* on brinjal. *National Symposium on Paradigms in Nematological Research for Biodynamic Farming*, Univ. of Agri. Sci., Bangalore, 80 pp.

Hashmi, S. and Hashmi, G. 1990. Crop rotation and *Meloidogyne javanica* in vegetable production. *Int. Nematol. Network Newsl.* **7** (1): 19–21.

Hassan, M.G. and Sobita Devi. 2004. Efficacy of bioagents against root-knot nematode, (*Meloidogyne incognita*) of tomato. *National Symposium on Paradigms in Nematological Research for Biodynamic Farming*, Univ. of Agri. Sci., Bangalore, 88 pp.

Hazarika, K. 1990. *Pathogenicity and Management of Meloidogyne incognita on Brinjal* (*Solanum melongena*). M. Sc. (Agri.) Thesis, Assam Agri. Univ., Jorhat.

Hemavathi, M.R. 2002. Evaluation of Bioagents, *Pasteuria penetrans* and *Glomus fasciculatum* in the Management of Root-Knot Nematode, *Meloidogyne incognita* Infesting Tomato. M. Sc. (Agri.) Thesis, Univ. of Agri. Sci., Bangalore, 102 pp.

Hide, G.A. and Corbett, D.C.M. 1974. Field experiments in the control of *Verticillium dahliae* and *Heterodera rostochiensis* on potatoes. *Ann Appl. Biol.* **78**: 295–307.

Jackai, L.E.N. and Singh, S.R. 1981. *Trop. Grain Legume Bull.* **22**: 3–6.

Jain, A.C. 1951. A bacterial disease (canker) of tomatoes, new to India. *Sci. Cult.* **17**: 46–47.

Jain, R.K. and Bhatti, D.S. 1978. Bare root treatment with systemics for controlling root-knot nematode in tomato transplants. *Indian J. Nematol.* **8**: 19–24.

Jain, R.K. and Bhatti, D.S. 1981. Systemics as seed treatment for the control of *Meloidogyne javanica* in okra. *Second Nematology Symposium*, Tamil Nadu Agri. Univ., Coimbatore, 44 pp.

Jain, R.K. and Bhatti, D.S. 1985. Effect of summer ploughing alone and in combination with some other effective practices on the incidence of root-knot nematode (*Meloidogyne javanica*) on tomato (Cultivar HS 101). *Indian J. Nematol.* **15**: 262.

Jain, R.K. and Bhatti, D.S. 1987. Population development of root-knot nematode (*Meloidogyne javanica*) and tomato yield as influenced by summer ploughing. *Trop. Pest Mangmt.* **33**: 122–25.

Jain, R.K. and Gupta, D.C. 1991. Integrated management of root-knot nematode, *Meloidogyne javanica* infecting *Solanum melongena*. *Afro-Asian J. Nematol.* **1**: 156–60.

Jain, R.K. and Hasan, N. 1984. Increased seed production of cowpea with low dose of phenamiphos. *Pesticides* **18**: 49.

Jain, R.K. and Hasan, N. 1986. Efficacy of "Neem" cake on fodder production, photosynthetic pigments and nematodes associated with oats and its residual effect on cowpea. *Indian J. Nematol.* **16**: 98–100.

Jain, R.K., Paruthi, I.J. and Gupta, D.C. 1988. Control of root-knot nematode (*Meloidogyne javanica*) in tomato through nursery bed treatment alone and in combination with application of çarbofuran at transplanting. *Indian J. Nematol.* **18**: 340–41.

Jain, R.K., Paruthi, I.J. and Gupta, D.C. 1990. Influence of intercropping on the incidence of root-knot nematode (*Meloidogyne javanica*) in brinjal. *Indian J. Nematol.* **20**: 236–38.

Jain, R.K., Paruthi, I.J., Gupta, D.C. and Dhankar, B.S. 1986. Appraisal of losses due to root-knot nematode, *Meloidogyne javanica* on okra under field conditions. *Trop. Pest Mangmt.* **32**: 341–42.

Jakobsen, I. 1999. Effects of *Pseudomonas fluorescens* DF 57 on growth and P uptake of two arbuscular mycorrhizal fungi in symbiosis with cucumber. *Mycorrhiza* **8**: 319–34.

Jatala, P., Kaltenbach, R., Devaux, A.J. and Campos, R. 1980. Field application of *Paecilomyces lilacinus* for controlling *Meloidogyne incognita* on potatoes. *J. Nematol.* **12**: 226–27.

Jatala, P., Martin, C., Mendoza, H.A., Hidalgo, O.A. and Rincon, R.H. (*eds*.). 1990. Role of nematodes in the expression of *Pseudomonas solanacearum* and strategies for screening and breeding for combined resistance. *Advances en el Mejoramiento Genetico de la Papa en los Paises del cono sur*., pp. 187–89.

Jayakumar, J., Rajendran, G. and Ramakrishnan, S. 2004. Management of reniform nematode, *Rotylenchulus reniformis* in okra through *Streptomyces avermitilis*. *National Symposium on Paradigms in Nematological Research for Biodynamic Farming,* Univ. of Agri. Sci., Bangalore, 98 pp.

Jayakumar, J., Rajendran, G. and Ramakrishnan, S. 2005. Bio-efficacy of avermectin against root-knot nematode, *Meloidogyne incognita* in tomato. *Indian J. Nematol.* **35**: 151–56.

Jenkins, W.R. and Coursen, B.W. 1957. The effect of root-knot nematodes, *Meloidogyne incognita acrita* and *M. hapla*, on *Fusarium* wilt of tomato. *Pl. Dis. Reptr.* **41**: 182–86.

Jones, F.C.E. 1961. The potato root eelworm *Heterodera rostochiensis* Woll. in India. *Curr. Sci.* **30**: 187.

Jones, F.C.E. 1969. Integrated control of the potato cyst nematode. *Proc. Fifth British Insecticide Fungicide Conf.* **3**: 646–56.

Jothi, G., Ramakrishnan, S. and Sundara Babu, R. 2004. Integrated management of *Pratylenchus delattrei* in crossandra. *National Symposium on Paradigms in Nematological Research for Biodynamic Farming,* Univ. of Agri. Sci., Bangalore, 97 pp.

Jothi, G. and Sundara Babu, R. 2000. Interaction effect of VAM and root-knot nematode on the growth and nutrient content of brinjal. *Int. J. Trop. Pl. Dis.* **18**: 91–100.

Kamra, A. and Gaur, H.S. 1998. Control of nematodes, fungi and weeds in nursery beds by soil solarization. *Indian J. Nematol.* **28**: 46–52.

Kandasamy, K. and Sivakumar, C.V. 1981. Effect of *Meloidogyne arenaria* on the germination of chilli (*Capsicum annum* Linn.) seeds. *Second Nematology Symposium*, Tamil Nadu Agri. Univ., Coimbatore, 16 pp.

Kantharaju, V., Prakash Babu, N., Krishnappa, K. and Reddy, B.M.R. 2002. Biodiversity of indigenous isolates of VAM, *Glomus fasciculatum* against *Meloidogyne incognita* race-1 on tomato (Abstr.). *National Symposium on Bio-diversity and Management of Nematodes in Cropping Systems for Sustainable Agriculture,* Jaipur, p. 3.

Kanwar, R.S. 1990. *Studies on Cropping Systems for Management of Root-knot Nematode, Meloidogyne javanica (Treub, 1985) Chitwood, 1949*. Ph. D. Thesis, Haryana Agri. Univ., Hisar, 118 pp.

Kanwar, R.S. and Bhatti, D.S. 1990. Yield performance of resistant and susceptible tomato cultivars in monoculture and related reproduction of *Meloidogyne javanica. Int. Nematol. Network Newsl.* **7** (3): 16–18.

Kanwar, R.S. and Bhatti, D.S. 1992. Controlling root-knot nematode, *Meloidogyne javanica* through crop rotation. *Curr. Nematol.* **3**: 65–70.

Karuna, K., Krishnappa, K. and Reddy, B.M.R. 2001. Utilization of *Pasteuria penetrans* in the management of root-knot nematode on brinjal. *National Conference on Centenary of Nematology in India–Appraisal and Future Plans,* Indian Agri. Res. Inst., New Delhi, 142 pp.

Karwasra, S.S., Chand, J.N. and Shaharan, G.S. 1981. Incidence, perpetuation and control of bacterial blight of guar. *Proc. 5th Intn. Conf. Pl. Path. Bact.*, California, pp. 143–47.

Karwasra, S.S., Chand, J.N. and Shaharan, G.S. 1984. Incidence, perpetuation and control of bacterial blight of guar. *Indian J. Mycol. Pl. Path.* **14**: 123–25.

Katan, J. 1981. Solar heating (solarization) of soil for control of soil borne pests. *Ann. Rev. Phytopath.* **19**: 211–36.

Katiyar, R.P. and Ram, R.S. 1987. *Indian J. Genetics Pl. Breeding* **47**: 46–48.

Kennedy, J.S., Day, M.F. and Estop, V.F. 1982. *A Conspectus of Aphids as Vectors of Plant Viruses.* Commonwealth Inst. Entomol., 114 pp.

Khan, A.M., Saxena, S.K. and Khan, M.W. 1971. Interaction of *Rhizoctonia solani* Kuhn. and *Tylenchorhynchus brassicae* Siddiqi, 1961 in pre-emergence damping-off of cauliflower seedlings. *Indian J. Nematol.* **1**: 85–86.

Khan, A.M., Saxena, S.K. and Siddiqi, Z.A. 1971. Efficacy of *Tagetes erecta* in reducing root infesting nematodes of tomato and okra. *Indian Phytopath.* **24**: 166–69.

Khan, A.M., Saxena, S.K., Siddiqi, Z.A. and Upadhyay, R.S. 1975. Control of nematodes by crop rotation. *Indian J. Nematol.* **5**: 214–21.

Khan, M.R. and Ghadipur, M.H. 2004. Management of root-knot disease of some solanaceous vegetables by soil application of fly ash. *National Symposium on Paradigms in Nematological Research for Biodynamic Farming*, Univ. of Agri. Sci., Bangalore, 92 pp.

Khan, M.R. and Tarannum, Z. 1999. Effects of field application of various micro-organisms on *Meloidogyne incognita* on tomato. *Nematol. Medit.* **27**: 233–38.

Khanna, A.S. and Sharma, N.K. 1994. Management of *Aphelenchoides composticola* infesting white button mushroom by neem cake. *Indian J. Hill Fmg.* **7**: 114–15.

Kiran Kumar, K.C., Krishnappa, K., Karuna, K., Ravichandra, N.G. and Sreenivas, K.R. 2004. Evaluation of proposed management package of growth, development and yield of tomato infested by root-knot nematode. *National Symposium on Paradigms in Nematological Research for Biodynamic Farming,* Univ. of Agri. Sci., Bangalore, 94 pp.

Kishun, R. 1981. Effect of different chemicals on black rot of cabbage (Abstr.). *Proc. 3rd Intn. Symp. Pl. Path.*, New Delhi, 250 pp.

Kishun, R. 1983. Soft rot of radish caused by *Erwinia* sp. and its control (Abstr.). *Indian J. Mycol. Pl. Path.* **13**: 3.

Kishun, R. 1984. Seed treatment for control of cabbage black rot. *J. Turkish Phytopath.* **13**: 81–88.

Kishun, R. 1989. Appraisal of loss in yield of cowpea due to *Xanthomonas campestris* pv. *vignicola*. *Indian Phytopath.* **42**: 241–46.

Kishun, R. and Chand, R. 1988. Epiphytotic survival of *Xanthomonas campestris* pv. *campestris* on *Centenella asiaticum* (L.) Orban. *Intn. J. Trop. Pl. Dis.* **6**: 189–93.

Kishun, R., Sohi, H.S. and Rawal, R.D. 1980. Screening of cowpea germplasm against bacterial blight. *Indian Phytopath.* **33**: 506–07.

Kishun, R. and Swarup, J. 1979. Further studies on bacterial stalk rot of onion. *Indian Phytopath.* **32**: 389–92.

Kishun, R. and Swarup, J. 1982. Studies on varietal resistance and chemical control of bacterial stalk rot of onion. *Veg. Sci.* **9**: 55–59.

Kore, S.S. and Dhaliwal, A.L. 1981. Epidemiology of leaf spot disease of chilli caused by *Xanthomonas campestris* pv. *vesicatoria. Indian Phytopath.* **34**: 113.

Krishnaiah, K., Jagan Mohan, N. and Prasad, V.G. 1981. Efficacy of *Bacillus thuringiensis* Ber. for the control of Lepidopterous pests of vegetable crops. *Entomon* **6**: 87–93.

Krishna Kumar, N.K. 1998. *Final report of ICAR Adhoc Scheme on Bioecology and Management of the Serpentine Leaf miner, Liriomyza trifolii (Burgess) on Vegetable Crops*. Indian Inst. of Hort. Res., Bangalore, 62 pp.

Krishna Kumar, N.K., Krishna Murthy, P.N., Easther, A.P. and Rama, N. 1997. Seasonality and management of the gall midge, *Aspondylia* sp. on brinjal and sweet peppers. *Proc. First Natl. Symp. Hortil. Crops*, Bangalore, pp. 98–101.

Krishna Kumar, N.K., Krishna Murthy, P.N., Rama, N. and Easther, A.P. 1997. Bitter gourd, an unsuitable host plant of serpentine leaf miner. *Insect Envt.* **2**: 125–26.

Krishna Kumar, N.K., Krishna Murthy, P.N., Rama, N., Easther, A.P. and Srinivasan, K. 1998. Assessment of crop loss and seasonality of the accidentally introduced pest, *Liriomyza trifolii* (Burgess), in India. *Proc. Int. Conf. Trop. Spices, Communities Ecosystems*, Tiruvananthapuram.

Krishna Kumar, N.K. and Srinivasan, K. 1988. Seasonality of *Eurytoma* sp. a parasitoid on okra petiole maggot, *Melanagromyza hibisci* Spencer. *Trop. Pest Mangmt.* **34**: 368–69.

Krishnamoorthy, A. and Mani, M. 1985. Investigations on *Telenomus remus* Nixon and *Apanteles marginiventris* Cress. against *Spodoptera litura* (F.) on cabbage. *Entomon* **10**: 277–80.

Krishnamoorthy, A. and Mani, M. 1996a. Suppression of brinjal mealybug *Coccidohystrix insolita* with *Cryptolaemus montrouzieri. Insect Envt.* **2**: 50.

Krishnamoorthy, A. and Mani, M. 1996b. Biosuppression of *Helicoverpa armigera* (Hubn.) on tomato using two egg parasitoids, *Trichogramma brasiliensis* (Ashm.) and *T. pretiosum* (Riley). *J. Ent. Res.* **20**: 37–41.

Krishnamoorthy, A. and Mani, M. 1997. The efficacy of green lacewing *Chrysoperla carnea* in controlling *Myzus persicae* on capsicum. *Proc. First Natl. Symp. Pest Mangmt. Hortil. Crops: Envtl. Implications and Thrusts*, Bangalore.

Krishnamoorthy, P.N. and Krishna Kumar, N.K. 2002. Advances in the use of botanicals in the IPM of major vegetable crops. *Proc. Intn. Conf. Vegetables,* Bangalore.

Krishna Moorthy, P.N., Krishna Kumar, N.K., Selvaraj, C. and Daniel, J.S. 1998. Neem seed kernel extract applications for diamondback moth management: Transfer of technology for mechanized farming. *Pest Mangmt. Hort. Ecosystems* **4**: 128–30.

Krishna Moorthy, P.N. and Srinivasan, K. 1989a. Petiole mining thresholds and potassium chloride sprays for management of bean fly (Diptera: Agromyzidae) in snap beans and cowpea. *J. Econ. Ent.* **82**: 246–50.

Krishna Moorthy, P.N. and Srinivasan, K. 1989b. Distribution and larval behaviour of *Lampides boeticus* in vegetable lablab. *Insect Sci. Applic.* **10**: 491–96.

Krishna Moorthy, P.N. and Srinivasan, K. 1990. Diagnosis and management of bean flies in beans. *In: Horticulture – New Technologies and Applications* (J. Prakash and R.L.M. Pierik, *eds*.).

Krishna Moorthy, P.N. and Srinivasan, K. 1997. Evaluation of neem products and standard chemicals for the management of bean fly, *Ophiomiya phaseoli* (Tryon) in French beans. *In: Neem and Environment, Vol. 1* (R.P. Singh, M.S. Chari, A.K. Raheja and W. Kraus, *eds*.), pp. 433–38. Oxford and IBH Publishing Co. (Pvt.) Ltd., New Delhi.

Krishna Moorthy, P.N., Tewari, G.C. and Srinivasan, K. 1988. Action thresholds for the management of pod borer, *Adisura atkinsoni* Moore on field bean, *Lablab purpureus*. *Insect Sci. Applic.* **9**: 489–92.

Krishna Rao, A.B.V., Padhi, N.N. and Acharya, A. 1987. Effect of different nematicides, oilcakes and urea in the control of *Rotylenchulus reniformis* in okra. *Indian J. Nematol.* **17**: 171–73.

Kumar, S. and Sivakumar, C.V. 1981. Disease complex involving *Rotylenchulus reniformis* and *Rhizoctonia solani* in okra. *Second Nematology Symposium*, Tamil Nadu Agri. Univ., Coimbatore, 12 pp.

Lakshmanan, P.L. and Sivakumar, C.V. 1981. Control of *Rotylenchulus reniformis* in Bhendi by integrated method. *Second Nematology Symposium*, Tamil Nadu Agri. Univ., Coimbatore, 54 pp.

Lal, B., Rana, B.P. and Dahia, R.S. 2002. Efficacy of antagonistic fungi on root-knot nematode, *Meloidogyne incognita* in okra. *Proceedings of the National Symposium on Biodiversity and Management of Nematodes in Cropping Systems for Sustainable Agriculture*, Rajasthan Agri. Univ., Bikaner.

Lal, B.S. and Ansari, M.N.A. 1960. Field studies on the root-knot nematodes (*Meloidogyne* spp.) (Nematoda: Heteroderidae). *Sci. Cult.* **26**: 279–81.

Laqman Khan, M. 2001. Dr. Y.S. Paramar University of Horticulture and Forestry, Solan, Himachal Pradesh. *In: Indian Nematology-Progress and Prospectives* (S.C. Dhawan *et al.*, *eds.*), pp. 153–58. Division of Nematology, Indian Agri. Res. Inst., New Delhi.

Lazarovits, G., Conn, K. and Tenuta, M. 2000. Control of *Verticillium dahliae* with soil amendments: Efficacy and mode of action. *In: Advances in Verticillium Research and Disease Management* (E.C. Tjamos, R.C. Rowe, J.B. Heale and D.R. Fravel, *eds.*), pp. 274–91. American Phytopathological Society, St. Paul, Minn.

Leonian, L. H. 1922. Stem and fruit blight of chillies caused by *Phytophthora capsici* sp. nov. *Phytopathology* **12**: 401–08.

Lodha, S. and Gupta, G.K. 1981. The relationship between bacterial blight and loss in yield components of guar (Abstr.). *Proc. 3rd Intn. Symp. Pl. Path.*, New Delhi, 94 pp.

Loganathan, M., Swarnakumari, N., Sivakumar, M., Prakasam, V., Mohan, L., Ramraj, B. and Samiyappan, R. 2001. Biological suppression of fungal nematode complex diseases of major cruciferous vegetables. *S. Indian Hort.* **49**: Special issue.

Machmud, 1986. *In: Bacterial Wilt Disease in Asia and the South Pacific* (G.J. Persley, *ed.*), pp. 30–34. ACIAR Proc. No. 13.

Maheswari, S.K., Gupta, P.C. and Suhag, L.S. 1988. Studies on chemical control of little leaf disease of brinjal. *Indian J. Mycol. Pl. Path.* **18**: 182.

Maheswari, T.U. and Mani, A. 1988. Combined efficacy of *Pasteuria penetrans* and *Paecilomyces lilacinus* on the biocontrol of *Meloidogyne javanica* on tomato. *Int. Nematol. Network Newsl.* **5** (3): 10–11.

Maheswari, T.U., Mani, A. and Rao, P.K. 1987. Combined efficacy of the bacterial spore parasite *Pasteuria penetrans* (Thorne, 1940) and nematicides in the control of *Meloidogyne javanica* on tomato. *J. Biol. Control* **1**: 53–57.

Mai, W.F., Brodie, B.B., Harrison, M.B. and Jatala, P. 1981. Nematodes. *In: Compendium of Potato Diseases* (W.J. Hooker, *ed.*), pp. 93–101. American Phytopathological Society.

Mani, M. and Krishnamoorthy, A. 1994. Selectivity of pesticides to the parasitoid *Aphelinus* sp. of the green peach aphid, *Myzus persicae* infesting sweet pepper. *J. Ahidology* **8**: 79–87.

Mani, M.P., Rajeswari, S. and Sivakumar, C.V. 1998. Management of the potato cyst nematodes, *Globodera* spp. through plant rhizosphere bacterium *Pseudomonas fluorescens* Migula. *J. Biol. Control* **12**: 131–34.

Martin, W.J. 1970. Elimination of root-knot and reniform nematodes and scurf infection from rootlets of sweet potato plants by hot water treatment. *Pl. Dis. Reptr.* **54**: 1056–58.

Matthews, G.A. 1984. *Pest Management*. Longman, 231 pp.

Mir, N.M. 1987. Screening of fungicides for field control of downy mildew (*Peronospora destructor*) on onions. *Indian J. Mycol. Pl. Path.* **17**: 321–22.

Mirakhur, R.K., Dhar, A.K. and Kaw, M.R. 1977. Downy mildew of *Allium cepa* and its control with fungicides in Kashmir valley. *Indian Phytopath.* **30**: 576–77.

Mishra and Shukla, B.N. 1997. Interaction between *Glomus fasciculatum* and *Meloidogyne incognita* on tomato. *Indian J. Mycol. Pl. Path.* **27**: 199–202.

Mondel, R. and Mukherjee, N. 1978. Sensitivity of plant bacteria to some antibiotics, antibactericides and other drugs. *Z. Pfenkh. Pflaschutz.* **85**: 607–16.

Nagarajan, K. and Ramakrishnan, K. 1971. Studies on cucurbit viruses in Madras State-I. A new virus disease in bittergourd. *Proc. Indian Acad. Sci. Section B* **73**: 30–35.

Nagarkatti, S. and Jayanth, K.P. 1982. Population dynamics of major pests of cabbage and their natural enemies in Bangalore district (India). *Proc. Int. Conf. Pl. Prot.*, Kaula Lumpur, Malaysia, pp. 325–47.

Nagesh, M., Hussaini, S.S., Ramanujam, B. and Chidanandaswamy, B.S. 2006. Management of *Meloidogyne incognita* and *Fusarium oxysporum* f. sp. *lycopersici* wilt complex using antagonistic fungi in tomato. *Nematol. Medit.* **34**: 63–68.

Nagesh, M., Parvatha Reddy, P., Vijaya Kumar, M.V. and Nagaraju, B.M. 1999. Studies on correlation between *Glomus fasciculatum* spore density, root colonization and *Meloidogyne incognita* infection on *Lycopersicon esculentum*. *J. Pl. Dis. Prot.* **106**: 82–87.

Naik, D. 2004. Biotechnological Approaches for the Management of Wilt Disease Complex in Capsicum (*Capsicum annum* L.) and Egg plant (*Solanum melongena*) with Special Emphasis on Biological Control. Ph. D. Thesis, Kuvempu Univ., Shimoga.

Nanjappa, C.K., Panda, M. and Seshadri, A.R. 1978. Effects of *Meloidogyne incognita* and *Rotylenchulus reniformis* on the emergence and survival of seedlings. *Third Int. Cong. Pl. Path.*, Germany.

Naqvi, Q.A. and Mahmood, K. 1975. Mosaic disease of spinach–A new record for India. *Indian Phytopath.* **28**: 268–71.

Narayanan, K. 1985. Control of *Heliothis armigera* through nuclear polyhedrosis virus. *In: Microbial Control and Pest Management* (S. Jayaraj, *ed.*). Tamil Nadu Agri. Univ. and ICAR Publication.

Narayanan, K. 1987. Field efficacy of nuclear polyhedrosis virus of *Adisura atkinsoni* Moore on field beans. *J. Biol. Control* **1**: 44–47.

Narayanan, K. and Gopalakrishnan, G.1988. Integrated control of *Heliothis armigera* (Hbn.) with their pathogens and pesticides in certain horticultural crops. *Proc. Natl. Workshop* Heliothis *Mangmt.* Tamil Nadu Agri. Univ., Coimbatore, pp. 165–76.

Narayanan, K., Jayaraj, S. and Subramanian, T.R. 1970. Control of three species of Lepidopteran insects with the pathogen, *Bacillus thuringiensis* Berliner. *Madras Agric. J.* **57**: 665–73.

Nath, A., Sharma, N.K., Bhardwaj, S. and Thapa, C.N. 1982. Nematicidal property of garlic. *Nematologica* **28**: 253–55.

Nath, R., Khan, M.N., Wanshi, R.S.K. and Dwivedi, R.P. 1984. Influence of root-knot nematode, *Meloidogyne javanica* on pre and post-emergence damping-off of tomato. *Indian J. Nematol.* **14**: 135–40.

Nath, R.P., Sinha, B.K. and Haider, M.G. 1976. Studies on the nematodes of vegetables in Bihar-II. Combined effect of *Meloidogyne incognita* and *Ozonium texanum* var. *parasiticum* on germination of eggplant. *Indian J. Nematol.* **6**: 177–79.

Nath, V.R., Vaheduddin, S. and Reddy, D.P. 1958. *Sci. Cult.* **23**: 605–06.

Nayar, K., Mathew, J., Kuriyan, J., Gnanamanickam, S.S. and Mahadevan, A. 1988. Role and association of root-knot nematode (*Meloidogyne incognita*) in the bacterial wilt of brinjal incited by *Pseudomonas solanacearum*. *In: Advances in Research on Pant Pathogenic Bacteria*. Based on the proceedings of the National Symposium on Phytobacteriology, University of Madras, Madras, India, pp. 45–48.

Nedumaran, S. and Vidyasekaran, P. 1982. Seed-borne infection of *Corynebacterium michiganense* in tomato. *Indian Phytopath.* **35**: 510–11.

Netscher, C. 1983. Control of *Meloidogyne incognita* in vegetable production by crop rotation in Ivory Coast. *Acta Hort.* **152**: 219–25.

Ngundo, B.W. 1977. Screening of bean cultivars for resistance to *Meloidogyne* spp. in Kenia. *Pl. Dis. Reptr.* **61**: 991–93.

Niknam, G.R. and Dhawan, S.C. 2001a. Induction of systemic resistance by *Bacillus subtilis* isolate Bs1 against *Rotylenchulus reniformis* in tomato. *Natl. Cong. on Centenary of Nematol. in India–Appraisal and Future Plans*, Indian Agri. Res. Inst., New Delhi, pp. 143–44.

Niknam, G.R. and Dhawan, S.C. 2001b. Effect of seed bacterization, soil drench and bare root-dip application methods of *Pseudomonas fluorescens* isolate Pf1 on the suppression of *Rotylenchulus reniformis* infecting tomato. *Natl. Cong. on Centenary of Nematol. in India–Appraisal and Future Plans*, Indian Agri. Res. Inst., New Delhi, 144 pp.

Nirula, K.K. 1963. Collateral host plants of root-knot nematodes. *Curr. Sci.* **32**: 25-221–22.

Nirula, K.K. and Bassi, K.K. 1965. Thermotherapy for root-knot nematode, *Meloidogyne incognita* in potato tubers. *Indian Potato J.* **7**: 9–11.

Nisha, M.S. and Sheela, M.S. 2004. Rhizome treatment–A low cost technology for the management of nematodes in Kacholam, an important medicinal plant. *National Symposium on Paradigms in Nematological Research for Biodynamic Farming,* Univ. of Agri. Sci., Bangalore, pp. 62–63.

Nisha, M.S. and Sheela, M.S. 2006. Bio-management of *Meloidogyne incognita* on Coleus, *Solenostemon rotundifolius* by integrating solarization, *Paecilomyces lilacinus, Bacillus macerans* and neem cake. *Indian J. Nematol.* **36**: 136–38.

Nisha, M.S., Sheela, M.S. and Beena, R. 2001. Effect of mulching for the management of plant parasitic nematodes associated with kacholam, *Kaempferia galangal* L. *Proc. Second National Symp. Integrated Pest Managmt. Hort. Crops: New Molecules, Biopesticides and Environment*, Bangalore, pp. 145–47.

Oostendorp, M. and Sikora, R.A. 1989. Seed treatment with antagonistic rhizobacteria for the suppression of *Heterodera schachtii* early root infection of sugar beet. *Rev. Nematol.* **12**: 77–83.

Paharia, K.D. 1963. *Proc. Second All India Potato Workers Conf.*, Central Potato Res. Inst., Shimla, India, pp. 198–202.

Palanisamy, S. and Sivakumar, C.V. 1981. Assessment of avoidable yield loss in cowpea, black gram, maize and finger millet. *Second Nematol. Symp.*, Tamil Nadu Agri. Univ., Coimbatore, 58 pp.

Pandey, K.K. and Pandey, P.K. 2001. Soil borne diseases in vegetable crops: Biological control approach. *Souvenir of AICRP on Vegetable Crops*, pp. 51–52.

Pandey, K.K. and Pandey, P.K. 2002a. Field evaluation of some fungal toxicants against fungal disease complex of cowpea. *Pestalogy* **24** (6): 20–22.

Pandey, K.K. and Pandey, P.K. 2002b. Study on soil solarization for nursery management of vegetable crops. *Asian Cong. Mycol. Pl. Path.*, Mysore.

Pandey, K.K., Pandey, P.K., Kalloo, G., Kumar, R. and Singh, B. 1999. Sources of resistance against powdery mildew (*Erysiphe pisi*) of pea and its pathogen reaction in natural and artificial conditions. *Veg. Sci.* **26**: 160–63.

Pandey, R.C. and Trivedi, P.C. 1992. Biological control of *Meloidogyne incognita* by *Paecilomyces lilacinus* in *Capsicum annum. Indian Phytopath.* **45**: 134–35.

Pani, A.K. and Das, S.N. 1972. Studies on etiological complexes in plant disease. I. Association of root-knot nematodes in bacterial wilt of tomato. *J. Res. Orissa Univ. Agri. Technol.* **2**: 54–59.

Parvatha Reddy, P. 1984. Efficacy of seed treatment with three nematicides for the control of *Meloidogyne incognita* infecting cowpea, French bean and peas. *Indian J. Nematol.* **14**: 39–40.

Parvatha Reddy, P. 1985a. Chemical control of root-knot nematodes infecting peas. *Indian J. Nematol.* **15**: 120–21.

Parvatha Reddy, P. 1985b. Estimation of crop losses in peas due to *Meloidogyne incognita. Indian J. Nematol.* **15**: 226.

Parvatha Reddy, P. 1988. *Nematode Management in Horticultural Crops.* Tech. Bull. No. 4, Indian Inst. of Hortil. Res., Bangalore, 52 pp.

Parvatha Reddy, P. and Khan, R.M. 1988. Evaluation of *Paecilomyces lilacinus* for the biological control of *Rotylenchulus reniformis* infecting tomato as compared with carbofuran. *Nematol. Medit.* **16**: 113–16..

Parvatha Reddy, P. and Khan, R.M. 1989. Evaluation of biocontrol agent *Paecilomyces lilacinus* and carbofuran for the management of *Rotylenchulus renfiromis* infecting brinjal. *Pakistan J. Nematol.* **7**: 55–59.

Parvatha Reddy, P. and Khan, R.M. 1991. Integrated management of root-knot nematodes infecting okra. *Curr. Nematol.* **2**: 115–16.

Parvatha Reddy, P. and Nagesh, M. 2002. A*vermectins–Isolation, Fermentation, Preliminary Characterization and Screening for Nematicidal Activity.* Tech. Bull. 17, Indian Inst. of Hort. Res., Bangalore, 28 pp.

Parvatha Reddy, P., Nagesh, M. and Devappa, V. 1997. Effect of integration of *Pasteuria penetrans, Paecilomyces lilacinus* and neem cake for the management of root-knot nematodes infecting tomato. *Pest Mangmt. Hort. Ecosystems* **3**: 100–04.

Parvatha Reddy, P., Nagesh, M. and Devappa, V. 1998. Management of *Meloidogyne incognita* on tomato by integration of *Trichoderma harzianum, Glomus fasciculatum* and neem cake. *In: Advances in IPM for Horticultural Crops* (P. Parvatha Reddy, N.K. Krishna Kumar and A. Verghese, *eds.*), pp. 349–52. Association for Advancement of Pest Management in Hortil. Ecosystems, Bangalore.

Parvatha Reddy, P., Nagesh, M., Devappa, V. and Vijaya Kumar, M.V. 1998. Management of *Meloidogyne incognita* on tomato by integrating endomycorrhiza, *Glomus mosseae* with oil cakes under nursery and field conditions. *J. Pl. Dis. Prot.* **105**: 53-57.

Parvatha Reddy, P., Rao, M.S. and Nagesh, M. 1998. Effect of bare-root dip treatment of tomato seedlings in plant leaf extracts mixed with *Paecilomyces lilacinus* spores for the management of *Meloidogyne incognita. In: Advances in IPM for Horticultural Crops* (P. Parvatha Reddy, N.K. Krishna Kumar and A. Verghese, eds.), pp. 334–38. Association for Advancement of Pest Management in Hortil. Ecosystems, Bangalore.

Parvatha Reddy, P., Rao, M.S. and Nagesh, M. 1999. Eco-friendly management of *Meloidogyne incognita* on tomato by integration of *Verticillium chlamydosporium* with neem and calotropis leaves. *J. Pl. Dis. Prot.* **106**: 530–33.

Parvatha Reddy, P. and Singh, D.B. 1979. Control of root-knot nematode, *Meloidogyne incognita* on tomato by chemical bare-root dips. *Indian J. Nematol.* **9**: 40–43.

Parvatha Reddy, P. and Singh, D.B. 1981. Assessment of avoidable yield loss in okra, brinjal, French bean and cowpea due to root-knot nematodes. *Third Int. Symp. Pl. Path.*, New Delhi, pp. 93–94.

Parvatha Reddy, P., Singh, D.B. and Ram Kishun. 1979. Effect of root-knot nematodes on the susceptibility of Pusa Purple Cluster brinjal to bacterial wilt. *Curr. Sci.* **48**: 915–16.

Parvatha Reddy, P., Singh, D.B. and Sharma, S.R. 1979. Interaction of *Meloidogyne incognita* and *Rhizoctonia solani* in a root rot disease complex of French bean. *Indian Phytopath.* **32**: 651–52.

Paruthi, L.J., Jain, R.K. and Gupta, D.C. 1987. Effect of different periods of degradation of subabool leaves alone and in combination with nematicides in okra. *Indian J. Nematol.* **17**: 30–32.

Patel, B.K. 1967. *The Bionomics of the Mustard Sawfly*, Athelia lugens proxima *Klug*. (Trenthredinae: Hymenoptera). M. Sc. Thesis, Sardar Patel Univ., Anand, 76 pp.

Patel, J.R.1980. *The Bionomics and Control Measures of the Potato Tuber Moth,* Phthorimaea (Gnorimoschema) operculella Zeller (Gelechidae: Lepidoptera). M. Sc. Thesis, Gujarat Agri. Univ., Anand Campus, Anand, 76 pp.

Patel, J.R., Patel, R.C. and Amin, P.R. 1983. Effect of systemic insecticides on thrips and mites and its relation to leaf curl disease of chilli. *Gujarat Agri. Univ. Res. J.* **8**: 129–33.

Patel, M.K., Abhyankar, S.G. and Kulkarni, Y.S. 1949. Black rot of cabbage. *Indian Phytopath.* **2**: 58–61.

Patel, N.C., Patel, J.J., Jayani, D.B. and Patel, J.R. 1997. Evaluation of insecticide schedules for controlling chilli thrip, *Scirtothrips dorsalis* Hood infesting chilli, *Capsicum annuum* L. *Indian J. Agric. Sci.*

Patel, P.N. *et al.* (*eds*.). 1972. *Plant Bacteriology Vol. I. Bacterial Diseases of Plants in India*. Indian Council of Agri. Res., New Delhi, 261 pp.

Pathak, K.N. and Keshari, N. 2004. Interaction of *Meloidogyne incognita* with *Fusarium oxysporum* f. sp. *conglutinans* on cauliflower. *Indian J. Nematol.* **34**: 85–87.

Patnaik, P.R., Mohanty, K.C., Mahapatra, S.N. and Sahoo, S. 2004. Evaluation of chilli varieties against root-knot nematode, *Meloidogyne incognita*. *National Symposium on Paradigms in Nematological Research for Biodynamic Farming,* Univ. of Agri. Sci., Bangalore, 100 pp.

Perveen, S., Ehteshamul-Haque, S. and Ghaffar, A. 1998. Efficacy of *Pseudomonas aeruginosa* and *Paecilomyces lilacinus* in the control of root rot-root knot disease complex on some vegetables. *Nematol. Medit.* **26**: 209–12.

Perwez, M.S., Rehman, M.F. and Haider, S.R. 1988. Effect of *Tagetes erecta* on *Meloidogyne javanica* infecting lettuce. *Int. Nematol. Network Newsl.* **5** (3): 18–19.

Prasad, H.H. 1932. A bacterial white soft rot of turnip. *Indian J. Agric. Sci.* **1**: 534–37.

Prasad, K.S.K. 1988. Nematology in 2000 AD as I see it in relation to potato cyst nematode. *National Nematology Symposium*, New Delhi, pp. 26–29.

Prasad, K.S.K. 1989. Nematological problems and progress of research on potato in India. *Fourth Group Meeting on Nematological Problems of Plantation Crops*, Univ. Agri. Sci., Bangalore.

Prasad, K.S.K. 1990. *Helicotylenchus dihystera*, a potential pest of potato. *J. Indian Potato Assn.* **17**: 204–05.

Prasad, K.S.K. and Krishnappa, K. 1981. Effect of bare-root dip treatment with pesticides on *Rotylenchulus reniformis* affecting brinjal. *Indian J. Nematol.* **11**: 154–58.

Prasad, K.S.K., Raj, D. and Sharma, R.K. 1983. Effect of planting time on *Meloidogyne incognita* infesting potato at Shimla Hills. *Third Nematol. Symp*, Y.S Parmar Univ. Hort. And Forestry, Solan, Himachal Pradesh, pp. 5–6.

Prasad, K.S.K. and Sharma, R.K. 1985. Parasitic nematodes associated with potato and the pathogenicity of *Quinisulcius capitatus* on variety Kufri Jyoti. *J. Indian Potato Assn.* **12**: 56–62.

Prasad, S.K. and Dasgupta, D.R. 1964. Plant parasitic nematodes associated with roses in Delhi. *Indian J. Ent.* **26**: 12–122.

Prasad, S.K., Dasgupta, D.R. and Mukhopadhyaya, M.C. 1964. Nematodes associated with commercial crops in Northern India. I. Uttar Pradesh. *Indian J. Nematol.* **3**: 8–23.

Raj, B.T. and Nirula, K.K. 1969. Trials on the eradication of root-knot nematodes from potato fields with DD (Abstr.). *All India Nematol. Symp.*, Indian Agri. Res. Inst., New Delhi, India.

Rajamohan, N. and Jayaraj, S. 1978. Field efficacy of *Bacillus thuringiensis* and some other insecticides against pests of cabbage. *Indian J. Agric. Sci.* **48**: 672–75.

Raju, K.S. and Rao, G.S. 1984. Studies to fix up optimum spray intervals against bacterial leaf spot and fruit rot disease of chilli. *South Indian Hort.* **32**: 319–20.

Ram, B. and Baheti, B.L. 2004. Management of reniform nematode, *Rotylenchulus reniformis* on cowpea through seed and soil treatment with plant products. *Indian J. Nematol.* **34**: 193–95.

Ram, S. and Gupta, D.C. 2001. Studies on the effect of intercropping of onion on the growth of tomato and population dynamics of root-knot nematode, *Meloidogyne javanica. National Conference on Centenary of Nematology in India–Appraisal and Future Plans*, Indian Agri. Res. Inst., New Delhi, p. 94.

Ramakrishnan, C. and Balasubramanian, P. 1981. Control of *Meloidogyne incognita* in chilli (*Capsicum annum* L.) nursery with nematicides. *Second Nematol. Symp.*, Tamil Nadu Agri. Univ, Coimbatore, 54 pp.

Ramakrishnan, G., Ranganathan, K., Murugesan, S., Damodaran, S.A.P. and Kandaswamy, T.K. 1971. A mosaic disease of *Amaranthus gangeticus. Madras Agric. J.* **58**: 679–83.

Ramana, K.V. and Mohandas, C. 1988. Occurrence of potato cyst nematode, *Globodera pallida* (Stone, 1973) in Kerala. *Indian J. Nematol.* **18**: 141.

Ramappa, H.K. 1988. *Studies on Nematode Parasites of French bean and their Control.* M. Sc. (Agri.) Thesis, Univ. of Agri. Sci., Bangalore.

Rameshchand and Gill, J.S. 2002. Evaluation of various application methods of *Pasteuria penetrans* against *Meloidogyne incognita* in tomato. *Indian J. Nematol.* **32**: 23–26.

Rangaswami, G. and Gowda, S.S. 1963. On some bacterial diseases of ornamentals and vegetables in Madras State. *Indian Phytopath.* **16**: 74–85.

Rangaswamy, S.D., Parvatha Reddy, P., Nagesh, M. and Nanje Gowda, D.N. 1999. Integrated management of *Meloidogyne incognita* using *Pasteuria penetrans, Trichoderma viride* and oil cakes in tomato. *Pest Mangmt. Hort. Ecosystems* **5**: 122–26.

Rangaswamy, S.D., Parvatha Reddy, P. and Nanje Gowda, D.N. 1999. Management of root-knot nematode, *Meloidogyne incognita* in tomato by intercropping with marigold and mustard. *Pest Mangmt. Hort. Ecosystems* **5**: 118–21.

Rangaswamy, S.D., Parvatha Reddy, P. and Narayana Swamy, B.C. 1995. Management of *Meloidogyne incognita* in tomato nursery by growing trap/antagonistic crops in rotation. *Curr. Nematol.* **6**: 9–12.

Rao, A.N.S. and Kishun, R. 1986. A note on the control of bacterial black rot [*Xanthomonas campestris* (Pammel) Dowson] of cabbage (*Brassica oleracea* var. *capitata* Linn.) by seed treatment. *Pesticides* **20**: 47.

Rao, M.S., Gowen, S.R., Pembroke, B. and Parvatha Reddy, P. 1997. Relationship of *Pasteuria penetrans* spore encumbrance on juveniles of *Meloidogyne incognita* and their infection in adults. *Nematol. Medit.* **25**: 129–31.

Rao, M.S., Kerry, B.R., Gowen, S.R., Bourne, J.M. and Parvatha Reddy, P. 1997. Management of *Meloidogyne incognita* in tomato nurseries by integration of *Glomus deserticola* with *Verticillium chlamydosporium*. *J. Plant Dis. Prot.* **104**: 419–22.

Rao, M.S., Mohandas, S., Parvatha Reddy, P. and Khan, R.M. 1993. Synergistic effect of endomycorrhizal fungus (*Glomus fasciculatum*) and biocontrol fungus (*Paecilomyces lilacinus*) in the management of root-knot nematode infecting tomato. *In: Soil Organisms and Sustainability* (D. Rajagopal, R.D. Kale and K. Bano, *eds.*), pp. 119–23. Indian Society of Soil Biology and Ecology, Univ. of Agri. Sci., Bangalore.

Rao, M.S., Naik, D., Shylaja, M. and Parvatha Reddy, P. 2002. Prospects for the management of the root-knot nematode and bacterial wilt disease complex in capsicum by integration of fungal and bacterial biocontrol agents. *Proc. Int. Conf. Vegetables,* Bangalore, pp. 347–51.

Rao, M.S. and Parvatha Reddy, P. 1992. Innovative approaches of utilization of biocontrol agents-*Paecilomyces lilacinus* and *Verticillium chlamydosporium* against root-knot nematode on tomato (Abstr.). *First Afro-Asian Nematol. Symp.*, Aligarh Muslim Univ., Aligarh, pp. 25–26.

Rao, M.S. and Parvatha Reddy, P. 1993a. Effective use of 'Welgro' and 'RD-9 Repelin' for the management of root-knot nematode on tomato. *In: Neem and Environment* (R.P. Singh, M.S. Chari, A.K. Raheja and W. Kraus, *eds.*), **2**: 651–56. Oxford & IBH Publishing Co. Pvt. Ltd., New Delhi.

Rao, M.S. and Parvatha Reddy, P. 1993b. Interactive effect of *Verticillium chlamydosporium* and castor cake on the control of root-knot nematode on eggplant. *In: Soil Organisms and Sustainability* (D. Rajagopal, R.D. Kale and K. Bano, *eds.*), pp. 114–18. Indian Society of Soil Biology and Ecology, Univ. of Agri. Sci., Bangalore.

Rao, M.S. and Parvatha Reddy, P. 1994. A method for conveying *Paecilomyces lilacinus* to soil for the management of root-knot nematodes on eggplant. *Nematol. Medit.* **22**: 265–67.

Rao, M.S. and Parvatha Reddy, P. 2001. Control of *Meloidogyne incognita* on eggplant using *Glomus mosseae* integrated with *Paecilomyces lilacinus* and neem cake. *Nematol. Medit.* **29**: 153–57.

Rao, M.S., Parvatha Reddy, P. and Mohandas, S. 1995. Studies on development of a biorational management strategy against root-knot nematode attacking tomato. *National Symposium on Nematode Problems of India–An Appraisal of the Nematode Management with Eco-friendly Approaches* (G. Swarup, D.R. Dasgupta and J.S. Gill, *eds.*), 14 pp. Nematol. Soc. of India, New Delhi.

Rao, M.S., Parvatha Reddy, P. and Mohandas, S. 1996. Effect of integration of *Calotropis procera* leaf and *Glomus fasciculatum* on the management of *Meloidogyne incognita* infesting tomato. *Nematol. Medit.* **24**: 59–61.

Rao, M.S., Parvatha Reddy, P. and Nagesh, M. 1997a. Integrated management of *Meloidogyne incognita* on okra by castor cake suspension and *Paecilomyces lilacinus*. *Nematol. Medit.* **25**: 17–19.

Rao, M.S., Parvatha Reddy, P. and Nagesh, M. 1997b. Integration of *Paecilomyces lilacinus* with neem leaf suspension for the management of root-knot nematodes on eggplant. *Nematol. Medit.* **25**: 249–52.

Rao, M.S., Parvatha Reddy, P. and Nagesh, M. 1997c. Management of root-knot nematode, *Meloidogyne incognita* on tomato by integration of *Trichoderma harzianum* with neem cake. *J. Pl. Dis. Prot.* **104**: 423–25.

Rao, M.S., Parvatha Reddy, P. and Nagesh, M. 1998a. Integrated management of *Meloidogyne incognita* on tomato using *Verticillium chlamydosporium* and *Pasteuria penetrans. Pest Mangmt. Hort. Ecosystems* **4**: 32–35.

Rao, M.S., Parvatha Reddy, P. and Nagesh, M. 1998b. Integrated management of *Meloidogyne incognita* on tomato using oil cakes and a bioagent, *Verticillium chlamydosporium. In: Advances in IPM for Horticultural Crops* (P. Parvatha Reddy, N.K. Krishna Kumar and A. Verghese, *eds.*), pp. 345–48. Association for Advancement of Pest Management in Hortil. Ecosystems, Bangalore.

Rao, M.S., Parvatha Reddy, P. and Nagesh, M. 1998c. Evaluation of plant based formulations of *Trichoderma harzianum* for the management of *Meloidogyne incognita* on eggplant. *Nematol. Medit.* **26**: 59–62.

Rao, M.S., Parvatha Reddy, P. and Nagesh, M. 1998d. Evaluation of *Paecilomyces lilacinus* cultured on neem cake extract for the management of root-knot nematodes on eggplant. *Pest Mangmt. Hort. Ecosystems* **4**: 116–19.

Rao, M.S., Parvatha Reddy, P. and Nagesh, M. 1999. Bare root-dip treatment of tomato seedlings in calotropis or castor leaf extracts mixed with *Paecilomyces lilacinus* spores for the management of *Meloidogyne incognita. Nematol. Medit.* **27**: 323–26.

Rao, M.S., Parvatha Reddy, P. and Nagesh, M. 2000. Management of *Meloidogyne incognita* on tomato by integrating *Glomus mosseae* with *Pasteuria penetrans* under field conditions. *Pest Mangmt. Hort. Ecosystems* **6**: 130–34.

Rao, M.S., Parvatha Reddy, P., Somasekhar, N. and Nagesh, M. 1997. Management of root-knot nematodes, *Meloidogyne incognita* in tomato nursery by integration of endomycorrhiza, *Glomus fasciculatum* with castor cake. *Pest Mangmt. Hort. Ecosystems* **3**: 31–35.

Rao, M.S., Parvatha Reddy, P. and Sukhada, M. 1998. Biointensive management of *Meloidogyne incognita* on eggplant by integrating *Paeciloymces lilacinus* and *Glomus mosseae. Nematol. Medit.* **26**: 213–16.

Rao, M.S., Parvatha Reddy, P., Sukhada, M., Nagesh, M. and Pankaj. 1998. Management of root-knot nematodes on eggplant by integrating endomycorrhiza (*Glomus fasciculatum*) and castor (*Ricinus communis*) cake. *Nematol. Medit.* **26**: 217–19.

Rao, M.S., Reddy, P.P. and Mohandas, S. 1995. Effect of integration of endomycorrhiza (*Glomus mosseae*) and neem cake on the control of root-knot nematode on tomato. *J. Plant Dis. Prot.* **102**: 526–29.

Rao, M.S. and Shylaja, M. 2004. Role of *Pseudomonas fluorescens* (Migula) in induction of systemic resistance (ISR) and managing *Rotylenchulus reniformis* (Linford and Oliveira) on carrot. *Pest Mangmt. Hort. Ecosystems* **10**: 87–93.

Rao, M.S., Tewari, R.P. and Pandey, M. 1992. Occurrence of mushroom nematodes in South India. *Mushroom Res.* **1**: 135–36.

Rao, V.R. and Singh, D.B. 1978. Chemical control of root-knot nematodes in okra with the aid of nematicides. *Indian J.Ent.* **35**: 151–54.

Rashid, A., Khan, F.A. and Khan, A.M. 1973. Plant parasitic nematodes associated with vegetables, fruits, cereals and other crops in north India. I. Uttar Pradesh. *Indian J. Nematol.* **3**: 8–23.

Ravi Dutt and Bhatti, D.S. 1986. Determination of effective doses and time of application of nematicides and castor leaves for controlling *Meloidogyne javanica* in tomato. *Indian J. Nematol.* **16**: 8–11.

Reddy, D.D.R. 1985. Analysis of crop losses in tomato due to *Meloidogyne incognita. Indian J. Nematol.* **15**: 55–59.

Reddy, D.D.R. and Seshadri, A.R. 1972. Studies on some systemic nematicides. II. Further studies on the action of thionazin and aldicarb on *Meloidogyne incognita* and *Rotylenchulus reniformis. Indian J. Nematol.* **2**: 182–90.

Reddy, D.D.R. and Seshadri, A.R. 1975. Elimination of root-knot nematode infestation from tomato seedlings by chemical bare-root dips or soil application. *Indian J. Nematol.* **5**: 170–75.

Reka Arya and Saxena, S.K. 1998. *Trichothecium roseum* together with *Rhizoctonia solani* and *Meloidogyne incognita* on germination of seeds of tomato *cv.* Pusa Ruby. *Indian J. Nematol.* **28**: 217–21.

Rhoade, R.A. and Jenkins, N.R. 1957. *Phytopathology* **47**: 29.

Ristaino, J.B. and Johnston, S.A. 1999. Ecologically based approaches to management of *Phytophthora* blight on bell pepper. *Plant Dis.* **83**: 1080–89.

Routaray, B.N. and Sahoo, H. 1985. Integrated control of root-knot nematode, *Meloidogyne incognita* with neem cake and granular nematicides on tomato. *Indian J. Nematol.* **15**: 261.

Rowe, R.C., Davis, J.R., Powelson, M.L. and Rouse, D.I. 1987. Potato early dying: causal agents and management strategies. *Plant Dis.* **71**: 482–89.

Sahoo, S., Mohanty, K.C., Sahoo, N.K., Mohapatra, A.K.B. and Mishra, H.N. 2004a. Effect of different cropping sequences on root-knot nematode (*Meloidogyne incognita*) and wilt incidence in tomato. *National Symposium on Paradigms in Nematological Research for Biodynamic Farming,* Univ. of Agri. Sci., Bangalore, 37 pp.

Sahoo, S., Mohanty, K.C., Sahoo, N.K., Mohapatra, A.K.B. and Mishra, H.N. 2004b. Comparative efficacy of different eco-friendly approaches for the management of root-knot nematode in okra. *National Symposium on Paradigms in Nematological Research for Biodynamic Farming,* Univ. of Agri. Sci., Bangalore, 79 pp.

Saifullah. 1996. Effect of nematophagous fungi on growth of potato and *Globodera rostochiensis* and their environmental tolerance. *Afro-Asian J. Nematol.* **6**: 128–32.

Saini, L.C. and Parashar, R.D. 1981. Comparative efficacy of stable bleaching powder with other antibacterial formulations in controlling black and soft rot of cauliflower. *Indian Phytopath.* **34**: 465–69.

Samiyappan, R. 2003. Biological control of fungal and nematode complex diseases by plant growth promoting rhizobacteria (PGPR). *Winter School on Biological Control of Plant Parasitic Nematodes*, Tamil Nadu Agri. Univ., Coimbatore, pp. 69–77.

Sasser, J.N. 1989. *The Plant-parasitic Nematodes, the Farmers Hidden Enemy.* Dept. of Plant Path., North Carolina State Univ. & Consortium for Intn. Crop Prot., 115 pp.

Sasser, J.N. 1989. *The Plant-parasitic Nematodes, The Farmers Hidden Enemy.* Dept. of Plant Path., North Carolina State Univ. & Consortium for Intnl. Crop Prot., 115 pp.

Sasser, J.N. and Freckman, D.W. 1987. A world perspective on Nematology: The role of the Society. *In: Vistas on Nematology* (J.A. Veech and D.W. Dickson, *eds.*), pp. 7–14. Society of Nematologists, Hyattsville, Maryland.

Sastry, K.S., Satyanarayana, A., Singh, S.J. and Rajendran, R. 1981. Screening of different bean cultivars against common bean mosaic virus. *Pulse Crops Newsl.* **1**: 30.

Sastry, K.S.M. and Singh, S.J. 1971. Effect of different insecticides on the control of whitefly (*Bemisia tabaci* Gen.) population in tomato crop and the incidence of tomato leaf curl virus. *Indian J. Hort.* **28**: 304–09.

Sastry, K.S.M. and Singh, S.J. 1973a. Field evaluation of insecticides for the control of whitefly (*Bemisia tabaci*) in relation to the incidence of yellow vein mosaic of okra (*Abelmoschus esculentus*). *Indian Phytopath.* **26**: 129–38.

Sastry, K.S.M. and Singh, S.J. 1973b. Assessment of losses in tomato by tomato leaf curl virus. *Indian J. Mycol. Pl. Path.* **3**: 50–54.

Sastry, K.S.M. and Singh, S.J. 1974. Effect of yellow vein mosaic virus infection on growth and yield of okra crop. *Indian Phytopath.* **27**: 316–17.

Sastry, K.S.M., Singh, S.J. and Sastry, K.S. 1977. Effect of border cropping and the use of insecticides in relation to the incidence of tomato leaf curl virus (TLCV). *Indian J. Hort.* **34**: 319–22.

Sastry, K.S.M., Singh, S.J. and Sastry, K.S. 1978. Studies on epidemiology of tomato leaf curl virus. *Indian J. Hort.* **35**: 269–77.

Saxena, S.K., Alam, M.M. and Khan, A.M. 1974. Chemotherapeutic control of nematodes with Vydate oxamyl, VC-13 and Dazomet on chilli plants. *Indian J. Nematol.* **4**: 235–38.

Scholte, K. 1990. Causes of differences in growth pattern, yield and quality of potatoes (*Solanum tuberosum* L.) in short rotations on sandy soil as affected by crop rotation, cultivar and application of granular nematicides. *Potato Res.* **33**: 181–90.

Seaby, D. 1989. *The Mushroom J.* **197**: 147–51.

Sen, B. and Kapoor, I.J. 1974. Chemical control of *Fusarium* wilt of tomato. *Pesticides* **8**: 40–42.

Sen, K. and Dasgupta, M.K. 1981. Effect of rice hull ash on root-knot nematodes in tomato. *Trop. Pest Mangmt.* **27**: 518–19.

Seth, A. 1984. Ph. D. Thesis, Dept. of Entomol. and Apiculture, Y.D. Parmar Univ. Hort. Forestry, Solan, Himachal Pradesh, India.

Seth, A. and Sharma, N.K. 1986. Five new species of genus *Aphelenchoides* (Nematoda: Aphelenchida) infesting mushroom in North India. *Indian J. Nematol.* **16**: 205–15.

Shah, R., Bhatnagar, M.K. and Mali, B.L. 1989. Epidemiology of bacterial blight of cowpea (Abstr.). *Proc. 7th Intn. Conf. Pl. Path.*, Budapest, 85 pp.

Shahzad, S. and Gaffer, A. 1989. Use of *Paecilomyces lilacinus* in the control of root rot and root-knot complex in okra and mungbean. *Pakistan J. Nematol.* **7**: 47–53.

Shanthi, A. and Sivakumar, C.V. 1995. Biocontrol potential of *Pseudomonas fluorescens* (Migula) against root-knot nematode, *Meloidogyne incognita* (Kofoid and White, 1919) Chitwood, 1949 on tomato. *J. Biol. Control* **9**: 113–15.

Sharma, A. and Trivedi, P.C. 1989. Influence of inoculum levels of fungus *Paecilomyces lilacinus* (Thom.) Samson on the biocontrol of root-knot nematode, *Meloidogyne incognita* (Chitwood). *Int. Nematol. Network Newsl.* **6** (2): 27–29.

Sharma, A. and Trivedi, P.C. 1994. Interaction of root-knot nematode with VAM on *Abelmoschus esculentus* and their effect on fruit yield. In *Vistas in Seed Biology* (T. Singh and P.C. Trivedi, eds.), Vol. II, pp. 61–81. Printwell, Jaipur.

Sharma, B.R., Swarup, V. and Chatterjee, S.S. 1977. Resistance to black rot disease [*Xanthomonas campestris* (Pam.) Dowson] in cauliflower. *Scientia Hort.* **7**: 1–7.

Sharma, G.L. 1989. Estimated losses due to root-knot nematodes, *Meloidogyne incognita* and *M. javanica* in pea crop. *Int. Nematol. Network Newsl.* **6** (1): 28–29.

Sharma, G.L., Trivedi, P.C., Sharma, M.K. and Tiagi, B. 1988. Fungus *Paeciloymces lilacinus* as a biocontrol agent of root-knot nematode, *Meloidogyne incognita*. *The Indian Zool.* **12**: 259–62.

Sharma, H.K. and Gill, J.S. 1992. Prevalence of races in root-knot nematode, *Meloidogyne incognita* in India. *Indian J. Nematol.* **22**: 43–49.

Sharma, H.K., Pankaj and Mishra, S.D. 2005. Polyethylene mulching in the management of plant parasitic nematodes. *Indian J. Nematol.* **35**: 82–85.

Sharma, N.K. and Sethi, C.L. 1976. Reaction of certain cowpea varieties to *Meloidogyne incognita* and *Heterodera cajani*. *Indian J. Nematol.* **6**: 99–102.

Sharma, N.K., Thapa, C.D. and Nath, A. 1981. Pathogenicity and identity of myceliophagous nematode infesting *Agaricus bisporus* (Lange) Sing. in Himachal Pradesh (India). *Indian J. Nematol.* **11**: 230–31.

Sharma, N.K., Thapa, C.D. and Damanjeet Kaur. 1984. Estimation of loss in mushroom yield due to myceliophagous nematode, *Aphelenchoides sacchari*. *Indian J. Nematol.* **14**: 121–24.

Sharma, N.K., Thapa, C.D. and Damanjeet Kaur. 1985. Pathogenicity and identity of *Ditylenchus mycetophagus* "Indian population" infesting *Agaricus bisporus* (Lange) Sing. in India. *Indian J. Nematol.* **15**: 233–34.

Sharma, R.C. and Sharma, S.L. 1985. Correlation of meteorological factors with stalk rot of cauliflower caused by *Sclerotinia sclerotiorum*. *Indian J. Ecol.* **12**: 327–30.

Sharma, S.B., Sharma, H.K. and Pankaj. 2002. Nematode problems in India. In *Crop Pest and Disease Management: Challenges for the Millennium* (D. Prasad and S.N. Puri, *eds.*), pp. 267–75. Jyoti Publishers, New Delhi, India.

Sharma, S.K., Singh, I. and Sakhuja, P.K. 1980. Influence of different cropping sequences on the population of root-knot nematode, *Meloidogyne incognita* and the performance of the subsequent mungbean crop. *Indian J. Nematol.* **10**: 53–58.

Sheela, M.S., Jiji, T. and Nisha, M.S. 2002. Evaluation of different control strategies for the management of nematodes associated with vegetables (brinjal) (Abstr.). *Int. conf. Vegetables,* Bangalore, 268 pp.

Sheela, M.S. and Nisha, M.S. 2004. Impact of biological control agents for the management of root-knot nematode in brinjal. *National Symposium on Paradigms in Nematological Research for Biodynamic Farming,* Univ. of Agri. Sci., Bangalore, pp. 63–64.

Sheela, M.S., Nisha, M.S. and Mohandas, C. 2002. Impact of trap cropping and crop rotation for the management of root-knot nematode in vegetables. *Proc. Int Conf. Vegetables,* Bangalore, pp. 344–46.

Shekhawat, P.S. and Chakravarti, B.P. 1976. Factors affecting development of bacterial leaf spot of chillies caused by *Xanthomonas vesicatoria*. *Indian Phytopath.* **29**: 392–97.

Shekhawat, P.S. and Chakravarti, B.P. 1977. Assessment of loss, symptomatology and occurrence of bacterial leaf spot of chillies caused by *Xanthomonas vesicatoria* in Rajasthan. *Indian J. Mycol. Pl. Path.*7: 1–14.

Shekhawat, P.S. and Chakravarti, B.P. 1978. Survival of *Xanthomonas vesicatoria* the incitant of chilli leaf spot and a medium for its isolation. *Indian J. Pl. Path.*14: 11–19.

Shekhawat, P.S., Jain, M.L. and Chakravarti, B.P. 1982. Detection and seed transmission of *Xanthomonas campestris* pv. *campestris* causing black rot of cabbage and cauliflower and its control by seed treatment. *Indian Phytopath.* **35**: 442–47.

Shekhawat, P.S. and Patel, P.N. 1977. Seed transmission and spread of bacterial blight of cowpea and leaf spot of green gram in summer and monsoon seasons. *Plant Dis. Reptr.* **61**: 390–92.

Siddiqi, M.A. and Alam, M.M. 1989a. Seed treatment with azadirachtin for the control of the stunt nematode attacking cabbage and cauliflower. *Ann. Appl. Biol.* **113**: 4–5.

Siddiqi, M.A. and Alam, M.M. 1989b. Control of stunt nematode by bare-root dip in leaf extracts of margosa and Persian lilac. *Pakistan J. Nematol.* **7**: 33–38.

Siddiqi, M.A. and Saxena, S.K. 1987a. Effect of interculture of margosa and Persian lilac with tomato and eggplant on root-knot and reniform nematodes. *Int. Nematol. Network Newsl.* **4** (2): 5–8.

Siddiqi, M.A. and Saxena, S.K. 1987b. Studies on the control of the stunt nematode, *Tylenchorhynchus brassicae* by interculture of margosa and Persian lilac. *Int. Nematol. Network Newsl.* **4** (4): 27–29.

Siddiqi, M.R. 1962. *Nematologica* **8**: 193–200.

Siddiqui, Z.A., Khan, A.M. and Saxena, S.K. 1972. Host range and varietal resistance of certain crucifers against *Tylenchorhynchus brassicae. Indian Phytopath.* **25**: 275–81.

Siddiqui, Z.A., Khan, A.M. and Saxena, S.K. 1973. Studies on *Tylenchorhynchus brassicae* II– Effect of temperature and moisture on multiplication of nematode. *Indian J.Ent.* **26**: 139–47.

Siddiqui, Z.A., Khan, A.M. and Saxena, S.K. 1976. Nematode population and yield of certain vegetables as influenced by oil-cake amendments. *Indian J. Nematol.* **6**: 179–82.

Singh, D. 1991. *Studies on the Effect of* Tagetes *Species on* Meloidogyne javanica *Infectivity in Brinjal* (Solanum melongena). Ph. D. thesis, Haryana Agri. Univ., Hisar, 83 pp.

Singh, D. and Patel, P.N. 1977. Studies on resistance in crops to bacterial diseases in India. VII. Inheritance of resistance to bacterial blight disease in cowpea. *Indian Phytopath.* **30**: 99–102.

Singh, D.B., Parvatha Reddy, P. and Sharma, S.R. 1981. Effect of the root-knot nematode, *Meloidogyne incognita* on *Fusarium* wilt of French beans. *Indian J. Nematol.* **11**: 84–85.

Singh, D.B., Rao, V.R. and Parvatha Reddy, P. 1978. Evaluation of nematicides for the control of root-knot nematodes on brinjal. *Indian J. Nematol.* **8**: 64–66.

Singh, D.V., Masoodi, S., Banerjee, A.K., Srivastava, S.S.L. and Rai, M. 1982. An outbreak of bacterial leaf spot of tomato in Uttar Pradesh. *Indian J. Mycol. Pl. Path.* **12**: 252–354.

Singh, H. and Singh, G. 1975. Biological studies on *Heliothis armigera* (Hubner) in Punjab. *Indian J. Ent.* **37**: 154–64.

Singh, J. and Swarup, J. 1986. Control of bacterial blight of cluster bean by seed treatment and foliar spray. *Pesticides* **20** (6): 23–24.

Singh, J. and Swarup, J. 1987. Epidemiology of bacterial blight of cluster bean caused by *Xanthomonas campestris* pv. *cyamopsidis*. *Farm Sci. J.* **2**: 91–92.

Singh, M., Samra, R. and Sharma, G.L. 1998. Effect of various soil types on the development of *Pasteuria penetrans* on *Meloidogyne incognita* in brinjal crop. *Indian J. Nematol.* **28**: 35–40.

Singh, N. and Gangopadhyay, S. 1984. Control of white rot of seed cauliflower. *Pesticides* **18** (12): 23–24.

Singh, R., Trivedi, B.M., Gill, H.S. and Sen, B. 1987. Breeding for resistance to black rot, downy mildew and curd blight in Indian cauliflower. *Cruciferae Newsl.* **12**: 96–97.

Singh, R.S. and Sitaramaiah, K. 1966. Incidence of root-knot of okra and tomato in oil-cake amended soil. *Pl. Dis. Reptr.* **50**: 668–72.

Singh, R.S. and Sitaramaiah, K. 1971. Control of root-knot through organic and inorganic amendments of soil: Effect of saw dust and inorganic nitrogen. *Indian J. Nematol.* **1**: 80–84.

Singh, R.S. and Sitaramaiah, K. 1973. *Control of Plant Parasitic Nematodes with Organic Amendments of Soil*. Final technical report on effect of organic amendments, green manuring and inorganic fertilizers on root-knot of vegetable crops, G.B. Pant Univ. of Agri. & Technol., Bull. No. 6, 289 pp.

Singh, R.V. and Gill, J.S. 1998. Important nematode problems on vegetables and their management. In *Pathological Problems of Economic Crop Plants and their Management* (S.M. Paul Khurana, *ed.*), pp. 557–70. Scientific Publishers (India), Jodhpur, India.

Singh, R.V., Midha, S.K. and Kumar, V. 2003. Major nematode pests of economically important crops in India and their management. In *Advances in Nematology* (P.C. Trivedi, ed.), pp. 233–56. Scientific Publishers (India), Jodhpur.

Singh, S.J. 1985. Bittergourd witches' broom–A disease in India. *Indian J. Virol.* **1**: 320–24.

Singh, S.J. 1990. Etiology and epidemiology of whitefly-transmitted virus diseases of okra in India. *Plant Dis. Res.* **5**: 64–70.

Singh, S.J. and Dutta, O.P. 1986. Enation leaf curl of okra, a new virus disease. *Indian J. Virol.* **2**: 114–17.

Singh, S.J. and Sastry, K.S. 1976. Tomato leaf curl virus and its control. *Indian Hort.* **21**: 9–10.

Singh, S.J., Sastry, K.S. and Sastry, K.S.M. 1977. Yellow vein mosaic virus of bhindi and its control. *Indian Hort.* **22**: 14–15.

Singh, S.J., Sastry, K.S. and Sastry, K.S.M. 1979. Combating leaf curl virus in chilli. *Indian Hort.* **24**: 9–10.

Singh, S.J., Sastry, K.S. and Sastry, K.S.M. 1981. Field tests with insecticides and mineral oil for the protection of French beans from yellow mosaic virus disease. *Gartenbauwissenchaft* **46**: 669–72.

Singh, S.J., Sastry, K.S. M. and Sastry, K.S. 1973. Effect of oil spray on the control of tomato leaf curl virus in field. *Indian J. Agric. Sci.* **43**: 669–72.

Sinha, P.P. 1989. Preliminary studies on bacterial leaf spot of cucumber. *Indian Phytopath.* **42**: 146–49.

Sitaramaiah, K. and Naidu, H. 2003. Plant parasitic nematodes-management opportunities and challenges. In *Nematode Management in Plants* (P.C. Trivedi, ed.), pp. 57–80. Scientific Publishers (India), Jodhpur.

Sitaramaiah, K. and Sikora, R.A. 1982. Effect of the mycorrhizal fungus, *Glomus fasciculatum* on the host-parasite relationship of *Rotylenchulus reniformis* in tomato. *Nematologica* **28**: 412–19.

Sivakumar, C.V. 1998. Bacterial antagonists for suppression of plant parasitic nematodes. In *Biological Suppression of Plant Diseases, Phytoparasitic Nematodes and Weeds* (S.P. Singh and S.S. Hussaini, *eds.*), pp. 128–37. Project Directorate of Biological Control, Bangalore.

Sivakumar, C.V. and Meerazainuddin, M. 1974. Influence of N, P and K on the reniform nematode, *Rotylenchulus reniformis* and its effect on the yield of okra. *Indian J. Nematol.* **4**: 243–44.

Sivakumar, C.V., Palanisamy, S. and Naganathan, T.G. 1976. Persistence of nematicidal activity in seed treatment of okra with carbofuran and aldicarb sulfone. *Indian J. Nematol.* **6**: 106–08.

Sivakumar, C.V. and Seshadri, A.R. 1972. Histopathology of infection by the reniform nematode, *Rotylenchulus reniformis* Linford and Oliveira, 1940 on castor, papaya and tomato. *Indian J. Nematol.* **2**: 173–81.

Smith, F.F., Webb, R.E. and Dusly, R.F. 1979. Greenhouse whitefly control by sticky yellow board traps. *9th Int. Cong. Pl. Prot.*, Washington DC, Abstr. 654.

Sobita Devi, L. and Dutta, U. 2002. Effect of *Pseudomonas fluorescens* on root-knot *(Meloidogyne incognita)* on okra plant. *Indian J. Nematol.* **32**: 215.

Somasekhar, N. and Gill, J.S. 1991. Efficacy of *Pasteuria penetrans* alone and in combination with carbofuran in controlling *Meloidogyne incognita. Indian J. Nematol.* **21**: 61–65.

Sreekantaradhya, R., Ravishankar, G. Chandrasekaraiah, S.C. and Seshadri, V.S. 1977. S-288, a promising cowpea variety resistant to bacterial blight. *Curr. Res.* **6**: 58–59.

Srihari, K. and Chakravarthy, A.K. 1998. Ecological approaches for the management of vertebrate pests in Western Ghats region of Karnataka. *Indian J. Ecol.* **25**: 55–61.

Srinivasan, K and Krishna Moorthy, P.N. 1991. Indian mustard as a trap crop for management of major lepidopterous pests on cabbage. *Trop. Pest Mangmt.* **37**: 26–32.

Srinivasan, K and Krishna Moorthy, P.N. 1993. Evaluation of neem products and other standard chemicals for the management of major pest complex on cabbage: Comparison between standard spray regime and IPM involving mustard as a trap crop. *In: Neem and Environment, Vol. I* (R.P. Singh, M.S. Chari, A.K. Raheja and W. Kraus, *eds.*), pp. 447–58. Oxford and IBH Publishing Co. (Pvt.) Ltd., New Delhi.

Srinivasan, P.M. 1958. Control of fruit borer, *Heliothis armigera* (Hub) on tomato. *Indian J. Hort.* **16**: 187–88.

Srinivasan, T.N. and Nariani, T.K. 1966. Studies on a mosaic disease of pea (*Pisum sativum* L.). *Indian Phytopath.* **19**: 189–93.

Srivastav, U.K. and Patel, N.T. 1990. Pesticide residues in foods – Present situation. *In: Pesticide Industry in India*. CMA Monograph No. 142, Oxford & IBH Publishing Co. (Pvt.) Ltd., New Delhi, pp. 89–105.

Srivastava, D.N. and Rao, Y.P. 1963. Bacterial blight of guar. *Indian Phytopath.* **16**: 69–73.

Srivastava, D.N. and Rao, Y.P. 1970. Epidemiology and control of bacterial blight of guar [*Cyamopsis tetragonoloba* (L.) Taub.]. *Indian Phytopath. Soc. Bull.* **6**: 1–4.

Subramanyam, S., Rajendran, G. and Vadivelu, S. 1990. Estimation of loss in tomato due to *Meloidogyne incognita* and *Rotylenchulus reniformis*. *Indian J. Nematol.* **20**: 239–40.

Sundarababu, P.C. 1968. Studies on the efficacy of Thuricide in the control of the leaf caterpillar, *Hymenia recurvalis* on amaranthus. *S. Indian Hort.* **16**: 17–19.

Sundarababu, P.C. 1969. Note on the possible use of *Bacillus thuringiensis* Berliner for the control of certain lepidopterous pests. *Madras Agric. J.* **56**: 824–26.

Sundarababu, P.C. 1970. Comparative efficacy of Thuricide against the spotted beetle, *Epilachna vignitioctopunctata* F. on brinjal (*Solanum melongena*). *S. Indian Hort.* **18**: 85–88.

Sundara Babu, R., Sankaranarayanan, C. and Shanthi, A.1993. Interaction between vesicular arbuscular mycorrhiza and *Meloidogyne javanica* on tomato as influenced by time of inoculation. *Indian J. Nematol.* **23**: 125–27.

Sundara Babu, R. and Suguna, N. 1998. Effective optimum dose of *Glomus mosseae* for the management of *Meloidogyne incognita* on tomato, chillies and bhendi. *Int. J. Trop. Pl. Dis.* **16**: 127–31.

Sundaram, H.D. and Arangarasan, V. 1996. Role of vesicular arbuscular mycorrhizal fungi on the growth, nematode (*Meloidogyne incognita*) disease incidence and fruit yield of tomato CO 3 grown in loamy soils of Arcot District (Tamil Nadu). *Mycorrhizae: Biofertilizer for the Future*, pp. 100–03.

Suresh, C.K. and Bagyaraj, D.J. 1984. Interaction between a vesicular arbuscular mycorrhiza and a root-knot nematode and its effect on growth and chemical composition of tomato. *Nematol. Medit.* **12**: 31–39.

Surulivelu, T., Chellaiah, S. and Balasubramanian, G. 1977. Efficacy of certain insecticides and *Bacillus thuringiensis* in the control of lab-lab pod borers. *Madras Agric. J.* **65**: 120–21.

Swamy, K.R.M. and Sadashiva, A.T. 2004. *Vegetable Varieties/Hybrids of IIHR*. Tech. Bull. 2, Indian Inst. of Hort. Res., Bangalore, 30 pp.

Swarup, G. and Seshadri, A.R. 1974. Nematology in India–Problems and Progress. *In: Current Trends in Plant Pathology* (S.P. Raychaudhuri and J.P. Verma, *eds*.), pp. 303–11. Dept. of Botany, Univ. of Lucknow, Lucknow.

Talukdar, Rita. 1993. *Management of* Meloidogyne incognita *(Kofoid and White, 1919) Chitwood 1949 on Okra* [*Abelmoschus esculentus* (L.) Moench]. M. Sc. (Agri.) thesis, Assam Agri. Univ., Jorhat.

Tanda, A.S. and Atwal, A.S. 1988. Effect of sesame intercropping against the root-knot nematode (*Meloidogyne incognita*). *Nematologica* **34**: 484–92.

Taylor, A.L. and Sasser, J.N. 1978. *Biology, Identification and Control of Root-knot Nematodes* (*Meloidogyne* spp.). North Carolina State University Graphics, Raleigh, 111 pp.

Tewari, G.C. and Krishna Murthy, P.N. 1984. Yield loss in tomato caused by fruit borer. *Indian J. Agric. Sci.* **54**: 341–43.

Tewari, G.C. and Krishna Murthy, P.N. 1986. *Eurytoma* sp. A new parasite on *Asphondylia* sp. infesting egg plant. *Entomon* **12**: 227–28.

Tewari, G.C., Krishna Murthy, P.N. and Sardana, H.R. 1987. Nature of damage and chemical control of gall midge, *Asphondylia* sp. infesting eggplant. *Indian J. Agric. Sci.* **57**: 745–48.

Thakar, N.A. and Patel, C.C. 1984. Screening of few cowpea varieties against reniform nematode, *Rotylenchulus reniformis* Linford and Oliveira, 1940. *Indian J. Nematol.* **14**: 204.

Thakur, M.S. and Dube, G.S. 1981. Studies on mosaic disease of radish (*Raphanus sativus*) (Abstr.). *Proc. 3rd Int. Symp. Pl. Path.*, New Delhi.

Thangaraju, D. 1983. Distribution of potato cyst nematodes in Kodaikanal Hills, Madurai district, Tamil Nadu. *Indian J. Nematol.* **13**: 222–23.

Thapa, C.D., Munjal, R.L. and Seth, P.K. 1987. The mushroom nematodes and their control. *Indian J. Mush.* **3**: 1–11.

Thapa, C.D., Jandaik, C.L. and Sharma, N.K 1987. Comparative resistance in some cultivated edible fungi against two mushroom nematodes occurring in India. *Proceedings of the International Conference on Science and Cultivation Technology of Edible Fungi*, pp. 169–72.

Thapa, C.D., Sharma, N.K and Nath, A. 1981. Mushroom nematodes and their source of infestation in mushroom beds in India. *Mushroom Sci.* **11**: 443–48.

Thirumalachar, M.J. 1951. Root-knot nematode on potato tubers in Shimla. *Curr. Sci.* **20**: 20, 104.

Thomas, K.M. and Krishnaswami, C.S. 1939. "Little leaf" a transmissible disease of brinjal. *Proc. Indian Acad. Sci.* **310**: 201–02.

Thorne, G. 1961. *Principles of Nematology*. McGraw-Hill, New York and London, 553 pp.

Tiyagi, S.A., Anver, S., Bano, M. and Siddiqi, M.A. 1986. Feasibility of growing zinnia as a 'mix-crop' along with tomato for control of root-knot and reniform nematodes. *Int. Nematol. Network Newsl.* **3** (3): 6–7.

Todd, J.M., Panniah, S. and Subramanyan, C.P. 1975. First record of tomato spotted wilt virus from the Nilgiris in India. *Madras Agric. J.* **62**: 162–63.

Trivedi, B.M. and Patel, P.N. 1972. *In: Plant Bacteriology Vol. I. Bacterial Diseases of Plants in India* (P.N. Patel, *ed.*), Indian Council of Agri. Res., New Delhi, 260 pp.

Trivedi, P.C. 1992. Evaluation of a fungus, *Paecilomyces lilacinus* for the biological control of root-knot nematode, *Meloidogyne incognita* on *Solanum melongena*. *In: Proc. of the 3rd*

Intn. Conf. on Plant Prot. in the Tropics (F.A.C. Ooi and G.S. Lim, *eds.*), Vol. 6, pp. 29–33. Genting Highlands, Malaysia.

Trivedi, P.C. and Tiagi, A. 1984. Effect of some plants on *Meloidogyne incognita* population in chilli soil. *Indian J. Bot. Sci.* **63**: 416–19.

Upadhyay, K.D. and Dwivedi, K. 1987. Analysis of crop losses in pea and gram due to *Meloidogyne incognita*. *Int. Nematol. Network Newsl.* **4** (4): 6–7.

Vadhera, I., Sheila, B.N. and Bhatt, J. 1995. Interaction between reniform nematode (*Rotylenchulus reniformis*) and *Fusarium solani* causing root rot of French bean. *Indian J. Agri. Sci.* **65**: 774–77.

Vadivelu, R.S. and Rajendran, G. 1986. Damage potential of the root-knot nematode, *Meloidogyne arenaria* to onion. *Int. Nematol. Network Newsl.* **4** (2): 5–8.

Van Berkum, J.A. and Seshadri, A.R. 1970. Some important nematode problems in India. *Proceedings of Xth International Symposium of European Society of Nematologists*, Pescara, pp. 135–37.

Varma, G.C. and Gill, G.S. 1976. Laboratory studies on the comparative efficacy of biotic insecticides for the control of *Plutella xylostella* (Linnaeus) (Plutellidae: Lepidoptera). *J. Res. PAU* **14**: 304–08.

Varma, M.K., Sharma, H.C. and Pathak, V.N. 1978. Evaluation of systemic granular nematicides against *Tylenchorhynchus* spp. parasitic on cabbage. *Indian J. Nematol.* **8**: 59–60.

Vats, R. and Dalal, M.R. 1997. Interaction between *Rotylenchulus reniformis* and *Fusarium oxysporum* f. sp. *pisi* on pea. *Ann. Biol. Ludhiana* **13**: 239–42.

Veere Gowda, R., Singh, T.H., Nagesh, M. and Rao, M.S. 2001. Evaluation of carrot (*Dacus carota* L.) genotypes for resistance to root-knot nematode (*Meloidogyne incognita*). *Second National Symposium on Integrated Pest Management in Horticultural Crops: New Molecules, Biopesticides and Environment*, Bangalore, 155 pp.

Verma, A.C. 2001. N.D. University of Agriculture and Technology, Kumarganj, Faizabad, Uttar Pradesh. *In: Indian Nematology-Progress and Perspectives* (S.C. Dhawan *et al., eds.*), pp. 121–25. Division of Nematology, Indian Agri. Res. Inst., New Delhi.

Verma, A.C., Singh, H.K. and Khan, N. 2005. Management of root-knot nematode, *Meloidogyne incognita* through antagonistic approaches in pointed gourd. *Indian J. Nematol.* **35**: 78–79.

Verma, K.K., Gupta, D.C. and Paruthi, I.J. 1999. Preliminary trial on the efficacy of *Pseudomonas fluorescens* as seed treatment against *Meloidogyne incognita* in tomato. *Proceedings of the National Symposium on Rational Approaches in Nematode Management for Sustainable Agriculture,* Indian Agri. Res. Inst., New Delhi, pp. 79–81.

Vetrivelkalai, P. and Subramanian, S. 2006. Plant parasitic nematode populations as affected by onion based cropping sequences. *Indian J. Nematol.* **36**: 10–14.

Virakthamath, C.A., Tewari, G.C., Srinivasan, K. and Gupta, M. 1993. American serpentine leaf miner is a new threat to crops. *Indian Fmg.* **43** (2): 10–12.

Walia, R.K., Bansal, R.K. and Bhatti, D.S. 1991. Effect of *Paecilomyces lilacinus* application time and method in controlling *Meloidogyne javanica* on okra. *Nematol. Medit.* **19**: 247–49.

Walia, R.K., Bansal, R.K. and Bhatti, D.S. 1992. Efficacy of bacterium (*Pasteuria penetrans*) and fungus (*Paecilomyces lilacinus*) alone and in combination against root-knot nematode (*Meloidogyne javanica*) infecting brinjal. *First Afro-Asian Nematol. Symp.*, Aligarh Muslim Univ., Aligarh, 38 pp.

Walia, R.K. and Dalal, M.R. 1994. Efficacy of bacterial parasite *Pasteuria penetrans* application as nursery soil treatment in controlling root-knot nematode *Meloidogyne javanica* on tomato. *Pest Mangmt. Econ. Zool.* **2**: 19–21.

Winslow, R.D. and Willis, R.J. 1972. Nematode diseases of potato. *In: Economic Nematology* (J.M. Webster, *ed.*), pp. 17–48. Academic Press, London.

Wood, F.C. and Goodey, J.B. 1957. Effects of gamma-ray irradiation on nematodes infesting cultivated mushroom beds. *Nature* **180**: 760–61.

Wyatt, J.E., Fassuliotis, G., Hoffman, J.C. and Deakin, J.R. 1983. "Nemasnap" snap bean. *Hort. Sci.* **18**: 776.

Zaki, F.A. 1998. Biological control of *Meloidogyne javanica* in tomato by *Paecilomyces lilacinus* and castor leaves. *Indian J. Nematol.* **28**: 132–39.

Zaki, F.A. and Bhatti, D.S. 1989. Effect of castor (*Ricinus communis*) leaves in combination with different fertilizer doses on *Meloidogyne javanica* infecting tomato. *Indian J. Nematol.* **19**: 171–76.

Zaki, F.A. and Bhatti, D.S. 1991. Effect of castor *Ricinus communis* and the biocontrol fungus, *Paecilomyces lilacinus* on *Meloidogyne javanica. Nematologica* **36**: 114–22.

Zuckerman, B.M., Dicklow, M.B. and Acosta, N. 1993. A strain of *Bacillus thuringiensis* for the control of plant parasitic nematodes. *Biocontrol Sci. Technol.* **3**: 41–46.

SUBJECT INDEX

D